ENVIRONMENTAL POLITICS IN SOUTHERN EUROPE
Actors, Institutions and Discourses in a Europeanizing Society

ENVIRONMENT & POLICY

VOLUME 29

Environmental Politics in Southern Europe

Actors, Institutions and Discourses in a Europeanizing Society

Edited by

Klaus Eder
Humboldt-Universität zu Berlin, Berlin, Germany

Maria Kousis
University of Crete, Rethimno, Greece

with contributions by

Susana Aguilar Fernández, Evi Arahoviti, Iñaki Barcena, Helen Briassoulis, George A. Daoutopoulos, Mario Diani, Teresa Fidélis, Elisabete Figueiredo, Joaquim Gil Nave, Pedro Ibarra, Manuel Jiménez, Leonidas Louloudis, Dimitris Á. Papadopoulos, Eugenia Petropoulou, Artur da Rosa Pires, Geoffrey Pridham, Myrto Pyrovetsi, Michael Redclift, Carlo Ruzza, Anna Triandafyllidou

KLUWER ACADEMIC PUBLISHERS
DORDRECHT / BOSTON / LONDON

A C.I.P. Catalogue record for this book is available from the Library of Congress.

ISBN 0-7923-6753-7

Published by Kluwer Academic Publishers,
P.O. Box 17, 3300 AA Dordrecht, The Netherlands.

Sold and distributed in North, Central and South America
by Kluwer Academic Publishers,
101 Philip Drive, Norwell, MA 02061, U.S.A.

In all other countries, sold and distributed
by Kluwer Academic Publishers,
P.O. Box 322, 3300 AH Dordrecht, The Netherlands.

Printed on acid-free paper

Printed in the Netherlands.

Contents

Contributors

Susana Aguilar Fernández is Senior Reader at the Faculty of Political Sciences and Sociology, Complutense University of Madrid, Spain.

Evi Arahoviti is a Msc. Researcher at the Department of Agricultural Economics and Rural Development of the Agricultural University of Athens, Greece.

Iñaki Barcena is Professor of Political Science at the Department of Political Science at the University of the Basque Country, Spain.

Helen Briassoulis is Associate Professor at the Department of Geography, University of the Aegean, Lesvos, Greece.

George A. Daoutopoulos is Professor of Agricultural Economics at the Department of Agricultural Economics at the Aristotelian University of Thessaloniki, Greece.

Mario Diani is Professor of Sociology at the Department of Government of the University of Strathclyde in Glasgow, Scotland, UK.

Klaus Eder is Professor of Sociology at the Institute for Social Sciences of the Humboldt-University Berlin, Germany.

Teresa Fidélis is Lecturer at the Department of Environment and Planning of the University of Aveiro, Portugal.

Elisabete Figueiredo is Lecturer at the Department of Environment and Planning of the University of Aveiro, Portugal.

Joaquim Gil Nave is Professor of Sociology at the Instituto Superior de Ciências do Trabalho e da Empresa (ISCTE), Lisbon, Portugal.

Pedro Ibarra is Professor of Political Science at the Department of Political Science of the University of the Basque Country, Spain.

Manuel Jiménez is a Doctoral Candidate and Researcher at the Center for Advanced Studies in Social Sciences, Juan March Institute, Madrid, Spain.

Marıa Kousis is Associate Professor of Sociology at the University of Crete, Rethimno, Greece.

Leonidas Louloudis is Assistant Professor at the Department of Agricultural Economics and Rural Development of the Agricultural University of Athens, Greece.

Dimitris Á. Papadopoulos is a PhD Researcher at the Department of Agricultural Economics and Rural Development of the Agricultural University of Athens, Greece.

Eugenia Petropoulou is Research Associate at the University of Crete Research Fund, Greece.

Artur da Rosa Pires is Professor of Development Planning at the University of Aveiro, Portugal.

Geoffrey Pridham is Professor of European Politics and Director of the Centre for Mediterranean Studies at the University of Bristol, UK.

Myrto Pyrovetsi is Assistant Professor at the Department of Ecology in the School of Biology of the Aristotelian University of Thessaloniki, Greece.

Michael Redclift is Professor of Rural Sociology in the Environment Section, Wye College of the University of London, UK.

Carlo Ruzza is PhD Lecturer in Sociology and European Studies at the University of Essex, UK.

Anna Triandafyllidou is a Jean Monnet Fellow at the European University Institute in Florence, Italy.

Preface

Environmental Politics in Southern Europe: Actors, Institutions, and Discourses in a Europeanizing Society addresses an alternative, more sociological view of the ways in which the North is intricately related to the environmental policies and politics of the South. The contributions focus on the structure of European political elites, on features of environment-related collective action, and on the institutional ordering of conflicts at national and supranational levels.

Putting this book together required substantial effort. In 1997 we established an informal research network on 'environmental discourses and policies in the EU' and in May 1998 the Department Sociology of the University of Crete hosted a conference on 'Environmental Movements, Discourses, and Policies in Southern Europe'. The conference was sponsored by the European Commission's DG XII, the Hellenic Secretariat for Research & Technology, and the University of Crete.

Several of the papers were revised on the basis comments made by conference panel leaders, discussants, and participants. The chapter by Gil Nave was added at a later stage. A proposal package for an edited volume was then submitted to the Environmental Sciences Division of Kluwer Academic Publishers. The package was reviewed by external referees and revisions were made prior to submitting the final manuscript for publication. To keep the volume 'fresh' the authors have incorporated in their final drafts the most recent developments in the field.

We are grateful to the European Commission (DG XII), the Hellenic Secretariat for Research & Technology, and the University of Crete for their financial support, to the reviewers of this volume for their comments, and to our publisher, Kluwer Academic Publishers, for making this work interna-

K. Eder and M. Kousis (eds.), Environmental Politics in Southern Europe, 1–2.

tionally visible. We also thank Eric J. Engstrom, for his efforts in polishing the English-language manuscript and making it as readable as possible, and Sibylle Scheipers, for her diligent engagement in editing the manuscript. The views presented in this book are the authors' alone and do not necessarily reflect those of the conference sponsors.

Klaus Eder and Maria Kousis

Rethimno, Crete, March 2000

Introduction

EU policy-making, local action, and the emergence of institutions of collective action

A theoretical perspective on Southern Europe

Maria Kousis and Klaus Eder

1. THE NORTH/SOUTH DIVIDE – A SOCIOLOGICAL PERSPECTIVE ON ENVIRONMENTALISM IN EUROPE

This volume presents a new way of looking at processes that result from environmental policy-making in Europe and affect environmental politics in its southern member-states. Instead of asking what Brussels is doing to harmonize environmental policies in Europe and how Brussels succeeds in implementing the idea of sustainability as a unifying policy frame in the member-states, this volume has a different aim. It argues that the 'new Europe' provides an opportunity structure for collective action in Europe. This, however, is only part of the story. The other part of the story is that those who act within this new opportunity structure do not act as monads, but as people shaped by common life experiences, i.e., by work and leisure, by production and consumption, and by common experiences of inclusion and exclusion. One major aim of this volume is to uncover such social factors and the driving forces which are fuelled by and react to environmental politics in Europe.

This tension between environmental policy-making, environmental politics, and the people has been a topic in both social-scientific and public discourse, where it is considered mainly in terms of an opposition between the EU-level policy-making process on the one hand, and the national member-states' reaction, i.e., compliance and non-compliance with the rules set by EU-authorities, on the other hand. Differential national reactions have then been used to structure the social field of collective action in Europe. Such differences have then been grouped together to identify different 'worlds'

K. Eder and M. Kousis (eds.), Environmental Politics in Southern Europe, 3–21.

within Europe (Pridham 1994; Baker, Milton and Yearley 1994; Ward, Lowe and Buller 1997), such as the North and the South. The very process of European integration seems to create a space with renewed political and cultural cleavages (Eatwell 1997).

The process of European environmental policy-making is not an exception to this rule. On the contrary, it even fosters the construction of difference. European policies produce different worlds of environmentalism, i.e., a heterogeneous sustainable Europe (Baker et al. 1997; Börzel 2000). Europe is – in a less harsh version – divided into leaders and laggards (Aguilar Fernández 1993; Ward, Lowe and Buller 1997; Pridham and Konstadakopulos 1997). The more skeptical studies[1] point to progress in the nineties and to a "less clear-cut divide between northern and southern members than is often supposed... Differences emerge between southern states, while some northern states have worse records than is generally believed" (Pridham 1996: 57). In a really harsh version, Europe is divided into a North-South divide, captured in the notion of a Mediterranean syndrome (La Spina and Sciortino 1993). Such constructions of difference resonate with the narrative of the good versus the bad – an elementary device used in making social differences.

This narrative has a political function. Environmental policy-making of the EU is considered in terms of a particular path toward European integration which assumes that Europe will converge on a form of environmentalism based on the principle of sustainability. At the same time, the Europeanization of environmental policies contributes to the making/reshaping of cleavages among the member-states of the EU. Framing these differences as the North-South divide offers a masterframe for a generalization of differences which are real.[2] However, what counts in public discourse are not the real differences, but their generalization in terms of categorizing such differences in well-resonating distinctions. Classifying real differences is thus part of a political game within Europe.[3] Such a theoretical perspective situates

[1] Such studies examine the transposition of environmental directives, registered environmental complaints, warning letters, reasoned opinions, and references to the European court by country (Pridham 1996).

[2] For a good discussion on the European cleavage structure see Flora (1999). Since the 1980s, southern European states have been distinguished from northern ones in a consistent and careful manner by many scholars of this area. The consolidation of the new democracies and the making of the 'new Southern Europe' (Gunther, Diamantouros and Puhle 1995) is now widely accepted.

[3] It has been argued that "a broad Southern European regional identity has been recently constructed in response to developments in northern Europe, and it will persist only so long as the EU provides a magnet of attraction and an incentive for co-operation between

European environmental policy-making within an emerging Europeanizing field of *political* struggles.

This way of looking at Europe leads us to conceive of Europe not in terms of a polity 'made in Brussels', but in terms of a society 'upset by Brussels'. We do not conceive of Europe as a space which, through an integrated market, transforms the old European cleavage system into a homogeneous social entity. Instead, we conceive of Europe as a social space which organizes itself within the institutional framework set by EU political institution building. This self-organization creates a social space often in opposition to, or in conflict with, the institutional framework from which it emerged. It becomes manifest in diverse forms of restructuring its cleavage structure. This is what we will call the Europeanization of society.

This implies empowering of a more sociological vision of European integration. European integration has changed the boundaries within which collective action takes place, thereby also dramatically changing the exit-options of collective action. This becomes particularly manifest in European environmental policy-making. The idea that southern and northern European countries follow different paths toward one 'sustainable' Europe defined in Brussels is a euphemism. What is true is that they are all in the same boat and are searching for their place in it. This search is accompanied by – usually contentious – public claim-making. Such unconventional political activities aim to create the pressure needed for countries' demands to be heard. When the pressure does not lead to a positive reaction on the part of the approached or challenged authorities, this indicates the countries' exclusion from power and the rigidity of political institutions. Europe emerges as a society by struggling with its newly institutionalized power structures. This relationship shapes movements, discourses, and policies in northern as well as in southern member-states.

There is a parallel European struggle with the environment which must be considered simultaneously. A brief examination of a few variations in European experiences with the environment reveals the following. In more industrialized northern areas, such as the United Kingdom, Germany, the Netherlands, Belgium, northern Italy, and northern Spain, higher concentrations of atmospheric pollutants have been recorded than in southern areas

---

these (and only these) political communities" (Whitehead 1996: 139). Discussions on Structural or Cohesion Funds to the 'poorer regions' of the EU support this statement. The 1988 reform aimed to use these funds to secure the political support of the southern states, while the 1993-99 Cohesion Fund – due to pressure from the Spanish government – targeted member states with a GDP of less than 90 per cent of the Community average, i.e., Greece, Spain, Portugal and Ireland (Bache 1998: 79-89).

(Eurostat 1996: 148-155). Biodiversity is more threatened (for birds, fish, and mammals) in Germany, Luxembourg, the Netherlands and UK, and less so in southern EU countries (Eurostat 1996: 327). Total consumption of pesticides in 1990 was highest for Spain, followed by France, Italy, and Germany (Eurostat 1996: 276). EU peripheral areas (from Ireland to Greece) contain regions of relatively undamaged habitat (Baker, Milton and Yearley 1994). Yet, at the same time, burned forested areas are largest for Spain, Italy, Portugal, France and Greece (in 1990) respectively (Eurostat 1996: 308). The generation of municipal waste (in kg/inhabitant) ranks highest in the Netherlands, Norway, Luxembourg, the UK, and Sweden, and is lowest for Portugal, Greece, Germany, Austria, and Spain (Redclift 1992: 82). Energy consumption per capita (kWh/person/year) is highest for Luxembourg, Belgium, France, Denmark, the UK, and Germany, while it is lowest for Portugal, Greece, Spain, Ireland, and Italy (Redclift 1992: 95). Energy production from nuclear power covers a considerable portion of this energy demand in France, Germany, Sweden, the UK, Belgium, Spain, and Finland (Eurostat 1996: 392). The export of hazardous and toxic waste to poor non-EU countries has been practiced by the UK, Germany, Italy, and Norway (Redclift 1992). In light of these distinctions, a serious evaluation of EU environmental politics and policies must take into account a wide repertoire of environmental issues before applying the leader-laggard labels to member states.

2. THE ENVIRONMENT AND THE MAKING OF EUROPE

The debate on environmental politics in Europe has been dominated by a theoretical perspective that looks at actors in Europe in terms of how they comply with European policy prescriptions (Liefferink, Lowe and Mol 1993; Baker, Milton and Yearley 1994). Movements are seen as pushers of EU policy-making, national governments and industrial actors as blockers, and the Commission as a particular actor making use of competing demands in this policy area. The goal of this system of interacting collective actors is taken for granted: environmental policy-making serves to enhance the environment. Research and theory have followed this orientation and identified implementation gaps, wrong means towards desired ends, or even the choice of wrong targets. This perspective is now institutionalized in an emerging professional expert culture in Europe.[4]

Our perspective takes a wider, more distanced view: we argue that environmental politics is action taken less on the environment than on society. Although it might also be good (or not) for the environment, environmental politics has ultimately to do with the good (or not) of society. Environmental politics (including environmental policy) shapes not simply the environment: above all, it shapes society. This perspective leads us away from the question of whether environmental politics in Europe succeeds in improving environmental conditions in Europe (and elsewhere).

What we propose is, ultimately, a reversal of the perspective. We start by asking to what extent environmental policy-making 'made in Brussels' provides an opportunity structure for environmental politics? European environmental policy based on sustainability principles provides an opportunity structure for environmental politics which tends to transgress its issue-specific boundaries. Since European environmental policy captures a wide range and intensity of environmental issues as elements of a policy of sustainability, it is necessarily associated with wider socio-economic and political issues. We therefore ask to what extent the environmental politics that emerges from such an opportunity structure shapes collective action aimed at the social organization of an emerging society in Europe.

We will examine collective action that turns the critique of environmental conditions into a critique of the social conditions of life in a Europeanizing society. In order to study empirically this function of environmental collective action in a Europeanizing social space, we have to identify and compare the respective collective actors, their political as well as their economic opportunities and constraints[5], the ways they organize themselves, and the way they view/frame sustainability.[6]

4 The term "epistemic communities" has been coined by Haas. For an application to Southern Europe see Haas (1989).

5 Advocates of the concept of political opportunity structure stress the relationship between environmental collective action and political cleavage structures, prevailing strategies and alliance structures (Kriesi 1995; Kriesi et al. 1995). The economic opportunity structure has only recently been pointed out as an important factor hindering or facilitating the emergence of such action; the nature of existing economic cleavages as well as the relationship to a given economic structure are important factors for the study of environmental collective action (Kousis 1998).

6 Three of the four dimensions have been considered factors that determine the development of collective action (McAdam, McCarthy and Zald 1996). The fourth, economic opportunity structure, has been investigated by students of local environmental mobilizations (Kousis 1998).

Our theoretical perspective on a European policy of sustainable development thus attributes to the issue of the environment more than the instrumental function of saving Europe from overexploitation or pollution of its natural resources, as laid out in specific, but poorly integrated policies. Environmental policy-making has – beyond this interest in a functioning natural environment – a social function (Eder 1996). The idea of sustainable development provides a frame that gives a 'transenvironmental' meaning to collective action.[7] In chapter 2, Michael Redclift discusses this issue in terms of how – albeit at times contradictory – environmental concerns are part of the European civil agenda on social justice and citizenship, in addition to being the outcome of an individual's experience with their immediate environment.

The relationship between environmental politics and the 'transenvironmental' meaning of collective action on the environment is, however, more complicated. This complexity is addressed in the three parts of the volume.

The instrumental function itself constitutes, as will be argued in the first part of this volume, a specific element in the analysis of the structure of an emerging European society, namely an analysis of its political elites that make Europe from above. The contributions to the first part of the book try to situate environmental action within this particular action space as defined and generated by European environmental policy-making. They look at socially embedded environmental discourses (Eder), at ecological modernization and global citizenship (Redclift), at informal activities and the quest for proper planning (Briassoulis), as well as the internal link connecting environmental problems, economics, and everyday activities such as tourism within EU policy-making (Ruzza).

The question underlying the second part of the volume is the self-organization of collective action proper in the social field constituted by environmental issues (Kousis 1999). This questioning is not restricted to a social-movement perspective (Daoutopoulos, Pyrovetsi and Petropoulou). It is, however, an important aspect of how social groups define the boundaries of acting together in a world of power. The term 'social mobilization' comprises a specific mode of collective action of people which is publicly visible and by definition conflictual. This term implies that structurally defined groups of people act contentiously in public, with a public interest, and on a public purpose which, if realized, would hurt the interests of the challenged group/s.

7 'Transenvironmental' implies that the meaning of collective action goes beyond the narrow environmental definition of the issue at stake. The environment carries with it more meanings than just the narrow one of doing something good to nature.

That social groups turn into movements is a special case of the general phenomenon of groups acting together. Movements are like the visible part of an iceberg. The social space opened up by environmental politics in Europe in general, EU policy-making in particular, is much more 'peopled' than a strong movement perspective might suggest. The theoretical idea is that social movements emerge when social groups succeed in adding a boundary condition to their collective action: a 'non-exit' condition in a given opportunity structure. Such boundaries can vary: they can be local, national, or transnational. Our hunch is that the local and the transnational will fuse at the expense of the national. The chapters in the second part address this issue. Within their given national or regional context, they identify the action space in which environmental NGOs as well as grassroots activists or local groups act collectively and intervene in environmental policies and discourses at the elite level. The topics analysed include the evolution and character of competing claims in southern European local environmental conflicts (Kousis), the establishment of ecosystem-destructive modern agricultural practices by powerful actors in Greek rural areas and the impacts on local groups (Daoutopoulos, Pyrovetsi and Petropoulou), the changing features, discourses, and political context of Basque environmentalism (Barcena and Ibarra), as well as the characteristics of grassroots environmental action in Portugal (Figueiredo, Fidélis and Rosa Pires).

To the extent that collective action is a latent possibility or even manifestly mobilized, it is forced to relate to given institutions and given powerful actors in the social field of environmental politics. We can observe new institutional forms that arise for organizing such a social action space and often challenge existing political institutions. Often, there is even a strong link with the issue of democratizing political institutions, an issue that seems to be all the stronger the more recent a non-democratic past may be (Gil Nave). This is the theme of the third part. Here we look at the changing relationship between the people, whose interests and identities are at stake in environmental politics, and the established national and supranational institutions dealing with the environment. The common theme of this part is the institutional ordering of conflicts arising from collective action in the name of the environment – both on the national and the supranational level. The topics covered range from institutionalizing participatory forms which join together environmental NGOs and environmental policy-making (Jiménez; Aguilar Fernández; Louloudis, Arahoviti and Papadopoulos; Gil Nave) or which implement sustainable agriculture and tourism (Louloudis, Arahoviti and Papadopoulos; Pridham) to the institutionalization of sustainability discourses among policy actors, business actors, and environmentalists (Diani; Triandafyllidou).

The three parts of this volume are organized not according to the logic of their distinctive environmental impact, but rather according to the logic of differential social impact. This social impact is addressed in terms of the public monitoring of the state of the environment, in terms of local protest against the social consequences of environmental politics, and finally in terms of institutional designs which allow for the coordination of conflicting actors through organizational and symbolic forms.

3. ACTORS

The theoretical assumption that environmental politics has a genuinely social impact can be specified analytically in three different ways: We can look first at the actors as carriers of environmental politics, secondly at the institutions that emerge in such action, and thirdly at the discourses through which these actors communicate with each other. This analytical perspective constitutes a second axis that links the contributions to this book and that will be explicated as follows.

How does one identify the actors who produce not only environmental impacts, but also social impacts? In the first chapter, Klaus Eder proposes a way of identifying the social location of different perspectives on environmental politics in Europe, namely, by speaking of different European publics observing EU environmental politics. He defines publics as the set of social actors who organize their action space in a way that takes into account Europe as a relevant environment for their action. The public at large, the European people, is however (still) an amorphous mass which exists in terms of 'citizens' who have the right freely to cross internal borders, to work within that space, and to move goods within that space as they like.

This description of the European people is an idealization or euphemism. Such a people does not exist as a homogeneous body, as a 'nation'.[8] Such rights of European citizens exist, but they are practiced in such a way as to allow boundaries to be drawn between people: national groups tighten their boundaries to make access to national labor markets difficult for non-nationals.[9] Consequently, the European citizen is an actor on paper, but not in reality.

8 The invention of the "First New Nation", as Lipset has called the American polity and its publics, would be an example of creating a homogeneous public out of heterogeneity. Such a process, however, is unlikely to take place in Europe in the foreseeable future.

9 This holds not only for workers, but also for professionals such as doctors.

A European people exists as a 'real' collective actor in at least three different modes. It exists as a *national actor*, and the question is whether this aggregate actor is still the central unit in the plurality of actors in the emerging European society.[10] Organized as 'national publics', such collective actors are defined as sharing a culture (a topic which is especially notable to those studying national stereotypes) and sharing institutions (e.g. the state) which 'represent' the people as a nation.[11]

In addition to national actors we can observe *elite actors* that crosscut national publics and transgress national boundaries. Organized as 'elite publics' they are composed of industrialists, NGOs, cultural entrepreneurs (intellectuals), and professional groups. These groups explicitly use Europe as an opportunity structure, be it by creating a European association, a European network, or a European image. The relationship of such elite publics to sustainability in Europe as a whole, and in southern Europe in particular, is studied in this volume by Redclift (chapter 2) and Ruzza (chapter 4). That this elite reference to sustainability is linked to a mode of organizing economic and social life in a formally rational way, is discussed under the heading of formal versus informal economies by Briassoulis (chapter 3).

Behind these publics, a third type of public is in the making. These are *popular actors* organized as 'popular publics'.[12] Popular publics are aggregates of people who respond to everyday experiences. Popular publics emerge in working class neighborhoods as well as in middle class suburbs; they emerge in both urban and agrarian communities. They are made of territorially fixed populations which share specific modes of living and of perceiving the world. Popular publics may be, but are not necessarily, class-related. They often constitute what has been called a 'milieu', a concept used as a substitute for waning classes. They tend to manifest the informal aspect of social life, outside the ordered and formal relationships defined by formal political and legal institutions (Briassoulis, chapter 3).

10 Evidence for the specificity of national actors on the periphery of Europe is provided by Kousis (1994). Note that here the notion of the national is already superseded by the notion of countries on the semi-periphery – a classic topos in the explanation of the European cleavage structure.

11 In a recent work on European political cultures, Eatwell (1997) argues that although the EU is still organized in terms of national cultures, changes are taking place at the macro (state) and meso (groups) levels, influenced by corporatist/economic, formal institutional, and cultural-historical linkages; these changes will offer the opportunity to achieve greater European convergence/unity.

12 We have avoided the term "mass public" because it was created in the fifties to denote an emerging media public in terms of national publics. Neither the mass, nor the nation, nor the media consumers are coextensive entities in present-day societies, thereby rendering this concept useless.

This popular public consists of networks of local or regional groups organizing themselves contentiously to protect local ecosystem, health, and economic interests, as shown in the chapters by Kousis (chapter 5), Figueiredo, Fidélis and Rosa Pires (chapter 8), as well as Barcena and Ibarra (chapter 7). These local groups represent a wide variety of informal as well as formal groups, such as residents and neighbors (usually leading groups), local governments, labor and trade unions, political party representatives, cultural activity clubs, environmental groups, and many others. Other popular publics may be less vocal in their responses to the socio-economic and political changes affecting their daily lives. One such public resides in rural areas and is engaged in the agricultural sector (Daoutopoulos, Pyrovetsi and Petropoulou). In this case, the destruction of a community mode of production and the imposition of chemical agriculture by powerful international and national actors has made local family-farmers[13] dependent on intensive cultivation and undermined local initiatives or institutions caring for the environment.

Popular publics have so far received less attention in research and public discourse on European environmental politics.[14] They easily disappear given the cultural dominance of elite publics interacting with national publics in European politics. A more sociological view of Europeanization can correct this bias.

4. INSTITUTIONS

Emerging institutional links between elite publics, national publics, and popular publics are addressed in chapters 9 through 15. Usually these participatory institutions involve environmental policy actors such as environmental state agencies, national or regional environmental organizations, political parties, groups of experts/professionals, local environmental groups, and organized economic interests (Jiménez; Aguilar Fernández; Gil Nave; Louloudis, Arahoviti and Papadopoulos). In addition, national or regional courts and the media play an increasingly important role in the institutionalization of the environmental field (Pridham). Branches of large international

[13] The adoption of alternative/environment-friendly methods necessitates technical as well as economic support which is still very limited for the small farmer who is not in a position to undertake economic risks that would endanger the survival of the family-farm (Kousis and Petropoulou, forthcoming).

[14] Only very recently have researchers become interested in the study of grassroots movements in a European perspective. This book is, in part, a reflection of this concern.

environmental associations, e.g. WWF or Greenpeace, have also played an important role in the past decade (Jiménez; Aguilar Fernández; Kousis). During the eighties, professional groups[15] held a strong position vis-à-vis other policy actors, while in the nineties, environmental associations – thanks to their increasing number, size, and importance – were able to enhance their status. Such emerging institutional forms of coordinating collective actors in the arena of environmental politics tend to have an intimate and intricately intertwined relationship with transnational actors on the European level, thus creating a transnational field of political action and policy-making.

The institutional change fostered by European institution building and advocated by transnational elite actors and national state actors has modified the opportunity structure for popular publics. The nation-state is no longer the only legitimate institution approached by popular mobilizers. The European Commission, the European Parliament, and the European Court of Justice have gained importance during the past decade, as the steady increase in the number of local contenders approaching them illustrates (Barcena and Ibarra; Kousis). Although at times successful, their demands tend to be rejected and thus, in most cases, the evidence suggests that they are more likely to be excluded. Paradoxically, since the late eighties, participation in European policy games has reduced participation in national policy games. Participation has become a two-tiered game in which the pay-off favors the dominant groups.

5. DISCOURSES

In these emerging institutional arenas, where the activities of elite and national actors are coordinated, while the people act as an observing and sometimes protesting public, metaphors symbolizing a shared transnational world become important. One of these metaphors is sustainable development. It can assume different political colors. It is either conservationist or ecologist in orientation, depending upon the actors, their various constellations, and their mediation in a particular media landscape (Diani; Trianda-

[15] For an interesting paper on the feasibility of environmental policy-making in the Mediterranean see Haas (1989). His explanation emphasizes the centrality of collective actors as epistemic communities. Thus, we again have an elite-centered explanation.

fyllidou).[16] At the same time, sustainability creates a sense of a common goal, a sense of commonness. Here the discourse of sustainability gains its eminently ideological function.

The sustainability discourse defines what is seen as a relevant 'problem' and defines the goals that have to be reached in order to solve a problem. It also serves to create a collective identity beyond naked interests among those involved in the dynamics of environmental politics. Sustainability is a narrative well suited to distinguishing the good from the bad guys in environmental matters. Since Europe has taken the lead in this discourse, it becomes a space of debate and conflict over this issue.

Yet, the question remains: What is the social function of sustainability discourses. Is sustainable development the dominant discourse, stabilized by European expert discourses? Is it a discourse driven by economic interests? Or might it even be a discourse imposed on elites by popular collective action?[17] The environment is certainly a conflictual discursive field in which sustainability discourses force other discourses to adapt. Whether such dominant discourses assimilate or exclude other discourses depends on the configuration of power in Europe.

An important part of the analysis presented in this book concerns the way in which movements and policy-makers make sense of themselves and produce an image that is communicated to other relevant actors in the field. The discourse of sustainability is not necessarily divisive. It is rather to be expected that it creates counter-discourses that follow local, regional, or national images of nature. Again, we are confronted with a paradoxical effect of Europeanization: the more the boundaries of legitimate discourse on the environment are defined by the dominant discourse, the more local, regional, or national counter-discourses are to be expected[18].

Sustainability discourse has both formal and informal aspects to it (Briassoulis). In its formal, institutionalized form, it is strongly linked with eco-

[16] The debate on sustainability is overwhelming. For some orientation see the contributions to Brand (1997), Baker et al. (1997), O'Riordan (forthcoming) or the classic by Redclift (1987).
See also his contribution to this volume as well as the different versions of what sustainability might mean presented in the other chapters of this volume. No attempt has been made to produce an artificial unity of this discourse. Its unity is the diversity of voices oriented toward a general normative idea.

[17] According to Golub "competitiveness objectives have delayed and in some cases prevented environmental regulation at the national level ... reconciling these conflicting objectives through collective international action has proven both politically and legally difficult" (Golub 1998: 25).

[18] This is not the only route that exists. Bottom up discourses may initiate sustainability discourses which in turn lead to counter-discourses from top groups.

nomic growth and the priorities of competition, thereby reflecting concerns of the North (Redclift; Aguilar Fernández; Daoutopoulos, Pyrovetsi and Petropoulou; Eder). Thus, tensions in the direction this discourse takes are evident when examining particular sectors, such as tourism. Business lobbies, environmentalists, and civil servants in Brussels adhere to different principles and targets within the sustainability discourse (Ruzza).

At a transnational/global level, this prevailing discourse includes the aim of maintaining or improving the economic wellbeing of the powerful North, while simultaneously enhancing the North's own competitive position vis-à-vis other economic blocs (Redclift). Simultaneously, since increasing competition on a global scale intensifies both formal and informal economic activities, the related sustainability discourse overlooks the real conditions of political economy (Briassoulis). For example, such discourses have not incorporated activities with serious impacts on local and global ecosystems – such as the illegal dumping of large amounts of toxic/hazardous waste.

The sustainability discourse at the nation-state level also reflects power relations, albeit in a geographically more limited and socially more restrained context. This discourse usually involves business interest groups, civil servants, the government, environmental organizations, a variety of publics, and, after entry into the EU, European Union agencies. Given the primacy of economic concerns and the target of economic harmonization with northern states, southern European governments, in collaboration with powerful economic interests in their countries, have in general opted for a more economically framed sustainability discourse (Aguilar Fernández; Gil Nave; Jiménez; Kousis; Figueiredo, Fidélis and Rosa Pires; Pridham 1994).[19] At the same time, focusing on smaller scale state and economic interests reveals – insofar as informal activities are concerned – the existence of a silent un/sustainability discourse, signaling tolerance of economic priorities (Briassoulis).

Simultaneously, signs of convergence have appeared especially in the past decade between state agencies and environmental organizations (Aguilar Fernández; Gil Nave; Diani; Triandafyllidou; Louloudis, Arahoviti and Papadopoulos; Jiménez; Kousis). However, the sustainability discourse of southern states does not appear to identify strongly with that of the northern ones, which have shaped the environmental policies of the EU (Aguilar

[19] In Spain and Greece, where the motives of economic growth prevail, this is more evident than in Italy, which modernized earlier. Nevertheless, this is not a discourse peculiar to the southern states. Even for countries such as Denmark, Germany, and the Netherlands, the economic costs of implementation have been stressed as an important factor leading to 'patchy' compliance (Pridham 1996: 54).

Fernández; Daoutopoulos, Pyrovetsi and Petropoulou; Jiménez; Gil Nave; Pridham).

Popular publics adhere to a succinct sustainability discourse. Although locals who protest for environmental protection do not usually refer to 'sustainable development' per se, they do demand the implementation of related policies and protective actions that are typical of sustainable development approaches. Their sustainability frames reflect interests focused mainly on family health, the preservation and protection of local ecosystems, and the maintenance of local economic resources (Kousis; Barcena and Ibarra; Daoutopoulos, Pyrovetsi and Petropoulou; Figueiredo, Fidélis and Rosa Pires). They are frequently, but not always, more concerned about local environmental issues – especially the residential groups. These pro-sustainability claims usually confront counter-claims from challenged groups, such as the state and producers, which emphasize economic values and downplay environmental ones (Kousis; Figueiredo, Fidélis and Rosa Pires).

Sustainability discourses, generated and fostered by constellations of Europeanizing actors, contribute to the symbolic representation of the structuring of an emerging European society. Local groups represent a class of people hit economically and socially by the consequences of economic change. This change expresses itself not so much in the loss of individual goods but rather in the loss of collective goods – as a result of pollution and other forms of environmental degradation –, coupled with the loss of social capital, i.e., of functioning social relations. National groups represent a class of people worried about its position in an increasingly competitive environment. This is the classic concern of the old middle classes which have always felt threatened in their relative position between the losers and the winners and which strive to protect their individual and, nowadays, even their collective goods. Finally, elite groups organize themselves increasingly as transnational, i.e., European groups. As epistemic communities or other carriers of an information and knowledge society, these groups take over the cultural and social hegemony that national elites are forfeiting in the course of Europeanization.

The environment – to repeat again – is not only an ecological problem; it is in the end a social (or even societal) problem. It is a collective good that serves as a new medium for rearranging social relations between groups, thus rearranging relations of power and restructuring forms of social inequality in an emerging European society.

6. THE REPRODUCTION OF CLEAVAGES: IS THERE A MEDITERRANEAN SYNDROME?

Finally, the contributions to this book share a common assumption that there is a difference between the North and the South which is reflected in different relationships either between the EU and its respective northern and southern member-states or between the EU and the South. The basic narrative regarding this cleavage is that the North is the leader and the South the laggard.

Given our model of a Europeanizing society, we can restate this narrative and invite the reader to consider an alternative narrative of the European South. Southern European discourses, as presented in the chapters of this volume, seem to follow a pattern according to which elite discourses on sustainability are coupled with local counter-discourses on how to improve it. What makes these discourses distinctive is the power relationship expressed through the adherence to such discourses. Sustainability is more costly for some and less so for others. Is the South paying for the North?

This alternative narrative can also be embedded in the long tradition of a system of cleavage structures in Europe. Since its modernization,[20] Europe has been torn by structural cleavages that either follow the East-West divide or the North-South divide. The East-West divide has involved mainly war, occupation, military confrontation, and power. The North-South divide is more cultural and, for the smaller southern states, more economic. It follows the religious differences separating Reformation from Counterreformation Europe. It separates republican Europe from authoritarian Europe, agrarian from industrial Europe. These cleavage structures have survived European integration and still influence the institutional forms of emergent European polity and society.

However, these cleavage structures are in constant flux. The factors that change them have to do with the effects of modernization: the shift in family structures toward single-child households, the rise of service industries, the expansion of secular culture, and last but not least changes in the relationship of man to nature. Such changes also affect the old cleavage structure of Europe. Behind the institutional changes which we associate with European integration and the new institutional forms of doing politics, there is an even more profound change: the remaking of the social basis of European economic and political institutions.

[20] Some – such as Mann (1986) – even claim that these cleavage structures existed prior to modernization.

The question that arises from such a perspective is to what extent the old cleavage structures will survive these changes? Do we reproduce the old cleavage of North and South, which in the dark ages was one of barbarism versus culture, tribes versus civilizations, and which in the course of nation-building was then was turned on its head as the South began to lag behind the North economically? Contemporary discourses still believe that such a difference shapes Europe and its economic and political integration.

The most explicit statement of this idea is the thesis of a 'Mediterranean syndrome' which – based on the differential compliance of member-states to EU-directives in environmental matters – reproduces the lag-theory used to understand modernized Europe. According to this theory, the early modern cleavage structure still holds. Such claims are supported by some empirical evidence. They will come up in nearly all the papers.

An alternative narrative is the counter-narrative of the South 'paying' for the North.[21] This narrative is based on the observation that the North is intricately intertwined in the environmental and economic politics of the South. Certain EU development policies in southern Europe not only fail to protect the environment, but also cause substantial environmental damage (Close 1995). This is not a peculiarity of the South. Liberatore (1997) argues that even though officially endorsed, the integration of environmental objectives in all EU policy-making areas "is facing various barriers that could lead to the dilution rather than the integration of sustainable development objectives into EU sectoral policies and across sectors" (Liberatore 1997: 124). She further argues that "...background conditions or vested interests can impede institutional development, the utilization of aspects of scientific evidence and the availability or otherwise of technological or financial resources" (Liberatore 1997: 121). Thus, although lack of capacity can be improved over time, it is the related problem of distribution between different groups and countries within and outside the EU that remains a decisive issue.

Northern European policies that have affected southern members include: the utilization of the natural resources of the South, the control of large-scale Mediterranean tourism activities by multinational corporations based in the North, the importation of northern European toxic waste into southern European countries (Jiménez, in this volume, Kousis 1994), the support of mega/infrastructural projects (e.g., the diversion of riverbeds; the

[21] Narratives as well as counter-narratives are interpretations of facts that orient the cognitive ordering of facts. This holds equally for public discourses as well as for scientific discourses. Scientific discourses add to this ordering the confrontation with facts which leads to a particular interaction of narrative interpretations and factual evidence. This is our way of navigating between the Skylla of interpretation and the Charybdis of empiricism.

building of highways), and the exportation of northern technologies (relating to environmental protection and ecological modernization) to southern Europe. Thus, we propose to take into account competing narratives in Europe that shape our political and cultural (including scientific) interpretations of the difference between the North and the South. Both narratives have some empirical evidence on their side.

We leave open – in this introduction – the question of whether the empirical evidence presented in this volume will corroborate one or the other narrative. An obvious empirical problem, however, can be stated at the outset: the differences between southern European countries are such that the claim of a unifying pattern will always run into problems. What one might conclude from such general observations and from the results presented in the papers to follow, will be discussed in the conclusion.

7. A REMINDER

We are not arguing that Europe is a bad invention. Nor are we Euro-skeptics, opposed to European integration. The Euro-skeptic discourse differs from ours. Given the analytical perspective presented above, one can assume either a skeptical or an optimistic attitude toward European integration. Our aim is to avoid a naive theory of social reality that identifies what happens with what political ideologies tell us happens. Our perspective rejects this naive view, trying instead to understand what happens and to identify the options (or non-options) available to actors in a Europeanizing society.

REFERENCES

Aguilar Fernández, S. (1993) Corporatist and statist designs in environmental policy: The opposing roles of Germany and Spain in the community scenario, *Environmental Politics* **2**, 223-247.

Bache, I. (1998) *The Politics of EU Regional Policy*, Sheffield Academic Press, Sheffield.

Baker, S., Kousis, M., Richardson, D., and Young, S. (eds.) (1997) *The Politics of Sustainable Development. Theory, Policy and Practice in the European Union*, Routledge, London.

Baker, S., Milton, K., and Yearley, S. (eds.) (1994) *Protecting the Periphery: Environmental Policy in Peripheral Regions of the European Union*, Frank Cass, Newbury Park, CA.

Börzel, T. (2000) Why there is no Southern Problem. On Environmental Leaders and Laggards in the European Union, *Journal of European Public Policy* **7**, 141-162.

Brand, K.-W. (ed.) (1997) *Nachhaltige Entwicklung - eine Herausforderung an die Soziologie*, Leske + Budrich, Opladen.

Briassoulis, H. (1999) Sustainable development and the informal sector: An uneasy relationship? *The Journal of Environment and Development* **8**, 213-247.

Close, D.H. (1995) Why the European Union's development policies threaten the environment: A Greek case study, in J. Perkins and J. Tampke, *Europe: Retrospects and Prospects*, Southern Highlands Publishers, Manly East.

Eatwell, R. (ed.) (1997) *European Political Cultures: Conflict or Convergence?* Routledge, London.

Eder, K. (1996) The institutionalization of environmentalism: Ecological discourse and the second transformation of the public sphere, in S. Lash, B.Szerszynski and B. Wynne, *Risk, Modernity and the Environment: Towards a New Ecology*, Sage, London, pp.203-223.

Eurostat (1996) *Environment Statistics 8A 1996*, Statistical Office of the European Communities, Luxembourg.

Flora, P. (ed.) (1999) *State Formation, Nation-Building, and Mass Politics in Europe: The Theory of Stein Rokkan*, Oxford University Press, Oxford.

Golub, J. (ed.) (1998) *Global Competition and EU Environmental Policy*, Routledge, London.

Gunther, R.P., Diamantouros, N., and Puhle, H.-J. (Eds., 1995) *The Politics of Democratic Consolidation: Southern Europe in Comparative Perspective*, The Johns Hopkins University Press, Baltimore and London.

Haas, P.M. (1989) Do regimes matter? Epistemic communities and Mediterranean pollution control, *International Organization* **43**, 377-403.

Kousis, M. (1994) Environment and the state in EU periphery: The case of Greece, *Regional Politics and Policy* **4**, 118-135.

Kousis, M. (1998) Ecological marginalization in rural areas: Actors, impacts, responses, Sociologia Ruralis **38**, 86-108.

Kousis, M. (1999) Sustaining local environmental mobilisations: Groups, actions and claims in Southern Europe, in C. Rootes, *Environmental Politics, Special issue: Environmental Movements: Local, National and Global*, pp.172-198.

Kousis, M., and Petropoulou, E. (in press) Local identity and survival in Greece, in T. O'Riordan, *Globalism, Localism and Identity: Perspectives on the Sustainability Transition in Europe*, Earthscan, London.

Kriesi, H.-P. (1995) The political opportunity structure of new social movements: Its impact on their mobilization, in J.C. Jenkins and B. Klandermans, *The Politics of Social Protest. Comparative Perspectives on States and Social Movements*, University of Minnesota Press, Minneapolis, MN, pp.167-198.

Kriesi, H.-P., Duyvendak, J.W., Giugni, M., and Koopmans, R. (1995) *New Social Movements in Western Europe. A Comparative Analysis*, University of Minnesota Press, Minneapolis, MN.

La Spina, A., and Sciortino, G. (1993) Common agenda, Southern rules: European integration and environmental change in the Mediterranean states, in J.D. Liefferink, P.D.Lowe and A.P.J. Moll, *European Integration and Environmental Policy*, Belhaven Press, London, pp.216-234.

Liberatore, A. (1997) The integration of sustainable development objectives into EU policy-making: Barriers and prospects, in S. Baker, M.Kousis, D.Richardson and S. Young, *The Politics of Sustainable Development: Theory, Policy and Practice within the European Union*, Routledge, London, pp.107-126.

Liefferink, J.D., Lowe, P., and A.rthur P. J. Mol, (Eds., 1993) *European Integration and Environmental Policy*, Belhaven Press, London.

Mann, M. (1986) *The Sources of Social Power. A History of Power from the Beginning to A.D. 1760 (Vol. I)*, Cambridge University Press, Cambridge, MA.

McAdam, D., McCarthy, J.D., and Zald, M.N. (eds.) (1996) *Comparative Perspectives on Social Movements: Political Opportunities, Mobilizing Structures and Cultural Framings*, Cambridge University Press, Cambridge, MA.

O'Riordan, T. (ed.) (in press) *Globalism, Localism and Identity: Perspectives on the Sustainability Transition in Europe*, Earthscan, London.

Pridham, G. (1994) National environmental policy-making in the European framework: Spain, Greece and Italy in comparison, in S. Baker, K.Milton and S. Yearly, *Protecting the Periphery: Environmental Policy in Peripheral Regions of the European Union*, Frank Cass, London, pp.80-101.

Pridham, G. (1996) Environmental policies and problems of European legislation in Southern Europe, *South European Society and Politics* **1**, 47-73.

Pridham, G., and Konstadakopulos, D. (1997) Sustainable development in Southern Europe? Interactions between European, national and sub-national levels, in S. Baker, M.Kousis, D.Richardson and S. Young, *The Politics of Sustainable Development: Theory, Policy and Practice within the European Union*, Routledge, London, pp.127-151.

Redclift, M. (1992) *Sustainable Development: Exploring the Contradictions*, Routledge, London.

Ward, N., Lowe, P., and Buller, H. (1997) Implementing European water quality directives: lessons for sustainable development, in S. Baker, M. Kousis, D.Richardson and S. Young, *The Politics of Sustainable Development: Theory, Policy and Practice within the European Union*, Routledge, London, pp.198-216.

Whitehead, L. (1996) Book Review on The Politics of Democratic Consolidation by Gunther et al. (1995), *South European Society & Politics* **1**, 139-141.

I

THE EUROPEANIZATION OF ENVIRONMENTAL POLITICS

Chapter 1

Sustainability as a Discursive Device for Mobilizing European Publics

Beyond the North-South Divide[1]

Klaus Eder

1. THE EUROPEANIZATION OF ENVIRONMENTALISM

European environmental policies are supposed to be good for the environment and should succeed in changing the environment in Europe (and elsewhere). This is of interest to policy makers and policy researchers who try to identify wrong policies, i.e., wrong means to realize (even good) ends, as well as gaps in the implementation of good (and bad) policies. To understand such effects it is necessary to go beyond the theoretical space accounted for in most research of the last ten years. This chapter will look at the embeddedness of such policy-making in a societal context that is itself in a process of transformation.

This societal transformation can be examined from two analytical perspectives: an action theory perspective and a structural perspective. The action theory argument states that more and more social actors enter the field of environmental struggle, including not only those who want to do something for the environment, but also those who want to take advantage of the fact that some actors care for the environment. The structural perspective refers to the claim that this social field of environmental struggle increasingly transgresses the national space and is transformed into a Europeanized space. Furthermore, environmental politics in Europe is a process of increasing mobilization of groups in the field and, as such, is part of the making of a European society. The action theory argument leads to the hypothesis that we will have increas-

[1] I have to thank Maria Kousis for critical comments on this paper which helped to improve it considerably. Nevertheless the responsibility for its content remains mine.

K. Eder and M. Kousis (eds.), Environmental Politics in Southern Europe, 25–52.

ing conflict over the environment in Europe. The structural argument engenders the hypothesis that Europe is confronting either a 'self-organizing European society' or a 'European society made in Brussels'.

Environmental policy-making is therefore considered in theoretical terms as a social field in which the structuration of Europe is at stake. The basic thesis is that environmental policy-making produces not only opportunity structures for collective action on the environment, but also new structural arrangements including new cleavage structures in an emerging European society Environmentalism is thus more than a particular policy issue. It is also a mechanism fostering the evolution of a European society.

The theoretical idea of Europe as an opportunity structure refers to a well-established concept in social movement research. It is used to explain the emergence of collective action and the conditions under which collective action can be sustained.[2] Such opportunity structures are constraints that foster or inhibit collective action. Thus, the task is to identify those opportunity structures which – depending on the question at hand – either foster or inhibit collective action. The assumption is that Europe provides a particular opportunity structure through its financial incentives as well as through an extended system of committees and consulting bodies.[3]

What is beyond the analytical reach of such concepts is the process of formation of collective action and the structuration of the space of collective action. This second theoretical perspective puts into perspective the self-organization of collective action within given social, political, and cultural contexts. Such self-organization not only solves the problem of collective action; it is also the basis for the structuration of the transnational space constituted in such collective action. This second theoretical perspective therefore goes beyond the questions of the origins of collective action toward the question of the structuration of the space opened up by such action. Collective action is therefore seen equally as a dependent variable (with regard to opportunity structures) and as an independent variable (with regard to the structuration of a social field).

This twofold analytical perspective locates Europe at the beginning (origins) and at the end (structuration) of collective action. Europe is to be seen not simply as determining environmental collective action; Europe is

[2] The political process model has made this rather static view more dynamic. For a good summary and presentation of these issues see Tarrow (1998).

[3] This system is due to the horizontal structures in the organization of the decision-making processes in the emerging European system of "governance" (Majone 1996; Kohler-Koch and Eising 1999).

also a means (or tool) of collective actors for organizing collective action on the natural environment.

Within this analytical model it is easier to situate those actors who make use of opportunity structures and those who produce structural effects which, in turn, constrain their action. In social movement research a distinction is made between activists, bystanders, and the concerned people. This classifies a particular type of social actor not according to the person, but according to the roles the person assumes. What all these roles have in common is that they involve actors in public. Of interest is not what they have in their minds as individuals, but what they communicate to others. They constitute what we will call 'publics'. These publics act in the social space of increasing interaction and communication which comprises emergent European society. Our question, therefore, is: In what way and to what extent do environmental issues lead to the structuration of European publics?

2. EUROPEAN PUBLICS – CONCEPTUAL AND HISTORICAL STARTING POINTS

2.1 A model of three publics

What are the European publics? These publics are defined as the set of social actors who organize their action space in a way that takes into account Europe as a relevant context for their action. The public at large is the European people which remain an amorphous mass that exists in terms of national citizens who have the right freely to cross the borders of the internal market, to work within that space, and to move goods within that space – ideally – as they like.[4] The European public is the aggregate of people sharing some (four) basic liberties.

There is little that binds this public together beyond such basic liberties: they do not watch the same TV programs (with the exception of Music channels which, nevertheless, are a youth phenomenon and therefore a transitory commonality); they do not speak the same language; they do not eat or drink the same things (Italian cooking in Germany is different from that done in Italy which, in turn, does not exist because cooking is highly regionalized).

[4] Such idealization presupposes a reality in need of such idealization, i.e., a system of internal closures.

The European public exists not as a homogeneous body, as a European 'nation'.[5] It exists as a plurality of publics. We will distinguish three types of publics within this amorphous mass of people. First, the European public is a series of *national publics*. National publics share a particular public space, which is reproduced by national media communication. The media (and related mechanisms of cultural reproduction such as the national army or the national school) make people into a national people. What such national publics finally share is that they read national newspapers and watch national television. The reference of such national publics to Europe is determined by a supposed national interest which sees Europe as a potential advantage to the national interest.[6]

Another type of public is the *elite public*. Elite publics are social groups often transgressing national boundaries and organizing themselves in functionally differentiated spaces. Such elite publics include industrialists, NGOs, cultural entrepreneurs (intellectuals), or professional groups. These publics come closest to the idea of a modern society. They construct functionally specific references to Europe. They organize their ties through networks, thus constituting professional publics. Central for such publics is a non-national, functional referent for identifying elite publics.

A third type of public is the *popular public*.[7] Popular publics are aggregates of people which follow interests related to concrete life experiences, i.e., in most cases (but not exclusively) work. Popular publics in working class neighborhoods differ from those in middle class suburbs; and the publics of urban communities differ from those in agrarian communities. Popular publics can be classes, but don't have to be. They can be less than classes – for example 'milieus' –, depending upon the degree of closure that such groups have developed.[8]

[5] The invention of the "First New Nation", as Lipset (1979) has called the American polity and its publics, would be an example for the creation of a homogeneous public out of heterogeneity, except for the fact that such a process is unlikely to take place in Europe in the foreseeable future.

[6] Such national interest is more than the interest of a utility maximizer. It is full of historical memories and carries with it a collective identity.

[7] I avoid the term 'mass public' which was created in the Fifties to denote an emerging media public in terms of national publics. See, e.g., the important research done by Converse (1964). Neither the mass nor the nation nor the media consumers are coextensive entities in present-day societies, thus limiting the usefulness of this concept.

[8] This is consistent with Max Weber's definition of social classes. Instead of arguing that classes are waning, we assume that there is an empirical variation between class and milieu depending upon the degree of closure.

2.2 European publics and environmental policy-making

These publics are exposed and react to (among many other things) European environmental policy-making. Instead of attributing the success of environmental policies to national characteristics, such an assumption allows us to account for varying reactions and modes of adapting to the institutional constraints imposed by the European Union.[9] The term institutional constraints denotes, firstly, substantive legal constraints (legal obligations). Secondly, it denotes procedural rules to comply with in order to profit from Community funds or access to power. Thirdly, it denotes ideas such as sustainability that work as strong symbolic constraints.[10]

To date research has concentrated on the institutional constraints imposed by policy actors. This is by definition an elitist perspective, involving Eurocrats and national governmental actors with a stake in transnational policy-making. At times, transnational corporations and even non-governmental organizations were added, thus generating an image of Europe as a corporatist regime (Ruzza, in this volume).

Within this regime an elite public is in the making. These elites judge the member-states according to their performance. European policy-making makes them comparable and fosters a process which draws invidious distinctions between the member-states. Research on the European Union itself has contributed to this game of making distinctions by producing a series of hypotheses regarding the comparative success of environmental policies in some member-states of the Union as opposed to others. In particular, by taking the lead in defining sustainability on the global level the Commission has defined an important parameter in this process of collective differentiation by separating the member-states into leaders or laggards in terms of national compliance with European legal norms. The most prominent hypothesis is that the Southern countries are particularly deficient regarding the implementation of European environmental norms. The explanation is that Southern countries share characteristics such as a civic culture based on familistic and/or clientelistic relationships, held together by corrupt and fragmented political parties engaged in eternal struggles between their frac-

9 The term 'institutional constraints' implies a double bonding device: organizational force and symbolic involvement. Institutions are the stronger the more they can exert constraints on both levels in a meaningfully coordinated way.

10 This notion of institutional environments is consistent with its use in neo-institutionalist theorizing. This approach is well expounded in March and Olsen (1989) and Powell and DiMaggio (1991). For its political implications in terms of democratic governance see March and Olsen (1995).

tions. Northern countries, on the contrary, are republican, based on non-familistic abstract relationships, and have rationally coordinated parties controlled by a strong independent legal system and corporatist arrangements which guarantee the compliance of major collective actors in society. But theories which draw on national cultures to explain differential compliance run into difficulties when confronted with comparative evidence. There is no such commonality as the 'South' or the 'Mediterranean syndrome', nor the 'North' or the 'Atlantic syndrome'.

The implementation of environmental policies in the different member-states varies in terms of national differences, as research into the compliance of member-states with European environmental regulation has shown (Börzel 2000). Germany is in some areas much less of a vanguard than a laggard.[11] Portugal has implemented to a higher degree than Britain. National publics behave selectively toward European rules, a selectivity which is due to national institutional ideas and organizational forms.

Popular publics, the third element, do not really enter into this type of reasoning. They are considered as irregularities in a regular pattern, as deviant cases (or as utopian cases – depending upon ideological preferences). They is the surplus that cannot be handled by national or elite publics. They are simply noise in terms of publics observing policy-making processes. Thus, there is no place for those publics which are defined neither by some elite connection nor by their nationality, but rather by their life experiences. Popular publics, be they lower or middle class, are a reality that do not count in the current models. This neglect has to be corrected in analyses of the dynamics of a (non)greening of Europe.

2.3 Sustainability in the discourse of the three publics

Instead of looking at the implementation of environmental directives we propose to address the use of the notion of sustainability by these different publics. This proposition is based on the assumption that institutional rules for complying with prescriptions are shaped less by national traditions than by specific modes of assimilating the rules of appropriate behavior in environmental matters. These rules are themselves institutionalized in the discourse on sustainability.

There is however no clear consensus among the different publics of what the notion of sustainability as a principle of environmental regulation really

[11] A good example is the EIA, the Environmental Impact Assessment Directive of 1985, which Germany still has not brought into formal compliance with its national law.

means.[12] Elites have developed a European interpretation of sustainability which is used as a 'European' position in global environmental negotiations. National publics adapt the elite discourse on sustainability to the needs of a nationally defined institutional system and legitimize member-state's behavior toward European rules.[13] Popular publics, by contrast, use sustainability in terms of local needs and experiences. This leads to a notion of sustainability that is much less technical and much more social, given the experience contained in local traditions and institutions of collective action (Kousis 1993).

Given the increasing interaction among these different publics through the medium of sustainability discourses, we would expect an increasing institutional isomorphism between these publics.[14] In spite of some empirical evidence for such isomorphism in Europe, an irritating observation remains: sharing sustainability discourses does not lead to similar outcomes in terms of implementing sustainable policies; nor does European rule-making lead to organizational isomorphism among the member-states, i.e., toward organizational convergence in the field of environmental action.

Sustainability – our argument will run – is a buzzword that has gained prominence across the different publics. Created as a device within elite circles, it has transgressed the elite boundaries and become a master frame for national publics as evidenced by the media attention which this concept has gained. Popular publics react to such frames within their own practical frames and attach a value to it, thereby orienting their environmental action. Thus, we have to expect three relatively autonomous reactions. European environmental legislation in the name of sustainability is the elite answer. National public monitoring (including the political representatives of people who depend upon such national monitoring) in the name of sustainability represents a second reaction. A differential mobilization of social classes which live in their own popular publics comprise a third reaction in the name of sustainability. The following argument proposes to embed the discourse

[12] This is even characteristic of academic debate. For an overview of the state of the debate see Zaccaï (1999). Historically, it was used by a tiny group of forest scientists before acquiring wider public resonance as an environmental bandwagon concept. It has close links to the idea of modernity as such and is therefore subject to the same criticisms as this concept (Sachs 1997; Conrad 1997).

[13] This is also emphasized by Richardson (1997). For its particular use in the Mediterranean see Chabason (1999).

[14] Institutional isomorphism, a theoretical argument proposed by DiMaggio and Powell (1983), is to be expected when organizational environments enter into increasing exchange relations among each other. Taking nation-states, popular life worlds, or elite publics as organizations, Europeanization would lead us to expect institutional isomorphism.

on sustainability in a social reality in which elites, nations and the people play different roles.

3. ELITE PUBLICS AND SUSTAINABILITY

Sustainability has become, despite and even because of the problem of translating the term into the national languages, a European concern. Elite publics have authored and propagated this term.[15] The social space which these elites provided was thus not confined to national spaces. It was disseminated in functionally defined action spaces, where experts meet experts, professionals other professionals, and intellectuals other intellectuals.

Such action spaces can coincide with national ones. Historically, this has been the case at the height of nationalism when the economy and high culture where seen to be in the service of the nation. Today, such coincidence is no longer normal. Economic elites define their action space in terms of their interest in finding consumers for the products they produce. Cultural elites define their action space in terms of literary resonance that mobilize the subcultures that crosscut national cultures. The effects of competition and the resulting coordination of these publics in more encompassing action spaces is becoming a central feature of the emerging European society.

The cultural cleavages that result from the attempts of these groups to influence EU policy-making are less known. We encounter observations of closure strategies on the part of professional groups against those who offer their services more cheaply. Such elite competition increases to the extent that neo-liberal rules and ideologies dominate the emerging European action space. This competition does not lead to a North-South divide. Even different national cognitive styles are replaced by shared interests and identities that emerge from extended markets for goods, knowledge, and taste. This hypothesis will be made plausible looking at some data regarding NGOs and the industrial actors in European society.[16]

The analysis of the institutional field in which the environment is an issue is characterized by the limited plurality of collective actors having different interests in this issue. The issue of the environment bundles different

[15] Regarding the role of economic elites in the construction of the sustainability concept see Donati (2000). The "integrative effort" of EU elites is well described and analysed by Liberatore (1997).

[16] The evidence will be taken from Eder (1995). Additional evidence is needed for professional and expert publics. An extensive discussion would, however, go beyond the limits of this paper.

interests, such as the private interest of the firm and the public interest of citizen groups, in a way that makes them dependent upon each other and that relates them to the interests of policy-makers. Beyond the specific interest in the environment, the diversity of interests is transformed into a general interest in not losing one's voice in the ongoing debate on the environment. The commonality of these interests cannot be kicked out of the game. This is the reason for the high politicization of the environmental issue in Western European countries.

The analysis of strategies of communication of consequential collective actors has enlarged the political field of environmental struggle beyond the traditional institutional boundaries of politics. Each actor has a stake in the communicative space, having invested in communicative practices which create an institutional order that can be called an emerging transnational 'postcorporatist order' (Eder 1996, part IV). This order is characterized by two elements: firstly, the inclusion of movement actors into the institutional framework; secondly, the perception – on the part of corporate actors – of a risky environment, requiring the investment of resources in non-economic institutions, i.e., in trust. These are the contours of a new elite-made and elite-dominated order in Europe.

These actors, bound together in a EU-specific institutional order, define the rules of the game for national actors. They reorganize and redistribute the symbolic capital in the field of environmental debate and politics. Elites create a social space in which the circulation of such symbolic capital is closed and tied to specific rules of access. Socializing into this elite social space means taking part in the definition and the practice of the rules of the game. Elite publics thus succeed in isolating themselves from the rest of society. They may even succeed in identifying themselves with society, transforming everyone else into individuals which either comply or do not comply.

Most studies on environmental policy analysis take this regime as given. In our view, it has succeeded in institutionalizing itself as a given. It has defined an institutional environment in which ideas of sustainability provide the symbolic basis of this institutional framework, defining its 'logic of appropriate action'. It has created organizational forms through which decisions have to pass. This makes it more than an epistemic community (Haas 1989). It makes it a policy network in which epistemic communities play an important role in its self-legitimation. Thus, the specific selectivities of epistemic communities are also part of the rules of the game in such regimes.

This amounts to saying that we have a European environmental regime that is more than a set of policies toward the environment. The environment does not affect the environment alone; it is also a medium for playing an elite power game. Its constraints have effects on what national actors can do.

Elite publics in Europe compete with national publics in Europe. Much of the national reaction to Europe is shaped by the reaction to those 'elites in Brussels'. This national backlash of Europeanization is part of the old antagonism between the people and power. National publics defends themselves against being assimilated into a European elite public. Since the national people do not act as such, those representing the people act for them. The proxies for the national people's society are their representatives, i.e., their governments. Governments thus act for and in the name of the people as 'national actors'. As long as the nation-state retains legitimacy in the eyes of the people, it will be able to act as a proxy for the people.[17]

For some governments, the environment is an important issue, for others less so. Comparative research on environmental policy-making is looking at the processes shaping such differential resonance – processes which are then used to explain the distinctive degrees of obedience to European rules. However, such comparative evidence remains descriptive. In order to explain national differences in terms of compliance with European rules, we have to identify the logic of the field in which national policy actors (we often speak also of 'countries') interact. 'Intergovernmental conferences' have added further options to the competitive game between national actors. Coalitions can be built, good guys and bad guys named and staged in order to strengthen national actors in a Europeanizing action.

An important – and now old – structure stabilizing this field of transnational competition and co-operation within Europe is the distinction between Northern and Southern European countries.[18] We are accustomed to the social representations of such differences in terms of the stereotypes we live by. The more European boundaries are drawn and structural differentiation takes place within the emerging European society, the more this difference becomes part of the rules of the political game between national actors. The Southern European countries become the European South. The stereotype for the other is transformed into a mechanism of assigning inferior status within an emerging society. The stereotype is transformed into a structural difference in transnational action space. The Greek or Italians or Portuguese

[17] This does not imply that the people of the nation-state agree. They often have different interests and seek to have their interests represented through non-institutional and non-governmental channels. This will be taken up by looking more closely into what we have called popular publics.

[18] It is part of the cleavage structure that Rokkan has identified as determinant for the historical dynamics if Europe (Rokkan 1975).

are no longer the poor other, the romantic outsider, but the bad environmental guy who is not willing to learn from the Northern good guys. The European rules of the game have thus found a particular and historically deep-rooted structural ground on which to play the game.

4. NATIONAL PUBLICS OF EUROPEAN ENVIRONMENTAL REGULATIONS

The discussion of elite publics has already shown that the European game cannot be understood without taking into account the national governments which act as the proxy of the people. This national perspective has to be extended because national governments are themselves bound by a national game in which the legitimacy or illegitimacy of those representing the people is defined. To really understand the national publics we have to take into account what gives a voice to the people: the national media publics.

Media publics are still defined by their national newspaper and TV-channel cultures. The national publics made by the mass media are undergoing profound changes because of the increasing competition between channels, which is spurred on by audience research (audience quota). These media publics in Europe develop isomorphic structures on the level of organizational forms (the new privatized media world). However, their institutionalized way of representing reality (which is their basic function) is still basically national.

Comparative research on national variation in the media coverage of environmental issues is rare. There are difficult problems to surmount when understanding and coding media messages in different languages. The media cultures differ widely as far as the messages sent and received are concerned. Drawing on some research done in the mid-nineties, we can garner some information to help answer these questions.

One good indicator of national differences in resonance of environmental issues are the narratives on the Chernobyl accident in the national media. These stories vary in a ways that can give us clues about the internal closure of national media publics and about the boundary construction built into these media communication practices.[19] Data taken from an analysis of the media coverage of environmental issues in the years after the Chernobyl

[19] The data is taken from the project funded by DG XII, Research Area III, Project Number PL 210493. See the report on this data (Eder 1995). An excellent analysis of the national resonance to Chernobyl can be found in Liberatore (1999).

accident in France, Britain, Germany, and Italy show a system of differences in the symbolic construction of the environment. In each case we find a particular combination of a metaphor with a series of moral prescriptions. Such a combination has itself an internal structure which can be called a national story line. These story lines explain not only the initial reaction to the Chernobyl accident, but also the particular value given to environmental issues in a national culture. Four such story lines will be presented to back up the argument that stereotypes underlie the way in which national cultures communicate environmental issues through their mass media – in this case, through, their newspapers.

(1) French concerns for the environment are expressed in French newspapers through the metaphor of the "jardin français" (Le Saout and Trom 1995). French media stories showed high degrees of de-personalization and de-politicization in the attribution of responsibility. These impersonal accounts removed the search for agents responsible for radiation effects.

(2) British concern for the landscape employed a metaphor similar to that of the French (Statham and Szerszinski 1995). The resonating theme of the 'landscape' in need of protection and particular care is linked to another British particularity, the centrality of societies for the protection of the Scottish or the English landscape. This also marks a difference with respect to France, where the central administration of the French garden made explicit that it was the garden of the state, a park rather than a landscape where people were walking or could walk.

(3) German concerns for the environment were mainly linked to the forest or the river as romantic metaphors for the good and unpolluted environment. Up until the early 1990s, such metaphors were linked to the ideological threats posed communism (Poferl and Brand 1995). The conservative German media exteriorized blame on to the Soviet Union for having caused the suffering of the victims. Thus, moralization was a predominant feature in the media coverage on the environment. The East-West opposition, the idea of superiority, and the politicization of risk frames were stable features in a media discourse where ideology was a crucial building block of collective identity. Furthermore, the prominence of discourses of technological superiority and technical efficiency shows that German pro-nuclear energy policies were supported by a strong expert culture. In other countries the role of expert discourses had been replaced by attention to public concerns. In the media discourse of the Frankfurter Allgemeine Zeitung such an alternative notion emerged only timidly in 1991. Chernobyl became a warning sign not

only for the sins of the Soviet system, but also for the internal risks of progress.[20]

(4) Italian concerns about life-style allowed risk discourses to be denied and rephrased in terms of control (Triandafyllidou, Donati and Diani 1995; Triandafyllidou 1995, 1996). This might be explained by the fact that Italy had decided to ban nuclear energy and thus had already taken on responsibility in this sense. On the other hand, in the Italian media discourse the moralization of environmental issues in terms of their posing a threat to the life-world was similar to the discourse of the Süddeutsche Zeitung in Germany in earlier phases. The Süddeutsche Zeitung used this pattern to impose a pro-environmental, moral, and ethical interpretation on the general risks of progress. By contrast, in Italy such moralization had the function of symbolizing both the devastating effects of nuclear disaster and a better Italian society, given the delusions with the political system which, with the 'mani pulite' affair, has started to dominate public discourse (Triandafyllidou, in this volume).

Such data and their theoretical analysis is also supported by more quantitative measures which – as usual – are never as stringent as well organized qualitative data.[21] The data show that German and UK media coverage was much more coherent than the Italian one. Coherence is necessary to provide convincing answers to pressing problems. Incoherence is possible when thought and action are loosely related, when no practical action results from thought. The rather large difference between attitude and action in Italy observed in good surveys supports this interpretation.

The national media – it can be argued – provides a field for dealing with environmental issues which is based on a short, specific history: it is a story that emerged with the making of the nation-state and which serves to reproduce history as a meaningful story for all. It is assimilated to new concerns such as the environment.[22]

[20] This alternative interpretation separated the risk and political system frames, and questioned the controllability of risk and the limits to progress. However, this admission of risk was misleading, since it was used as an apparent concession to the critical discourse of opponents to nuclear power, while in reality the aim was to legitimate its use.

[21] Qualitative data, treated with care and according to the logic of scientific inference, produce much more valid data than any quantitative study. Quantitative data therefore are more of an additional device to ensure the reliability of qualitative data by reflecting it in quantitative results. Even this has limits, however, given the fact that the object of study changes constantly, not only by virtue of it being studied and observed, but also because of it being the object of reflexive action by actors. But in practical research this is a question of degree.

[22] Even the discourse of sustainability seems to reproduce this story line. The difficulty of translating the term into German, French, Italian and other languages is a formal indicator.

This comparative analysis of differences can be pushed further to help distinguish between national political cultures competing over the distribution of symbolic power. This symbolic power differentiation articulates itself in the form of 'stereotypes' for naming the typological particularity of a national actor. Such a 'stereotypological' approach (stereotypes understood as attempts to derogate the other thus exerting symbolic power) will normally end up with as many types as there are (national) cases. We will discuss shortly four cases of stereotyping: the French, English, German and the Italian case.

The French case concerns the stereotype of 'French modernism'. The elitist culture of French political institutions, together with the image of a technologically advancing power, has led to the image of an institutional order in which expert culture is far ahead of popular culture on any complex issue of regulating modern societies. The environment is a field in which this institutional order has produced the most impressive symbolic expressions, such as the French nuclear industry. However, this modernism rests on a tiny cultural base which periodically erupts into French institutional life. Such stereotypes lie at the base of a *technological regime*. In its hierarchical organization, its use of careerism as a mechanism to bind diverse collective actors into the game, and its retention of real power at top of the system, this regime leaves little space for collective action from below. High thresholds of collective mobilization are thus established and hence, any mobilization, when it occurs, will be simultaneously an attack on this institutional regime. Reference to technology as an elite knowledge guarantees the stability of its selective mechanisms.

The stereotype in the British case is 'English individualism' (which is mainly individualist utilitarianism). This individualism favors solutions to environmental problems by calculating the gains and costs of decisions. This stereotype has been fostered by the image of a country of environmental free-riders fostered during the Thatcher era. The institutional regime resulting from such a configuration may be called a *pragmatist regime*. British protest groups and movements as well as business and political actors have increasingly engaged in what can be called pragmatic co-operation, trying to maximize the interests of all. Consequently, conflictual styles of dealing with environmental issues are rare.[23]

The stereotype in the German case is 'German avant-gardism'. Such avant-gardism has been claimed by German collective actors and oftentimes

---

That it serves as an umbrella for diverging practices points to the multi-user friendliness of this term.

[23] Sellafield is probably the exception to the rule.

is even accepted by others as an adequate self-description (except by the English, who are the closest in terms of the type of institutional order dealing with the environment). This avant-gardism is often regarded as being linked to the German past and the historical quest for a new basis on which to construct a national identity. The environment as a substitute for traditional collective identity has been a widespread topos in political debate. What is true about it is that experimenting with 'discursive' forms of creating networks of decision-making is a German specialty. The institutional regime of environmentalism characterizing the German case can be called a *dialogical regime*. It moves problems and their solutions from one body to the next, thus trying to accumulate democratic legitimacy for whatever decision finally is taken. Dialogues bind the actors into the game of environmental politics.

The stereotype in the case of Italy is 'Italian exceptionalism'. Italy is considered and describes itself as different, as an exception to the pattern followed by other European countries because of the specific mode of functioning of its institutions. The permanent discourse on institutional reforms (which rarely happen) and the cumulative production of institutional rules without assimilating the new to the old ones, creates a Byzantine culture of public administration which can only be managed through the ingenious and playful competence of collective actors which are consequently regarded more as players than as institutionally committed actors. The regime characterizing the Italian case is a '*Popperian' regime* (in the metaphorical sense of following Popper's precept that complex societies cannot be planned and therefore there is no way other than 'muddling through'). However, the transaction costs of this type of regime are extremely high and generate its characteristic cyclical convulsions (the last one being the mani pulite phase of Italian institutional life).

This analysis leads us first to a deconstructive statement. National particularities are linked so closely to national stereotypes that any generalization is excluded in principle. Thus, there is no Southern style because national styles cannot be generalized. The constructive statement is that we have a convergence between national countries in terms of a transnational regime which is the unintended result of the interaction of national cultures. Even if there is a relationship of domination[24], the transnational regime of co-operating national publics produces its own institutional logic, i.e., the mobilization of national discourses for the transnational regulation of the

[24] It is difficult to answer the question of who dominates. The French? The Germans? The English? Or the Belgians? Since supranational legal forms force them to co-operate, one is well-advised to explain the forms of co-ordination of national publics as the unintended consequence of imposing a national style on the others.

environment. In such transnational forms the national story lines are retold. The environment is well-suited for retelling such stories because they refer to nature and culture, the body and the mind, the good and the bad, the dangerous and the familiar, pollution and purity. Such stories are always packed with moral messages, here with the message of a world based on trust and rationality fighting against a world of distrust and irrationality, a world based on cooperative political games against a world of cheating and self-interest.

European society emerges as a society which uses its national cultures to reconstruct a cleavage among themselves. This story then tells us about 'Southern' and 'Northern' cultures, which for the sake of their sharp boundaries are sometimes said to have strange cases at their (normally outer!) borders.

Thinking of environmental politics in terms of national publics leads us to a theory which sees environmental politics more as a form of symbolic politics than a practical effort to solve environmental problems. National publics talk the environment into being, and such discourses help to legitimate purposes as diverse as selling a national actor to his electorate or serving a nation's interest in the face of encroachment by other nations. Whatever the concrete symbolic use, environmental politics is driven more by symbolic factors than by practical necessities. For national publics an identity is at issue – this explains the dynamics in both the North and the South.

5. POPULAR PUBLICS

How do popular publics contribute to the structuring of environmental politics in Europe? We are accustomed to viewing popular classes in national terms. This is the historical model inherited from the heyday of the nation-state, when the nation-state was made identical with society. In describing the nation-state as the society in which class compromise is achieved under the shield of the (national!) welfare state, environmentalism becomes the property of a nation. This identification is, however, nothing but a consequence of the power game between political actors and social classes: classes have to be the nation. This model no longer holds. The middle classes transgress national boundaries to the extent that new markets open new fields of action for them. Agrarian classes discover common interests and common enemies beyond the national confines and within the national confines, thus creating a specific popular realm that follows its own logic and creates its on specific boundaries.

Thus, we are back to classes instead of nations. There is a first and classic hypothesis explaining the rules of the game involving popular publics in terms of classes. Petty bourgeois groups (the new and the old middle class) are attracted to environmental issues; environmentalism has even been called a petty bourgeois world view (a view which even has good historical arguments to back it up). The new urban middle classes are probably the best predictor of long-term environmental consciousness and readiness for action.[25] On the basis of this observation, it could be concluded that where we have more new middle classes, we will have more environmental politics. Since new middle classes are more developed in Northern European countries than in Southern European ones, we could take this as a structural indicator of a North-South divide.

The role of the middle classes in environmental matters is an hypothesis that is still open to debate. The hypothesis assumes that the middle classes have a stronger life-world orientation than the lower classes (who are, as Bourdieu once argued, characterized by the force of 'necessity'). The inclination of middle classes toward environmental concerns can, however, be neutralized in diverse ways. One is their increasing dependence on market forces in a Europeanizing and globalizing economy. Another is the marketization of consumption patterns which are in part the unintended result (a Pyrrhic victory) of the middle-class dominated environmental movement itself. More doubtful are causal inferences concerning the higher potential mobilization of these groups for such concerns. For our discussion, it is important to stress that the middle classes have used the environment for reorganizing their class position. The new petty bourgeois way of life thus appropriates the working class's action repertoire of collective action. Moreover, fighting pollution marks a symbolic boundary of purity vis-à-vis those who pollute – it secures symbolic victories not only against big industry (which pollutes, above all the nuclear industry), but also against the working class. This new middle class society is spreading across Europe (Sulkunen 1992), distancing itself from the lower classes in terms of income and property and legitimizing itself as the savior of nature. Thus, the growth of the middle classes might be a pan-European phenomenon that undermines and traditionalizes the North-South divide.

The middle-class hypothesis excludes the lower classes as relevant actors in environmental politics. The role of working and especially agrarian classes also has to be reassessed. Research in environmental politics has observed that an agrarian society is a good bulwark against environmentalism (Osti

[25] This is a classic argument. See as an overview with additional, often critical references Eder (1996).

1990). This observation requires qualification. In many rural areas of Southern Europe and Northern areas as well, populations are pluri-active – they work on different activities in different seasons to achieve the desired income e.g. tourism and agriculture/fishing. Thus, it's not surprising that the observed link is most likely valid for agri-environmental measures which impinge upon economic survival. If we look at other forms of environmentalism, i.e., organizing for the protection of the local ecosystem and against various forms of ecological marginalization (Kousis 1998), we find an increase of local environmental mobilization in rural areas since the 1980s, especially in Southern Europe (Kousis 1999). When looking not just at agricultural activities that damage the environment, we find rural populations intensively organizing against the imposition of toxic waste facilities in rural areas (Kousis 1997)[26], against the construction of large infrastructure projects such as anti-road protest (Barcena and Ibarra, in this volume), or against nuclear energy installations, such as the nuclear waste siting in Sweden or the protest against river diversion in Greece. Thus, agrarian protest extends into rural protest, mobilizing a wider array of interests among the old agrarian classes.[27]

There is a handy structural explanation for this: The more agrarian a society, the stronger we can expect this effect to be. From this the following conclusion could be drawn: Since the South is more agrarian than the North, the agrarian problem might foster, via environmentalism, a growing North-South cleavage in terms of more protest in the South than in the North. A first caveat to such a conclusion is that rural social life over the past two decades has come to involve activities and policies that go far beyond simple agriculture, both in the South (especially with the spread of tourism in rural regions since the seventies) and in the North (Kousis 1998). The agrarian problem is a general problem throughout Europe and increases to the extent that agrarian interests feel threatened. There is no difference per se between the North and the South. Whether agrarian classes are more inclined to environmental concerns in the North than in the South depends on factors other than simply agrarian ones. One of them concerns the form in which rural life develops in different regions of Europe. Since rural life in the South develops more complex social forms, we would arrive at the paradox that the more rural an area is, the more environmentally oriented it is and, conversely, the more merely agrarian it is, the less environmentally oriented it

[26] This phenomenon not only exists in Europe. It applies to rural populations in other parts of the world as well. See the protest against extracting activities such as the current Papua revolt against US copper companies in New Guinea.

[27] One is tempted to call them the 'new agrarian classes'.

is. Since some regions of the North are developing this kind of agrarian mono-culture more than regions in the South, the prospects of environmentalism turn out – according to such theoretical reasoning – to be less auspicious in the rural classes in Northern regions. This however is a hypothesis in need of testing.

Existing research on this link, however, points to a further caveat and a paradox, namely the question of the capacity of rural life to organize itself politically. When looking at rural non-agrarian interests which call for environmental protection, even though they do not have as strong ties to political organizations as urban groups do, rural environmental mobilizations find and use a variety of other networks (Kousis 1999). This holds for the North and the South equally, even slightly more so for the North. This has to be qualified with respect to the framing of the protest. What this difference tells us is that agrarian interests will be more inclined toward disembedded interest politics, whereas rural interests will be more inclined toward life world concerns which favor environmental attitudes. Paradoxically, this might even be fostered by observations made in a growing literature on the environmental attitudes of farmers that small farmers who are dependent on chemical agriculture rarely have alternative options for economic survival. The more people are forced into such negative options, the more they might realize alternative options.

Thus, there is no clear connection between the number of those still in the agrarian sector and the environmental concerns of these classes. The idea that strong agrarian interest implies less environmental concern loses much of its initial plausibility. This initial plausibility was based on the projection of a high-culture notion of environmental concerns onto these classes, thus generating misconceptions of the other. This is a continuation of symbolic 'class fare'.[28] Evidence to the contrary suggests an alternative hypothesis: popular environmentalism couched in class terms might even be an effect of environmental concerns on the part of European elites pushing environmental concerns on the discursive level. The staging of sustainability discourses by European elite publics unites agrarian classes against agri-environmental policies and middle classes against food policies.

A further hypothesis leading us away from the simple model of a North-South divide is the emerging relationship between rural classes and middle classes in Europe as a new conflictual class relationship. According to this hypothesis, middle class environmentalism is a mechanism for creating an economic cleavage which results from the poor having to pay the costs of environmentalism (the costs of the new purity lived by the new middle

[28] Some new empirical evidence is presented below in Part II dealing with local actors.

classes) while themselves being forced to produce environmental pollution for survival. In such terms ecology is linked to economy in a way normally neglected in research: ecology contributes to the reproduction of economic inequality between Europeanizing classes.

The environmental orientation and action potential of popular publics are therefore not determined by their nationality or by any supposed national character of these classes, but rather by their position in an emerging system of social relations in Europe which defines the dependence of agrarian and industrial classes, of petty bourgeois and new middle classes on national and transnational actors who act – this is the implication of a democratic and institutionalized class struggle – as their proxies.

6. THREE WORLDS OF SUSTAINABILITY IN EUROPE

The tentative result of this argument is that environmental policy-making in Europe is shaped by an economically divided Europe, a division that is not between the North and the South, but that is increasingly decoupled from spatial and geographical fixations. Cleavages related to environmental politics increasingly depend on deterritorialized class relations. Cleavages develop not between nation-states; instead, they depend on one's position in a system of economic dependence in European society. We do not claim that Europe is a prolongation of the old, nationally defined, industrial class society or of the new, nationally defined, post-industrial class society.[29] We are certainly confronted with changing classes, but not with the end of classes.[30]

[29] These have been the debates of the sixties and seventies (Touraine 1971; Giddens 1973; Gouldner 1979) and their attempts to reformulate class theory by adapting it to the emergence of new post-industrial classes. The central question however remained which was to which extent the concept of class itself had to be changed in the course of the development of a post-industrial society. Since social inequality did not disappear and even increased where welfare state tradition lost ground, the class concept still describes the fact that people are different (Hout, Brooks and Manza 1993). The most radical solution is the one by Bourdieu: defining classes as groups constructed on the basis of statistical differences that have social effects in terms of boundary construction of social groups (through constructing "distinctiveness"). See Bourdieu (1986, 1987). The alternative solution of Touraine to replace the concept of social class by the notion of social movement did finally not work. This theoretical conceptualization holds probably only in times of rapid social change (Touraine 1988).

[30] For theoretical and empirical attempts to come to terms with such changes see Eder (1993) and Esping-Andersen (1993).

Our proposal therefore is to see environmental policy-making not so much embedded in national or regional[31] characteristics, as in a system of European class structuration which determines the outcome of such policies. Class explains the patterns of popular reactions, and this impinges upon defensive national reactions or transnational pressure.

Sustainability is the strategy of transnational elite actors to impose a regime of environmental control on European society which is – as is sometimes the case with action by dominant classes and, even more so, with democratic societies – good for the environment. National elite actors are caught in a no-exit situation. They continue – with national coloring – the discourse of sustainable development. Thus, a complex system of elite action emerges which is not only good for the environment, but which also fosters a system of class relations in Europe. This system puts agrarian people, middle classes, the old industrial labor class and the new lower service classes in a position which shapes their reaction to the issue of environment. The Greek small farmer does not behave as he does because he is a Greek, but because he acts in a social position in which he has to live, which means either to give up and leave behind what has been his life-world or to defend a life-world that secures his economic survival.

The tentative proposal is to talk about '*Three Worlds of Sustainability*'. The three worlds of sustainability are compounds of shared interests and identities which shape the ways of dealing with environmental issues.[32] Environmental politics is a politics in which actors participate not with a view toward winning or losing material assets, but rather toward establishing a better world for themselves and others. The basic problem of environmental politics is that here the free-rider problem is more acute here than in any other policy field: any betterment of the environment simultaneously betters the conditions both for those striving and struggling for it, and for those who do nothing.

The different worlds of sustainability provide specific structures and resources to overcome this problem of environmental politics. The incentives to do something about the environment have to be built into constraints exerted by the institutional structure of a society – a society which is supported by a cultural orientation which legitimates this institutional framework. We

[31] Regional meaning here a group of nationally defined entities such as the European 'South'.

[32] They differ from *The Three Worlds of Welfare* (Esping-Andersen 1990) by their focus on the common good problem of the "environment". Whereas social policy regimes are regimes to redistribute the wealth of a nation, environmental policy regimes are regimes to create a better collective state of affairs.

hypothesize that the worlds of sustainability are based on highly demanding structural constellations which do not develop without strong pressure from the outside world.

We suggest the following 'Three Worlds of Sustainability': the *transnational elite world*, the *nation-state world*, and the *popular world*:

- The (new) 'transnational elite world' abounds with deliberative bodies and councils, green images of business, and sustainability discourses which translate economic interests into metaphors of a good life. The consequences of such institutional changes are not yet clear – they will be more than ideological 'Überbau'.
- The (old) 'nation-state world' comprises corporatist arrangements in the environmental field, state responsibility with strong temporary anti-state mobilizations, and national economic growth minimizing ecological transaction costs (sometimes couched in terms of 'ecological modernization'), all of which are couched in metaphors of a good life.
- In the 'popular world' we have strong localism with scattered mobilizations (depending in scope upon highly varying degrees of policing), informal economies embedded in life-world, i.e., moral economies based on specific relations of trust, and class-specific conceptions of nature as metaphors of a good life.

These worlds provide specific solutions to the problem of collective action in environmental politics. This problem is resolved in these worlds in the following ways:

- In the transnational elite world, the involvement of collective actors in transnational negotiation turn environmental issues into public issues which allow policy outcomes to be attributed to these actors, thus creating particular constraints on them.
- In the nation-state world, a strong identification of environmental politics with national interests forces collective actors to act when these interests are at stake, an often irrational constraint of collective action.
- In the popular world, the idea of bad forces coming from the outside world (upper classes, the state, Europe, globalization) produces collective action, i.e., local protest and grassroots mobilization, the logic of which is closer to that of class conflict than that of lobbying strategies.

The interaction of three worlds of sustainability is structured in time and space. Here, the limits of a merely national perspective become evident. In terms of time, we have a space that, with the end of the Second World War and of the Cold War, marks periods that separate these worlds from our own. In terms of space, the world is divided ever less between national societies and evermore between organized blocks of national societies. Europe defines a space that is distinctive from that defined by the US or Japan.[33] Thus, the spatial and temporal structuration of the interaction of these three worlds constrains transnational actors, national actors, as well as those acting in the name of some people (or class) in the making of the environment. This structuration has surprising implications. The spatial metaphor is no longer linked to a strict geographical location. These different worlds are scattered across national boundaries within Europe. Neither the elite world nor the popular world can be grasped adequately in terms of nationally defined spaces. The transnational elite world in Europe is identified with Brussels, but is spatially highly volatile because it exists in a deterritorialized communicative space; these groups need only space to meet. Such space is provided by a European public space which has been built (in stone and glass and behind it) between Strasbourg and Brussels. This is not to say that the national space in Europe no longer plays an important role; on the contrary, it still does. But even this space is increasingly bound to the first world, which has created a non-exit situation for actors in the national space. Loyalty of national actors is the only alternative left, as more transnational actors succeed in closing off exit-options for national actors.[34] To conclude from Hirschman's theory of the relation between 'voice', 'exit' and 'loyalty': if there is no exit and no loyalty, 'voice' is the only strategy left for the people. It is not national membership but class location which shapes their voice on environmental problems. The people can express themselves through different forms of collective mobilizations[35], be it through elite attacks, retreat in informal economies, self-help, or NIMBY actions. They are functionally equivalent (although they certainly have different impacts on the environment).

The temporal and spatial structuration of Europe has also made the field of environmental politics a unified field of action. Here transnational actors have succeeded in providing a common discourse for this unified field: the discourse on sustainability, produced and continuously reproduced in and by expert circles[36], taken up by committees and eventually transformed in some legislative act. To stabilize this space, a public monitoring system for the

[33] This is well reflected in the market competition of currencies: we compare the Dollar with the Yen and the Euro. This defines the objective and symbolic boundaries of the space.

[34] For an interesting use of Hirschman's categories in analysing European society see Bartolini (1997).

[35] When such collective action gets started against centralized states they often become violent. France has experienced the most violent mobilization in the anti-nuclear campaigns, and Italy has seen an especially strong terrorist mobilization.

[36] In fact, originally, the new political meaning of the concept was invented in business circles in preparation for the Rio summit (Donati 2000).

national actors in the field is installed which involves national publics in a common, yet often conflictual public space – conflictual because national interests is often at stake. This spatial organization is defined by clear boundaries: a field of discursive relations forces the different publics to establish links via this field to other publics. Each public has to situate its strategic actions and the pursuit of its interests in this context, with the effect that these strategies become more and more isomorphic.

However, this system does not control collective action anchored in different class positions. This is the final result of our analysis: the field of environmental politics has a third European dimension characterized by isomorphic class positions within the European space. The Irish and the Greek farmer, the French and the German steel worker, the Belgian and Italian lorry driver, they all denote class positions that do not coincide with national class locations. To put it in other words: class locations are nationally deterritorialized. Old and new environmental problems are transformed into social problems by such transnationalization and deterritorialization of the class structure of modern societies.

This has important implications for an understanding of difference in environmental politics. The idea of a North-South divide presupposes the congruence of national actors and popular actors which react to transnational policy-making. This coupling is unwarranted. Rather, a coupling of transnational and national actors is taking place, which is accompanied by a decoupling of national actors and popular actors. The key to an understanding of environmental policy-making by transnational and national actors is therefore not the *national compliance* with EU regulations, but the *popular reaction* to the politics of environmentalism imposed on the life-world of people. The determinant of these life-worlds are not – to repeat – given by a national character or a national way of life, but by a class position which exists equally in Western Ireland and Southern Greece, in Milan, Paris or Berlin, in the old industrial lands of Birmingham or Bitterfeld or Southern Belgium, and through which the environment is articulated and represented as a problem.

This argument also sheds some new light on the role of the middle classes throughout Europe in environmental politics. With the institutionalization of environmental policies based in the hypercode of sustainability and a legal apparatus fostering environmental legislation on the European and national level, the privilege of the middle classes as the central class in environmental politics has waned. Farmers, workers, the unemployed are made part of the game. Paradoxically enough, this has created cleavages that cut across the North-South divide. This forces us to find new analytical tools in order to grasp divisions that lack the neat and clear spatial location of a 'country'. The new cleavages of European societies will no longer be those dividing 'countries', but rather those separating the life-world of the people in Europe and the policy action of European elites on these life-worlds. In this emerging cleavage structure, national actors will turn – in a situation of decreasing exit-options – from voice to loyalty. This does not imply that the people will follow the path of loyalty taken by these national actors. They might do it. But they also have another option: voice, which, in turn, opens up still further options: anything from NIMBY to discursive democracy, from environmental carelessness to fundamentalist ecology is possible. Enhancing our knowledge means understanding such diversity of reactions instead of simply repeating the stereotype of a North-South divide.

7. CONCLUSION

Popular publics point to a social reality in Europe beyond elites and nations. Environmental politics in Europe is a good example of this argument. It is not only made by Brussels and its policy makers and experts. It is also made by popular groups which occupy diverse, but isomorphic positions in Europe. This is linked to changes in the role of national actors and national publics. They are increasingly forced into loyal alliances with emerging transnational European institutions – alliances which at the same time reduce the strong link between the nation and the people. Europe seems to open a new space for the self-organization of people by providing a space beyond the nation.

These people turn into a new type of public for environmental politics. They observe within their life-worlds and from their class position what is going on; they monitor effects and draw their conclusions regarding their particular interests and ways of life. Europe – due to the homogenizing capacity of its elites – has provided a large space for the articulation of these interests and ways of life which no longer allows these groups to escape the consequences of their actions. There is no exit as soon as they are in the European boat.

It is through such processes that a European society emerges, not as the ideal project of policy makers, but as the reality generated through collective action with all its cleavages, conflicts, and coalitions. Europe is made and reproduced through the structuration of action spaces as described above. In this sense, an analysis of environmentalism in Europe – its actors, its discourses, its policies – might reveal to us something that goes beyond the policy field of the environment: the emergence of European social structures, the making of a distinctly European society.

The North-South divide then disappears as a relevant parameter. It appears merely as a device for legitimating national publics and rationalizing the loyalty game they play, the parameters of which are set by European institutions. This game fails to including the people. Beyond the North-South divide there reappear social classes in Europe which are defined by their market position in a Common Market. Their interests and identities enter the process of environmental politics to the extent that elites, supported to different degrees by national actors, engage in changing the environmental regime through policies which use the regulative idea of sustainability to do so. This conflict between people and political elites is a rather normal phenomenon as such. It comes, however, as a surprise to those adhering to a harmonious image of Europe. This image is a delusion. Instead, the emergence of transnational social classes appears to become the most profound structuring principle of emergent European society. Environmental policy-making is – perhaps more so than social policy - a medium for this social change, thus unintentionally fostering the social articulation of and the collective mobilization for environmental problems.

REFERENCES

Bartolini, S. (1997) *Exit Options, Boundary Building, Political Structuring*, Unpublished Manuscript, European University Institute, Florence.

Börzel, T. (2000) Why there is no Southern Problem. On Environmental Leaders and Laggards in the European Union, *Journal of European Public Policy* **7**, 141-162.

Bourdieu, P. (1986) Forms of capital, in J.G. Richardson, *Handbook of Theory and Research for the Sociology of Education*, Greenwood, Westport, CT, pp.241-258.

Bourdieu, P. (1987) What makes a social class? On the theoretical and practical existence of groups, *Berkeley Journal of Sociology* **32**, 1-17.

Chabason, L. (1999) The concept of sustainable development in the Mediterranean: Emergence and recurrence, *Geographica Helvetica* **2**, 105-112.

Conrad, J. (1997) Nachhaltige Entwicklung - ein ökologisch modernisiertes Modell der Moderne? in K.-W. Brand, *Nachhaltige Entwicklung. Eine Herausforderung an die Soziologie*, Leske + Budrich, Opladen, pp.51-70.

Converse, P.E. (1964) The nature of belief systems in mass publics, in D.E. Apter, *Ideology and Discontent*, Free Press, New York, NY, pp.207-261.

DiMaggio, P.J., and Powell, W.W. (1983) The iron cage revisited: Institutional isomorphism and collective rationality in organizational fields, *American Sociological Review* **48**, 147-160.

Donati, P.R. (2000) *The Industrial Class in Post-Industrial Age: Corporate Political Strategies in the Environmental Issue*, Ph.D. Thesis, European University Institute, Florence.

Eder, K. (1993) *The New Politics of Class. Social Movements and Cultural Dynamics in Advanced Societies*, Sage, London.

Eder, K. (1995) Framing and Communicating Environmental Issues. Final Report to the Commission of the European Communities, DG XII, Research Area III, Project Number PL 210493, European University Institute, Florence.

Eder, K. (1996) *The Social Construction of Nature. A Sociology of Ecological Enlightenment*, Sage, London.

Esping-Andersen, G. (1990) *The Three Worlds of Welfare Capitalism*, Princeton University Press, Princeton, NJ.

Esping-Andersen, G. (ed.) (1993) *Changing Classes. Stratification and Mobility in Post-Industrial Societies*, Sage, London.

Giddens, A. (1973) *The Class Structure of Advanced Societies*, Hutchinson, London.

Gouldner, A.W. (1979) *The Future of the Intellectuals and the Rise of the New Class. A Frame of Reference, Theses, Conjectures, Arguments and a Historical Perspective on the Role of Intellectuals and Intelligentsia in the International Class Contest of the Modern Era*, Seabury, New York, NY.

Haas, P.M. (1989) Do regimes matter? Epistemic communities and Mediterranean pollution control, *International Organization* **43**, 377-403.

Hout, M., Brooks, C., and Manza, J. (1993) The persistence of classes in post-industrial societies, *International Sociology* **8**, 259-277.

Kohler-Koch, B., and Eising, R. (eds.) (1999) *The Transformation of Governance in the European Union*, Routledge, London.

Kousis, M. (1993) Collective resistance and sustainable development in rural Greece: The case of geothermal energy on the island of Milos, *Sociologia Ruralis* **33**, 3-24.

Kousis, M. (1997) Grassroots environmental movements in rural Greece: effectiveness, success and the quest for sustainable development, in S. Baker, M. Kousis, D.Richardson and S. Young, *The Politics of Sustainable Development: Theory, Policy and Practice within the European Union*, Routledge, London, pp.237-258.

Kousis, M. (1998) Ecological marginalization in rural areas: Actors, impacts, responses, *Sociologia Ruralis* **38**, 86-108.

Kousis, M. (1999) Sustaining local environmental mobilisations: Groups, actions and claims in Southern Europe, in C. Rootes, *Environmental Politics, Special issue: Environmental Movements: Local, National and Global*, pp.172-198.

Le Saout, D., and Trom, D. (1995) *Framing and Communicating Environmental Issues. The French Case*, Project Report, European University Institute, Florence.

Liberatore, A. (1997) The integration of sustainable development objectives into EU policy-making: Barriers and prospects, in S. Baker and M. Kousis and D. Richardson and S. Young, *The Politics of Sustainable Development: Theory, Policy and Practice within the European Union*, Routledge, London, pp.107-126.

Liberatore, A. (1999) *The Management of Uncertainty. Learning from Chernobyl*, Gordon and Breach, Amsterdam.

Lipset, S.M. (1979 [1963]) *The First New Nation*, Norton, New York, NY.

Majone, G. (ed.) (1996) *Regulating Europe*, Routledge, London.

March, J.G., and Olsen, J.P. (1989) *Rediscovering Institutions: The Organizational Basis of Politics*, Free Press, New York, NY.

March, J.G., and Olsen, J.P. (1995) *Democratic Governance*, Free Press, New York, NY.

Osti, G. (1990) Società rurale e voto ambientalista nell' Italia nord-orientale, *Quaderni di Sociologia* **13**, 123-140.

Poferl, A., and Brand, K.-W. (1995) *Framing and Communicating Environmental Issues. The German Case*, Project report, European University Institute, Florence.

Powell, W.W., and DiMaggio, P.J. (eds.) (1991) *The New Institutionalism in Organizational Analysis*, University of Chicago Press, Chicago, IL.

Richardson, D. (1997) The politics of sustainable development, in S. Baker and M. Kousis and D. Richardson and S. Young, *The Politics of Sustainable Development: Theory, Policy and Practice within the European Union*, Routledge, London, pp.43-60.

Rokkan, S. (1975) Dimensions of state formation and nation-building, in C. Tilly, *The Formation of National States in Western Europe*, Princeton University Press, Princeton, NJ, pp.562-600.

Sachs, W. (1997) Sustainable Development. Zur politischen Anatomie eines internationalen Leitbilds, in K.-W. Brand, *Nachhaltige Entwicklung. Eine Herausforderung an die Soziologie*, Leske + Budrich, Opladen, pp.93-110.

Statham, P., and Szerszinski, B. (1995) *Framing and Communicating Environmental Issues. The British Case*, Project report, European University Institute, Florence.

Sulkunen, P. (1992) *The European New Middle Class. Individuality and Tribalism in Mass Society*, Avebury, Aldershot.

Tarrow, S. (1998) *Power in Movement. Social Movements, Collective Action and Politics*, Cambridge University Press, Cambridge, MA.

Touraine, A. (1971) *The Postindustrial Society. Tomorrow's Social History: Classes, Conflicts and Culture in Programmed Society*, Random House, New York, NY.

Touraine, A. (1988) *Return of the Actor*, University of Minnesota Press, Minneapolis, MI.

Triandafyllidou, A. (1995) The Chernobyl accident in the Italian press: A 'media story-line', *Discourse and Society* **6**, 517-536.

Triandafyllidou, A. (1996) 'Green' corruption in the Italian press: Does political culture matter? *European Journal of Communication* **11**, 371-391.

Triandafyllidou, A., Donati, P.R., and Diani, M. (1995) *Framing and Communicating Environmental Issues. The Italian Case*, Project report, European University Institute, Florence.

Zaccaï, E. (1999) Sustainable development: characteristics and interpretations, *Geographica Helvetica* **2**, 73-80.

Chapter 2

Sustainability and the North/South Divide

Global and European Dimensions

Michael Redclift

1. INTRODUCTION

In most of the countries in the European Union environmental policy questions are largely dictated by public policy in a number of separate sectors: health; food; transport; education and land use among them. The direction that environmental policy is taking in Europe is thus closely linked to the 'balance' that societies wish to achieve between economic growth (or 'wealth creation') and their quality of life. Environmental concerns are often the outcome of the individual's immediate experience of production and consumption, but also (and sometimes, contradictorily) part of the civil society agenda, including elements such as social justice, citizenship and governance. It follows that the way that environmental problems and policies are constructed in Europe influences both the scope for achieving sustainability, and the way in which it is achieved. Sustainable development is a declared goal of policy, but it is not clear who 'owns' the policy discourse surrounding it.

In this chapter attention is paid to the ways in which thinking about the North/South divide carries implications at the European level, as well as that of the globe. Consideration is given to aspects of global modeling, and the profile attached to policy issues, both of which assume largely 'northern' proportions. The environmental policy agenda – at the global level – largely reflects Northern issues, such as carbon emissions as a factor in determining the scale and effect of global climate change. Relatively little attention is given, in this 'global' discourse, to issues which arise from Southern preoccupations: such as infant mortality levels attributable to environmental con-

K. Eder and M. Kousis (eds.), Environmental Politics in Southern Europe, 53–72.

ditions, particularly clean water, or the effect of debt rescheduling on the environmental health of both human populations, and that of other species. This concern with what are referred to as 'global' issues, in effect prioritizes the policy concerns of the Northern, industrialized countries, at the expense of the less developed countries of the South. It is the contention of this chapter that these structural tensions, and discontinuities at the global level are reproduced, albeit less sharply, within Europe. In many respects 'northern' Europe exhibits some of the same environmental concerns as the *generic* North – waste disposal, climate change, recycling, concerns over nuclear energy. At the same time 'Southern' Europe exhibits many of the same environmental concerns (and discourses) as those of the *generic* South: such as, aridity, water shortages, decline in land quality and degradation, and the erosion of forests. The distinction between North and South is, therefore, of some metaphorical force, carrying cultural and political meaning. At the same time the substantive characteristics of Northern and Southern European *environments* reflect ecological and geophysical, as well as cultural, differences. Even the question of what constitutes an 'environmental' problem (as opposed, to a problem of poor livelihoods, or poverty, or distribution) is subject to different interpretation, from within the discourses of Northern and Southern social science. In this chapter some of the ways of identifying sustainable development discourses, and in particular the positioning of these discourses around the North/South distinction, are considered . It is suggested that some of the environmental policy preoccupations of *Northern* Europe, notably that of ecological modernization, lose much of their 'global' relevance, if placed within the contours of Europe as a whole. They remain important concerns, but need to be identified with specific projects of economic and commercial convergence, such as that embarked upon by the European Union, in its much-vaunted Action Plans.

This chapter begins by examining the concept of sustainable development in relation to two key – and interrelated – questions. How might it be possible to secure improved sustainability at the level of the macro-economy while, at the same time, ensuring that these policy goals correspond with the aspirations and values of multiple 'publics'? It takes the debate from one of conceptual clarification towards a focus on policy 'ownership'. The four main component 'dimensions' of sustainable development are reviewed: ecology, equity, modernity and globalization. After considering some of the conceptual and ontological questions surrounding sustainable development, each of these dimensions is considered in turn.

Later, as we have seen, the chapter turns to consider 'ecological modernization', as the latest manifestation of public policy in Europe, and considers attempts to integrate sustainability and economic development through internalizing environmental costs. It is suggested that this policy emphasis does not reflect the principal environmental concerns of Southern European areas. It is suggested that the European debate should not be divorced from the wider global issues – and those affecting modernity. Sustainable development, it is argued, might be seen as part of a new definition of global citizenship, which builds on European experience, rather than simply as a strategy for making European business and technology more competitive.

2. SUSTAINABLE DEVELOPMENT

Each scientific problem which is resolved by human intervention, using fossil fuels and manufactured materials, is viewed as a triumph of management, and a contribution to economic good, when it might also represent a future threat to sustainability. Having jettisoned the fear that resources themselves were limited, in the 1970s, we are today faced by the prospect that the means we have used to overcome resource scarcity, substitution and increased levels of industrial metabolism, contribute to the next generation of problems which are associated with global environmental problems. This realization provides an enormous challenge to conventional social science thinking, encapsulated in the term 'sustainable development'.

Sustainable development was defined by the Brundtland Commission in the following way: "development that meets the needs of the present without compromising the ability of future generations to meet their own needs" (Brundtland Commission 1987).

This definition has been brought into service in the absence of agreement about a process which almost everybody thinks is desirable. However, the simplicity of this approach obscures underlying complexities and contradictions. Before exploring whether we can establish indicators of sustainability, it is worth pausing to examine the apparent consensus that reigns over sustainable development.

First, following the Brundtland definition, it is clear that 'needs' themselves change, so it is unlikely (as the definition implies) that those of future generations will be the same as those of the present generation. The question then is, where does 'development' come into the picture? Obviously development itself contributes to 'needs', helping to define them differently for each generation, and for different cultures.

This raised the second question, not covered by the definition, of how needs are defined in different cultures. Most of the 'consensus' surrounding sustainable development has involved a syllogism: sustainable development is necessary for all of us, but it may be defined differently in terms of each and every culture. This is superficially convenient, until we begin to ask how these different definitions match up. If in one society it is agreed that fresh air and open spaces are necessary before development can be sustainable, it will be increasingly difficult to marry this definition of 'needs' with those of other societies seeking more material wealth, even at the cost of increased pollution. And how do we establish which course of action is *more* sustainable? Recourse to the view that societies must decide for themselves is not very helpful. (Who decides? On what basis are the decisions made?). At the same time there are problems in ignoring culturally specific definitions in the interest of a more inclusive ontology.

There is also considerable confusion surrounding *what* is to be sustained. One of the reasons why there are so many contradictory approaches to sustainable development (although not the only reason) is that different people identify the objects of sustainability differently.

For those whose primary interest is in ecological systems and the conservation of natural resources, it is the natural resource base which needs to be sustained. The key question that is usually posed is the following: how can development activities be designed which help to maintain ecological processes, such as soil fertility, the assimilation of wastes, and water and nutrient recycling? Another, related, issue is the conservation of genetic materials, both in themselves and (perhaps more importantly) as part of complex, and vulnerable systems of biodiversity. The natural resource base needs to be conserved because of its intrinsic value.

There are other approaches. Some environmental economists argue that the natural stock of resources, or 'critical natural capital', needs to be given priority over the flows of income which depend upon it. They make the point that human-made capital cannot be an effective substitute for natural capital. If our objective is the sustainable yield of renewable resources, then sustainable development implies the management of these resources in the interest of the natural capital stock. This raises a number of issues which are both political and distributive: who owns and control genetic materials, and manages the environment? At what point does the conservation of natural capital unnecessarily inhibit the sustainable flows of resources?

Second, according to what principles are the social institutions governing the use of resources, organized? What systems of tenure dictate the ownership and management of the natural resource base? What institutions do we bequeath, together with the environment, to future generations? Far from taking us away from issues of distributive politics, and political economy, a concern with sustainable development inevitably raises such issues more forcefully than ever.

The question 'what is to be sustained?' can also be answered in another way. Some writers argue that it is present (or future) levels of production (or consumption) that need to be sustained. The argument is that the growth of global population will lead to increased demands on the environment, and our definition of sustainable development should incorporate this fact. At the same time, the consumption practices of individuals will change too. Given the choice, most people in India or China might want a television or an automobile of their own, like households in the industrialized North. What prevents them from acquiring one is their poverty, their inability to consume, and the relatively 'undeveloped' infrastructure of poor countries.

Is there anything inherently unsustainable in broadening the market for TV set or cars? If the answer is 'yes', then those of us who possess these goods need to be clear about why we consume goods unavailable to others. The response is usually that it is difficult, or even impossible, to function in our society without information or private motorized mobility. But, this is to evade the question of underlying social commitments. We define our needs in ways which effectively exclude others meeting theirs, and in the process increase the long-term risks for the sustainability of their livelihoods. Most important, however, the process through which we enlarge our choices, and reduce those of others, is largely invisible to us.

If we concentrate our attention on our own societies in Europe, we can begin by identifying aspects of our management of the environment that are unsustainable. However, we are socialized into behavior which is unsustainable. Everyday practices have served to institutionalize unsustainable practices which reflect the way that industrialized societies in Europe are currently organized. We, in Europe, rarely imagine that environmental management extends to aspects of behavior that are not presented as personal choices.

Economics developed, historically, around the idea of scarcity. The role of technology was principally that of raising output from scarce resources. Among other benefits of economic growth was the political legitimacy it conferred, within a dynamic economy, on those who could successfully overcome the obstacles to more spending, and wealth was usually regarded as a good thing, in itself. This assumption of scarce resources and technological benefits sits uneasily with sustainability in the industrial North today and underlines the difficulty in reconciling 'development' with 'sustainability'. It strikes at the legitimization of only one form of 'value' , albeit the principal one, within capitalist, industrial societies. Habermas expressed his criticism of this view forcefully, in the following way:

> can civilization afford to surrender itself entirely to the … driving force of just one of its subsystems – namely, the pull of a dynamic … recursively closed, economic system which can only function and remain stable by taking all relevant information, translating it into, and processing it in, the language of economic value … (Habermas 1992: 134)

There is another dimension to the problem of diminished sustainability, which is relatively recent. This is the extent to which, at the end of the twentieth century, we need to refer to genuinely global processes. As Miller has argued, global consumption "provides a new egalitarianism between subject and subject" as Central Africans wear suits, Indonesians and Brazilians produce soap operas, and branded commodities acquire general importance (Miller 1994: 3). At the superficial level of dress conventions the ethnography of consumption has the same referents, and the commodities consumed can be construed as great 'levellers' (Brewer and Porter 1993).

A major obstacle to the 'leveling' effects of global consumption is the way we measure wealth. William Cobbett, the early nineteenth century English Radical, referred to these over one hundred and sixty years ago, in supporting the reduction of factory hours for children from twelve to ten hours a day, in the textile mills. Speaking in support of Lord Shaftesbury's bill, Cobbett said: "it is interesting to learn that all of Britain's wealth, power and security ... lay not in her virility, nor in her agriculture, banking or merchandize, but ... in the labor of three hundred thousand little girls in Lancashire". If two hours of their daily work were deducted, "away goes the wealth, away goes the capital, away go the resources, the power and the glory of England" (Cobbett 1842: 26).

The question today, as in the 1830s, is 'what is it *worth* to measure wealth in this way?' It is often assumed that increasing 'sustainability' jeopardizes the creation of wealth, but unless we are clear about how we measure wealth, it is difficult to assert that the creation of wealth is necessary for improvements in the quality of life. The U.K. White Paper on Science, *Realising Our Potential*, (HMSO 1993) tries to side-step this problem by referring to wealth creation *and* the quality of life, as if they are mutually compatible goals. In practice they often pull against each other. It depends on where we measure wealth creation, at the national, global, or local level? In many countries balance-of-payments surpluses are the result of decreases in national wealth, if by 'wealth' we mean infant mortality rates, or female and child nutrition. The creation of wealth, as a policy objective, tends to confine environmental factors to the closet, enabling politicians to wring their hands over the supposed high levels of unemployment that higher environmental standards herald, or the dangers of interfering with market forces which are assumed to work best when they are free from government control.

Similarly, within the European Union, a recent White Paper on *Growth, Competitiveness and Employment* places emphasis on economic growth and increased employment intensity as vehicles for economic recovery (European Commission 1993). The familiar argument is that we need to increase both growth and employment, to generate the means to deal with environmental problems, before sustainability can be achieved. In practice, increased private consumption is seen as the key policy lever. If we increase the consuming capability of the household within the European Union, we can invest the benefits in employment-creating activities such as child care, education, vocational training, and better facilities for the old and handicapped. However, widening the tax base depends, in turn, on freeing labor for productive purposes. For some groups – such as female single parents – to effectively enter the workforce requires improvements in social services, including child care, before their earned income can be enhanced. It is difficult to increase 'consuming capacity' without increasing disposable resources, a key element of which is the generation of income through women's paid employment...

There is another problem in the orthodox argument, that is rarely exposed by either Left or Right of the political spectrum. The social policy agenda that is supposed to be the beneficiary of increased growth carries environmental implications, both in the goods and services to be provided in the social field, and the means of achieving them. If we pursue the creation of needs, as a means of lifting overall consumption, and enhancing current production, we are unlikely to identify the needs which our economic system currently does *not* meet. There is a considerable risk that we will create more casual employment rather than more socially useful employment and, in the process of raising personal consumption and personal welfare, place environmental standards in greater, rather than less, jeopardy.

The alternative path to follow is a radical one. It means pursuing better environmental standards – in energy production and conservation, in more efficient transport, better air and water quality – as the first objectives of policy, rather than the supposed beneficiaries of more economic growth. At the moment the European Union is setting 'environmental targets' around 'what we can afford', from the wealth created by unsustainable levels of production and consumption. The alternative is to make environmental targets the *instruments* for improving the quality of our lives. As Fleming (1995) has argued we need to bring sustainability out of the environmental closet and start applying it to the economy at large. The task of the social sciences in Europe, is to assist in the redefinition of needs, and the ways in which they may be satisfied: to place the different *dimensions* of sustainability at the top of our agenda. In exploring these dimensions within the context of modern Europe we can begin with the physical limits on sustainability represented by ecology.

2.1 The ecological dimension of sustainability

Chapter Four of *Agenda 21* examines the idea of sustainable development at the international level, by considering consumption patterns within a global context. Clearly the most obvious, and pressing, dimension of sustainability is represented by our inability to live within ecological limits, and the environmental costs of increased pressures from population and economic development. Two concepts which have been used to represent these limits are environmental space and ecological footprints. The concept of environmental space expresses each individual's entitlement to a share of the world's resources, and sink capacities. It seeks to define acceptable levels of waste and pollution. The concept of ecological footprints measures the effect of individual consumption in one country on the environment of others. Consumption patterns draw on resources, and the capacity for waste assimilation, from a much wider area than that occupied by a given population. For example, Japan has access to 'invisible acres' through its extensive fishing of the world's oceans; and large cities like New York, Mexico City or Los Angeles source water and energy from vast hinterlands, and dispose of waste far outside their local political boundaries.

Pressures on these ecological limits have, in turn, led to the gradual loss, or increased vulnerability, of sustainable social practices, and our knowledge of sustainable resource uses. Existing social practices define the social limits to which human populations can adjust, or accommodate, to the physical constraints represented by the environment. These social practices, into which we are socialized throughout our lives, reflect the way societies are currently organized. To accommodate successfully to ecological limits we have, in effect, to begin to re-invent our social practices.

The bounds of sustainability, then, are set by cultural and historical factors, as well as by material, physical limits. In poorer societies where resources are particularly scarce it is material limits – the absence of groundwater, for example, or of forest cover – on which the largely poor populations depend. However, the current concern with global sustainability takes us beyond traditional conceptions of physical limits, to embrace the social mechanisms on which our appetites for resources depends. Beyond a consideration of physical limits, and ingrained social practices, sustainable development needs to address questions, like these, of equity.

2.2 Sustainable Development: The dimension of equity

The second dimension of sustainable development is that of *equity*. To understand why equity lies at the heart of the sustainable development discourse we must consider the competing claims of 'wealth creation' and the 'quality of life', the two goals which are routinely incorporated in the corporate plans of companies, and governments, in the European Union. There are broadly three discernible intellectual and policy positions on the relationship between economic growth (and wealth creation) and sustainability:

2.2.1 First perspective: Sustainability and economic optimality are incompatible

This is the view that, for business to expand, and to optimize its potential, the costs incurred by the environment need to be acknowledged as serious, and negative. By 2030, it is suggested, global Gross Domestic Product (GDP) will be over three times as large as it is today. It follows that pursuing sustainable development in its own right would put this economic achievement in jeopardy. Environmental policy objectives should be set in such a way that no serious limits are placed on economic growth: unsustainable practices should be avoided only when they are likely to reduce the net economic benefits of growth. Under these circumstances, then, if sustainability is pursued, especially at the international level, it will be because it is deemed important in its own right, and not because it is an essential condition for more economic growth. This is the perspective of most neo-classical economists working in government departments, and international bodies.

2.2.2 Second perspective: The pursuit of economic growth and sustainability is made compatible by measuring economic growth to make allowance for environmental costs

The second position on wealth creation and sustainability argues that the problem we face at the moment is not that increased sustainability will impede more economic growth, but that currently economic growth is measured in ways that do less than justice to the environmental costs of that growth. This view is espoused by ecological economists, and many other more orthodox environmental economists working in government departments, and for private companies. Essentially we are measuring 'wealth' in the wrong ways; for example, by failing to include pollution and resource depletion on the negative side of our balance sheet. This is the 'sustainability gap' represented by the difference between monetary indicators of current and future (more sustainable) levels of resource use. It is possible, say the advocates of this second position, to achieve *cleaner* growth, through the use of cleaner technologies, together with better ways of dealing with waste. Economic growth and development should be targeted on more sustainable wealth creation: the only kind of growth to make sense in the long run.

2.2.3 Third perspective: Sustainability will improve with faster economic growth

In contrast to the first position, which argues that economic growth and enhanced sustainability need to be pursued separately because sustainability and economic optimality are usually incompatible, this position argues that rapid economic growth is a prerequisite in achieving sustainability. It is argued that above a certain level of income some aspects of environmental quality improve with further economic growth. This was traditionally the view espoused by many economists working for the multilateral development banks, and other lending institutions.

There are clear difficulties with this position, which carry implications for global equity between the richer Northern countries and the developing countries of the South. This third prescription would appear to carry heavy environmental costs for the South during the period of economic transition – and this is borne out, for example, by much higher than average levels of air pollution in cities like Bangkok and Jakarta. Environmental costs are, on the whole, irreversible, so achieving economic targets may actually be prejudiced, at least in the longer term, by short-term economic gains associated with high levels of 'externality'.

If the developing countries were to emulate the standard of living in the North, then the environmental costs of such a transformation would be – globally at least – unsustainable, if measured by global effects such as carbon emissions. At the same time it is not clear that economies that are already rich – those of the North – are necessarily more sustainable, in most cases, than those of the South. Much of their economic growth is sourced from the South, and much of their waste and pollution is transferred there. As argued earlier, these 'dispersal effects' of sourcing and dumping, at several removes from the original site, need to be built into our thinking. They are the invisible dimensions of apparent growth, and apparent equilibrium (Redclift 1996).

The problem that lies at the heart of sustainable development, especially on the global scale, is that as long as most of the world persistently *under*-consumes critical goods and services (such as education, clean water and good housing) they remain suspicious of the minority who *over*-consume, less essential goods to sure-up their relatively privileged lifestyles.

The threat represented by global inequality is therefore not merely an abstract ethical one; it is a political one, manifest in the inability to reach binding and adequate global agreements over Greenhouse gas emissions. Still more remote is a global compact for future economic development based on securing improved living standards for the majority of the world's population (the original objective of the Bretton Woods institutions). In the face of these failures increased economic growth and consumption serves to fuel inequalities, both between countries, and within them. The price paid for the neglect of global equity is increased global insecurity. Difficulties in achieving sustainable development reflect both global inequalities and the process of globalization itself.

2.3 The dimension of globalization

The third dimension of sustainable development is that of globalization. There are several aspects of economic liberalization that are often linked with globalization, among them: increasing convergence in economic goals, increased emulation in patterns of consumption, and attempts to 'manage' the consequences of globalized markets through agreements on trade, capital markets and the environment.

The difficulty underlying the convergence of economic trajectories, which has characterized the last fifteen years, is that although there is more agreement internationally on the goals of the international economic system, there is also more competition to achieve them. Tensions emerge between economic competitors over trade, capital flows and access to 'natural' resources, which can quickly become intractable problems for policy. On the one hand the deregulation of markets is taken as the guiding principle of international agreements to liberalize trade, and investment. On the other hand, the consequences of economic liberalization include enhanced risk, and environmental damage, which, in turn become the subject of new regulations, aimed at establishing global standards. It also becomes clear that a tighter regulatory framework needs to be matched by precautionary action to prevent, rather than react to, environmental problems. A marriage of precautionary measures and re-regulation is not impossible to achieve, and may indeed be a *sine qua non* for progress in achieving sustainable development, but it also requires high levels of trust between states, and verifiable ways of measuring performance.

Finally, global management of the environment, combined with better global economic management, encounters problems at the sub-global level. Two such problem areas have become evident in the last decade, in particular.

First, economic restructuring has led to political, as well as economic resistance, in many parts of the world where the benefits of globalization are not universally agreed. There is social, cultural and political resistance to aspects of globalization in many Islamic states, where it is equated with 'westernization'. Similarly, where severe political conflicts, and structural power, is distributed very unequally, opposition to economic restructuring has frequently become the gathering point for ethnic and political dissent. An example of the latter process is the effective opposition mounted by the *zapatista* rebels in Chiapas, to the demands of the Mexican state, and its place within the wider constellation represented by NAFTA (North American Free Trade Area).

Second, the effectiveness and authority of international action, to achieve 'global' goals, such as averting climate change, is undermined by the neglect of the domestic environment and livelihood agendas of many governments in the South. Even for relatively large economies, like China and India, the effect of their economic behavior on global environmental change, such as biodiversity loss or climate change, is overshadowed by the need to meet their own policy priorities in areas such as health, education and immediate urban environment. The assertion that there is an overarching moral imperative to protect the biosphere's integrity is used to justify the creation of a new global regulatory order, but this new order is undermined by the 'real world' political economy that exists between the wealthiest countries (including those of the European Union) and the poorest countries.

Globalization, then, is not merely an inexorable process, but also one that carries the seeds of its own undoing by underlining the differences between 'global' criteria for sustainability, and the actual conditions under which most people live. Globalization is also a highly normative concept, whose very use changes the terms of international debate. It evokes political and ethical responses from different regions, religious groups and cultures, which differentiate themselves from processes of global economic and cultural convergence It is to this normative, representational question that we turn next.

The way we represent the global environment and population is the discursive terrain of 'development'. This discourse carries the preoccupations of the North, including the countries of the European Union. These include the aspiration to maintain, or improve, the economic well-being of its own populations, and to enhance its own competitive position *vis-à-vis* that of other economic blocs.

The discourse surrounding 'development' and 'wealth creation' also serves to obscure issues, by proposing materially verifiable data on global relations, and correlations between 'development' options, rather than the analysis of the social relations which underpin these structures. In place of an analysis of the responsibilities for global resource transfers, or global pollution, we are offered 'global data' and 'global models', some of which, (like Global Circulation Models for climate) take existing circumstances as the given. The point from which such 'global' models begin, and the direction of globalization 'must' take, are determined by existing economic strengths and resource inequalities. Structures are sacrosanct, while the processes determining them historically are seldom explored.

On the whole this global 'development' discourse takes place in a geopolitical space where popular demands for access to natural resources are also ignored. Given the primacy of the Northern agenda, which drives the globalization imperative, and which proposes geopolitical space as the canvas for discussion of the environment, opposing interpretations of development sometimes acquire political visibility, there is little room for opposing interpretations. The trajectory for development aspirations – patterns of consumption and resource use – is assumed to be that mapped out by the industrial world. Although in some respects developing countries become the imitator of these tendencies, in other ways they exhibit cultural distinctions driven, in part by distinct political agendas.

The limitations of 'global' environmental action, consistent with only one view of development, and thus lacking local legitimacy, are evident in the South. We can take the Amazon as an example. As Carvalho Teixeira states, in her discussion of the vast Tucurui dam:

> hydroelectric projects in the Brazilian Amazon ... aim to explore certain natural resources and geopolitical spaces, and to mobilize certain territories with a specific objective in mind: the production of energy in the name of economic growth. (Carvalho Teixeira 1994)

Within any territorial space groups of people are regarded, and routinely treated, as if they contribute to this end: the use of natural resources for the larger purpose of, (in this case Brazilian) energy policy and economic 'development'. Local populations in the Brazilian Amazon who oppose the development of large dams for energy production, are deemed part of the 'environmental problem'. Their protection by the Brazilian state, like that of the flora and fauna of the region, is essentially a guarantee of their role.

Ideologically speaking they can be represented as 'preserving nature', while they remain quiescent, but their political role changes when they fail to endorse the larger project. The idealized role that indigenous people play as 'jungle Indians' is much less important than the purchase this gives the Brazilian state in *interpreting* the interests of indigenous people. Globalization confers ideological power and legitimacy on those who utilize it for their own ends.

Part of the significance of the global development discourse, then, is that it is both hegemonic *and* contested. By challenging the representations of their place within the new order of globalization, and seeking to illuminate the relationship between global economic interests and its hold on our consciousness, environmental and political protests effectively challenge the assumptions of the global model. There are several areas – of which climate change is the most obvious – in which dependence on global modeling has matched that of global rhetoric. Not surprisingly, dependence on global modeling devices does not serve to increase peoples' responsibilities for the global environment, but serves to diminish them, by accelerating the processes through which global economic interests come to control local markets and resources.

By establishing a technical plane of 'expert' opinion, linked to inevitable economic processes with expected benefits for everybody, the aspirations of individuals are only partially accommodated. Globalization and global management can be used to undermine locally sustainable practices and fit awkwardly with global development trajectories. At the same time, the very existence of global consciousness has prompted local and national consciousness.

Less than a decade since the Earth Summit in 1992, global environmental management is still at a relatively early stage in its development.. However, our reliance on global modeling might eventually displace more conventional diagnostic tools for measuring welfare and livelihood benefits, rather as our material production of nature, (via genetics) has begun to displace 'nature' itself. The effect of globalization is thus intimately linked with the role of modernity, in the way 'sustainable development' is achieved.

2.4 The dimension of modernity

According to Giddens (1984) there are three elements that help us define modernity: First, social interaction takes place across space, with people we hardly know, through the medium of new information technologies, the media and 'virtual' communication. Second, people are increasingly forced to place their trust in 'expert' systems of knowledge, which they feel unable to evaluate or understand. Third, both these changes pose problems for human 'reflexivity', the process through which human behavior becomes self-conscious, and people express this awareness through human agency.

In the context of sustainable development the significance of modernity is that it poses questions for the way sustainable development is realized: it needs to be 'willed' by human agents, who may feel powerless. For societies the problem becomes: how to manage the transition to greater sustainability, to manage the environment better, without prejudicing 'human rights' and by making more effective use of 'human choices' (Rayner and Mallone 1998).

The discourse surrounding environmental policy in the European Union has tended to explore policy issues, such as the effects of consumption, and human welfare on sustainability, by articulating the need to develop a broad basis of consensus for environmental measures. It is assumed that 'the public' will lend support to measures if they are provided with the mantle of objective science – they are imperatives for human action. Thus we are told that there needs to be *more* public understanding of environmental issues, *better ways* of enrolling the 'public' in their implementation, and estimates of their *willingness to pay* for environmental improvements. Bodies of scientific knowledge are taken as givens; the task of public policy is to ensure that the vocabulary of choice is used to ensure that the right decisions are made, and that environmental policy acquires legitimacy. Clearly the agreement of human actors is an essential complement to the regulations, and market instruments available to the European Union, and its component national governments in undertaking sustainability planning.

At this point, however, sustainable development, as a policy commitment, confronts some of the problems associated with modernity discussed above. First, the rights and responsibilities of citizens in Europe are changing, with new systems of communication and new authoritative practices. Areas such as genetics have opened up new rights to consideration: those of the unborn child, or of tropical forest peoples' to 'their' natural forests. The recognition that these rights and responsibilities transcend environmental space, in a way hitherto largely unknown, adds another complication. Our behavior in Europe carries important consequences for the environment elsewhere in the world. Second, as we have seen, new forms of knowledge and communication have served to transform existing social bonds, and social relations. People in Europe today meet via the Internet, as well as in the street; they communicate virtually through telecommunications, as well as in person. The basis on which political consensus might be based, and authority given for public policy, has been transformed by human interaction itself.

This does not mean that structural power no longer exists, or that equity is no longer a facet of sustainable development. However, the manifestations of structural power are different in our societies, and power is more often wielded at a distance, and held to be dependent on impersonal forces, like the international market, or information held on computers.

One example of the way in which sustainability is being pursued in Europe, which embodies elements of the kind of 'expert knowledge' discussed above, and which seeks to incorporate the advances of science through technological change, is that of 'ecological modernization'. In other respects, too, 'ecological modernization' reflects other dimensions of sustainable development: the pressure to live within ecological limits, and in a context increasingly constrained by the experience of global markets and cultural processes.

The European Union's policy response to many of the problems in achieving greater sustainability has been to seek to 'internalize' environmental costs within processes and products, leaving European countries in a competitive position on global markets. This, in a nutshell, is the basis of ecological modernization.

2.5 Ecological modernization, as the 'internalized product'

Ecological modernization has been reviewed in a variety of publications (Simonis 1989; Spaargaren and Mol 1991; Weale 1992; Jänicke 1990; Hajer 1994; Gouldson and Murphy 1998). The central proposition behind ecological modernization is that economic growth can be adapted to meet environmental goals.

As Gouldson has expressed it: "[it] assumes that there can be a synergy between environmental protection and economic development, where in the past there has been conflict" (Gouldson and Murphy 1998: 5). The prime mover is government, which helps to provide a broader context than is usually provided by environmental policy alone. In specific terms this means the creation of new products and services that demonstrate improved environmental and economic performance. Essentially, ecological modernization proposes to internalize 'externalities', designing cleaner, more sustainable goods, which meet clear environmental standards.

In seeking greater integration of environmental policy goals with those of other sectors, ecological modernization seeks to accommodate late industrial society. It seeks to redefine international competitiveness in such a way that early technological innovators reap market advantages. It does not represent a threat to capitalist development, however, and those who argue for ecological modernization do not challenge the logic of international capital. As Gouldson and Murphy put it: "ecological modernization can be viewed as very selective in just where it apportions blame for environmental degradation" (1998: 8).

It is assumed that advanced industrial societies can shift their technologies, and patterns of production while leaving the structures of private capital accumulation fundamentally intact. There are a number of problems with this approach on the global scale. First, it is insufficiently grounded in international political economy, where recent debates have focused on 'flexible specialization' in production, the primacy of information and associated technology, and internationally differentiated labor markets. Ecological modernization suggests that economic restructuring can be modified to incorporate environmental goals of society. These Green goals serve to act upon the 'real world' of contemporary capitalism, enabling new environmental values to penetrate the very heart of the industrial process. The results is that companies, and governments, will be more competitive in the longer term within the global system. The economic restructuring of global capital is a reality, but in some of the most dynamic economies environmental 'externalities' remain just that, *external*.

As yet, however, there is little evidence that economic competitiveness has been refashioned to reflect more sustainable objectives. For example, a recent survey, conducted by the United Nations, of 794 leading transnational corporations with sales over $1 billion (US) per annum shows that most large companies attach relatively little importance to any environmental considerations likely to reduce their profitability. The conviction that ecological modernization represents a way forward for business, in no way suggests agreement with higher levels of external regulation, or commitment to longer-term environmental objectives. As with other concepts, including sustainable development, different writers have found different things in ecological modernization. Gouldson argues that the concept represents a challenge to the nation state, and to national regulation (Gouldson and Murphy 1998). Fleming questions whether ecological modernization is an effective way of addressing the problem of economic growth, in economies where growth is beginning to flounder (Fleming 1995). He draws attention to the contradictions between the European Union's goal of increased employment, and that of 'labor-saving' ecological modernization.

The problems are more severe if we look outside Northern Europe. The White Paper from the European Commission, on Growth, Competitiveness and Employment (1993) states that extrapolating current consumption and production patterns within the European Union to the entire world would require a ten-fold increase in resources. Europe's environmental protection industries, the nub of ecological modernization, are currently incapable of shouldering the burden of growth within Europe. It remains to be seen whether 'social coupling', the organization of the workplace around best environmental practice, can work in Germany, Scandinavia or the Netherlands, where it is advocated most strongly. To 'globalize' from European experience would require not merely major shifts in global economies, but also exacerbate divisions and distributional problems. Where does competitive advantage take you, if everybody gains from it?

There are other problems too. At the moment ecological modernization is largely confined to 'end of pipe' technologies, where environmental regulation is usually operative. It is significant, then, that those who favor environmental regulation usually see ecological modernization as a facet of business development, rather than a means of raising environmental standards. It is argued that business will take ecological modernization seriously, once it benefits financially from doing so.

The real challenge, however, as Herman Daly noted some time ago is to reduce energy throughout in the economy, rather than in the production of a limited range of 'greener' goods and services (Daly 1992). What is required is not the creation of 'greener' management accounting and environmental regulation, but a shift towards the wider recognition that sustainability might drive the economy. Until the globe's sink capacities have been assessed, and production modified to reflect these capacities, we will not have turned the corner to greater sustainability.

Finally, as Spaargaren and Mol, among others, have shown, ecological modernization does not extend environmental protection to many global environmental problems and risks (Spaargaren and Mol 1991). Such risks tend to be what Ulrich Beck has called risks of 'high consequence' but 'low probability', such as those of nuclear accidents or chemical warfare (Beck 1992). Ecological modernization, by contrast, seeks to allay uncertainties in policy formulation. Risk is materialized, in production, as part of a long-term strategy of economic harmonization.

The universality of high consequence risks makes management responses, such as those of ecological modernization, an irrelevance. Even if one dissents from Beck's view that the 'positive logic' of wealth distribution has been overshadowed by the 'negative logic' of risk distribution, it remains clear that only preventative action on the global scale will enable us to deal with global risks of this kind. This has served to redefine distributive problems; but they have not disappeared. Economic harmonization around products and markets exposes the poor to exploitation as a cheap resource. The poorer you are, the less effective is preventative action.

In practice, effective international action to address environmental problems is not amenable to technical solutions alone. It requires agreement about both means and ends, in which the internalization of environmental costs (through ecological modernization) can represent a market advantage from which the rich reap most of the benefits. Economic convergence towards 'greener' production, measured by indicators of sustainability is envisaged within the industrialized countries as a substitute for restructured economies and restructured international institutions (MacGillivray and Zadek 1995).

3. CONCLUSION

This chapter has argued that many of the environmental policy concerns of the European Union are, essentially, those of *Northern* Europe, and that there are parallels with the situation at the global level. A policy discourse is not a culturally objective phenomenon: it carries cultural and political bias, preoccupations and interests. Even a debate like that surrounding sustainability carries considerable cultural and political baggage. Perhaps the key question for environmental politics in Southern Europe, the focus of this volume, is the way in which sustainable development, like other policy discourses, has particularistic value while, at the same time, being clothed in the language of universals.

REFERENCES

Beck, U. (1992) Risk Society: Towards a New Modernity, Sage, London.

Brewer, J., and Porter, R. (1993) Consumption and the World of Goods, Routledge, New York, NY.

Brundtland Commission (World Commission for Environment and Development) (1987) Our Common Future, Oxford University Press, Oxford.

Carvalho Teixeira, M.G. (1994) Hydropower Development in Amazonia, Ph.D. Thesis, University of East Anglia.

Cobbett, W. (1842) Advice to Young Men and (Incidentally) to Young Women, in the Middle and Higher Ranks of Life. In a Series of Letters Addressed to a Youth, a Bachelor, a Lover, a Husband, a Father, a Citizen, or a Subject., Anne Cobbett, London.

Daly, H.E. (1992) Steady-State Economics, Earthscan, London.

European Commission (1993) Growth, Competitiveness and Employment: The Challenges and Ways forward into the 21st Century, White Paper Luxemburg, Office of Publications.

Fleming, D. (1995) Towards the low-output economy: The future that the Delors White Paper tries not to face, European Environment July 1994, 11-16.

Giddens, A. (1984) The Constitution of Society, Polity Press, Oxford.

Gouldson, A., and Murphy, J. (1998) Regulatory Realities: The Implementation and Impact of Industrial Environmental Regulation, Earthscan, London.

Habermas, J. (1992) Postmetaphysical Thinking, MIT Press, Cambridge, MA.

Hajer, M. (1994) Ecological modernisation and social change, in S. Lash, S. Szerszinski and B. Wynne, Risk, Environment and Modernity. Towards a New Ecology, Sage, London.

HMSO (Her Majesty's Stationary Office) (1993) Realising our Potential, London.

Jänicke, M. (1990) State Failure: The Impotence of Politics in Industrial Society, Polity Press, Oxford.

MacGillivray, M, and Zadek, S. (1995) Sustainability Indicators, New Economics Foundation, London.

Miller, D. (ed.) (1994) Acknowledging Consumption, Routledge, London.

Rayner, S., and Mallone, E. (eds.) (1998) Human Choice and Climate Change, Vol.3, Battelle Press, Columbus.

Redclift, M. (1996) Wasted: Counting the Costs of Global Consumption, Earthscan, London.

Simonis, G. (1989) Technikinnovation im ökonomischen Konkurrenzsystem, in U. Alemann, H. Schatz and G. Simonis, Gesellschaft - Technik - Politik. Perspektiven der Technikgesellschaft, Leske + Budrich, Opladen, pp.37-73.

Spaargaren, G., and Mol, A. (1991) Sociology, Environment, and Modernity: Ecological Modernisation as a Theory of Social Change, Wageningen University, Netherlands.

Weale, A. (1992) The New Politics of Pollution, Manchester University Press, Manchester.

Chapter 3

Sustainable Development – The Formal Or Informal Way?

The Case of Southern Europe

Helen Briassoulis

1. INTRODUCTION – WHY THE INFORMAL SECTOR?

The voluminous literature on sustainable development rests on an almost unquestionable assumption; namely, that sustainable development can be achieved by a variety of actions directed towards the *formal sector* – known, explicit, overt, and legitimate social groups and economic activities – that can be regulated, controlled and coordinated by the state. The *informal sector* – covert, invisible, non-institutionalized and frequently illegitimate social groups and economic activities – is ignored although considerable evidence exists to date that it constitutes a separate system running parallel to the official one (De Soto 1989; Roberts 1994) and influences significantly spatial and socio-economic development world-wide.

Considering the formal sector only, the achievement of sustainable development does not appear to be an easy task. Critics have already questioned its meaningfulness and feasibility (Beckerman 1994; Lele 1991; Toman 1992). But, if it is difficult to achieve sustainable development with known, formal actors, how this can be done with unknown, informal actors? Is it possible to plan, design and implement sustainable development in the presence of a sector that is currently outside the range of vision of those involved in this task? The literature has not touched this question yet, despite its significant practical implications. This author has made a first attempt to intro-

K. Eder and M. Kousis (eds.), Environmental Politics in Southern Europe, 73–99.

duce the informal sector into the current sustainable development debate at a general level (Briassoulis 1999).

The present paper focuses on Southern Europe as one of the regions where an extensive literature documents the presence and impacts of informal activities on local and regional development (Cappecchi 1989; Georgakopoulou 1988; Lobo 1990; Mingione 1990; Vaiou and Hadjimichalis 1997). It offers a first background analysis of potential influences of informal activities on the sustainability of development to set the stage for a search for policy approaches that take into account both the formal and the informal sectors in making development sustainable in the European South. It must be stressed that the 'European South' is not an undifferentiated whole, as differences exist among the Southern European countries (Hadjimichalis 1995). For the present purposes, this implies that the origins, causes and impacts of the informal sector will differ among these countries although basic commonalities do exist.

The discussion starts in the following section with a brief presentation of the informal sector, both world-wide and in Southern Europe. The third section examines how the formal relates to the informal sector and how the latter may impact on the sustainability of development. The last section suggests some principles for policy choices to manage informal activities on the way to sustainable development and indicates directions for future research.

2. THE INFORMAL SECTOR WORLD-WIDE AND IN SOUTHERN EUROPE

In 1972, the report of the ILO Employment Mission in Kenya brought the concept of the informal sector into international usage (Lubell 1991). Most studies identify informality with *lack of regulation*, and, consequently, with *non-institutionalization*, *non-registration*, and *illegality*. Informal activities are persistent, universal, regular. They are not random events (Ybarra 1989) as frequently has been assumed because they do not fit the model of formal (institutionalized) activities (Georgakopoulou 1988). They constitute part of the daily life of individuals and households and they have increased during the last decades (Castells and Portes 1989; Vaiou and Hadjimichalis 1997). The informal sector is an integral component of national economies, defined with reference to the formal sector:

> Any change in the institutional boundaries of regulation of economic activities produces a parallel realignment of the formal-informal relation-

> ship. In fact, it is because there is a formal economy ... that we can speak of the 'informal' one. In an ideal market economy, with no regulation of any kind, the distinction between formal and informal would loose meaning since all activities would be performed in the manner we now call informal ... The more a society institutionalizes its economic activities following collectively defined power relationships and the more individual actors try to escape this institutionalized logic, the sharper the divide between the two sectors. The informal sector grows, even in highly institutionalized economies, at the expense of the already formalized work relationships. .... The informal economy is ... a process of income generation characterized by one central feature: *it is unregulated by the institutions of society, in a legal and social environment in which similar activities are regulated.* (Portes, Castells and Benton 1989: 12-13)

A host of diverse and variegated activities, covering the same range as formal, regulated ones and aiming to satisfy various human needs, are carried out in an unregulated manner, thus generally falling into the informal category. They include trade and commercial activities, manufacturing (usually subcontracted), piece work at home, handicrafts, artistry, squatter housing, second home and tourist development, transportation, solid waste collection, disposal and recycling, water provision, urban agriculture, social services (e.g. baby and elderly care). The basic difficulty encountered in their study is that they are well concealed from public view (Portes, Castells and Benton 1989; Vaiou and Hadjimichalis 1997). It is no accident that, instead of the term 'informal', alternative expressions are used such as: non-standard, underground, illegal, hidden, shadow, unobserved, submerged, irregular, invisible, unofficial, parallel, silent, alternative, world underneath, shadow state. Informal activities are mostly desirable productive activities that differ from formal activities with respect to the *manner* by which a final product is produced and exchanged, namely, licit goods and services are produced and sold outside the regulatory apparatus (Castells and Portes 1989; De Soto 1989; Roberts 1994; Sassen-Koob 1989; Williams and Windebank 1995). However, undesirable, counter-productive, and criminal activities (e.g. black market, drug trafficking, illegal dumping of hazardous wastes) cannot be excluded. Informal activities may develop in close relation to formal activities, as in the case of large firms subcontracting certain parts of the production process to small, informal ones (Roberts 1994; Vaiou and Hadjimichalis 1997).

Available estimates of the magnitude of the informal sector differ depending on the assessment techniques used (Kanellopoulos 1992). The table below indicates the relative size of this sector in the member-states of the EU assessed as percent of GDP (Williams and Windebank 1995).

Table 1. The relative size of the informal sector in the member-states of the EU assessed as percent of GDP

	Smallest	Highest	Average
Ireland	0.5	7.2	3.9
Britain	1	34.3	6.8
Germany	3.4	15	8.7
Netherlands	9.6	9.6	9.6
Denmark	6	12.4	10.1
Belgium	2.1	20.8	10.9
Spain	1	22.9	11.1
France	6	23.2	11.4
Portugal	11.2	20	15.6
Italy	7.5	30.1	17.4
Greece	28.6	30.2	29.4

Various causes trigger and underlie the emergence of informal activities that are specific to the societies and time periods where informalization takes place (Castells and Portes 1989). Among them, the literature emphasizes as most important general causes: (a) the restructuring of the world economy since the 1970s – intensification of international competition, integration of previously protected national markets and decentralization of production, (b) the process of industrialization in many third world countries where social and economic conditions forbid much enforcement of state regulations, (c) improvements in technology and communications, (d) the 'retreat' of the state from economic and social intervention as a strategy to promote market-led growth, (e) the reaction by both firms and individual workers to the power of organized labor, (f) the reaction against the state's regulation of the economy (Castells and Portes 1989; Leontidou 1990; Roberts 1994; Williams and Windebank 1995). Overall, structural problems in a particular area of the formal socio-economic system cause the creation of a *parallel socio-economic system* that more effectively satisfies social needs (Sassen-Koob 1989; Ybarra 1989).

In Southern Europe, the growth of informal activities in the 'golden' 1968-1990 period is attributed to: (a) the structure of the labor force – low rates of economically active population, high ratios of self-employed, pluri-employment, low female participation rates, (b) the high unemployment and underemployment rates, (c) the direct and indirect cost of labor, (d) the structural characteristics of enterprises (magnitude, modernity, labor intensity, seasonality) – existence of many delinquent small and medium enterprises, (e) the tax burden and the tax 'ethos', (f) lax state regulatory systems – especially, as regards spatial development and land use control, (g) the structure of the land market and high urban and rural land subdivision, (h) the decrease of outmigration combined with repatriation and the revival of the countryside, (i) foreign immigration from the Third World (especially in Italy), former socialist countries (especially in Greece) and former colonies (especially in Portugal) and (j) the creation of 'intermediate regions' caused by the decentralization of production and population towards smaller urban centers and the periphery of big cities (Briassoulis 1992; Economou 1995; Leontidou 1990; Roberts 1994; Vaiou and Hadjimichalis 1997). The 'Third Italy' and the old industrial triangle Milan-Torino-Genoa, Braga, the axis along the river Duro in Portugal, Madrid and the coast of Alicante up to Barcelona in Spain, the regions of Attica and Macedonia in Greece exemplify successful, dynamic, endogenous economic development, mainly in the 1970s to mid-1980s period, based upon diffused industrialization and flexible specialization (Cappecchi 1989; Piore and Sabel 1984; Vaiou and Hadjimichalis 1997). Other areas have also exhibited endogenous, though less successful development based on informal activities such as Central and North Portugal, Valencia, Catalonia, Madrid and the Basque region in Spain, the Languedoc in France, Mezzogiorno in Italy, certain regions in Macedonia and Thrace and tourist islands in Greece (Vaiou and Hadjimichalis 1997).

Deeper, socio-psychological, cultural and political forces, however, underlie the emergence and impacts of informal activities in the countries of the European South. Familism and the strong value attached to the private sphere (atomism/individualism) give preponderance to the family as a basic productive and reproductive unit that serves also as a pool of cheap female labor. A traditional mode of direct communication – without the agency of third parties – gives priority to personal ties and favoritism that foster and maintain a tradition of a relatively atrophic civil society. Combined with state corruption and lack of satisfaction with the (usually inefficient) state, these traits create a political culture of clientelistic statism and a deeply rooted mistrust of and cynicism towards the state and its rules (Demertzis 1997; Griffin 1997; Pinto and Nunez 1997). To these, add a Southern European and Balkan tradition of 'cleverness' and a desire to make quick money that combine with efforts to survive working on wage labor (Vaiou and Hadjimichalis 1997). Particular types of farmer-businessmen and various forms of pluri-employment, thus, arise as well as particular modes and codes of social conduct. In a nutshell, when it comes to meeting social needs (housing, employment, extra income) the formal system is frequently ignored (even if it happens to be efficient). This is simply because, traditionally, formal (state) intervention has been out of question for certain (though not for all) aspects of socio-economic life. Hence, the precedence of the informal over the formal in most Southern European countries to date.

Vaiou and Hadjimichalis (1997) classify informal (economic) activities in Southern Europe into: (a) criminal activities (gambling, drug trafficking, illegal arms trade, illicit dealing in antiquities), (b) activities exploiting the profitable 'gap' of the formal system of control/regulation (tax evasion, illegal banking, favoritism, evasion of labor legislation), (c) traditional production activities (construction, tourism, handicrafts, petty trade, fruit collection) benefiting from international immigration since the 1980s, (d) activities generated by the restructuring strategies of certain businesses in agriculture, manufacturing, trade and services (flexible specialization).

The state has a dual role in this context. On the one hand, the inadequate state provision of goods and services, lack of coordination between state agencies and lax enforcement give rise to informal activities. On the other, the state is a principal, conspicuous or inconspicuous, participant in the formation and functioning of that system (Pichon 1992, 1996, Williams and Windebank 1995). Governments world-wide tolerate or even stimulate informal activities as a way to resolve potential social conflicts (e.g. housing crises), promote political patronage, protect their national territory, reduce unemployment and spur economic development (Castells and Portes 1989; Pichon 1996; Williams and Windebank 1995). In Greece and other Southern European societies, "informal activities constitute a … form of control from the dominant social classes and sex and nationality groups – frequently with the semi-suppressed support of the state, the local government and the political parties" (Vaiou and Hadjimichalis 1997: 202). In Italy, for example, the Communist party in the government of the central regions and the Christian Democrats in the Northeast have both collaborated in the informal practices of small and medium-sized enterprises through their influence on industrial relations (Vinay 1987). Similarly, in poor agricultural areas, the textile industry around Barcelona and in the footwear industry in Alicante, Spain local authorities protect black market businesses. In Greece and in Portugal, the state does not (among other things) enforce industrial homeworking legislation (Lobo 1990; Mingione 1990).

In summary, informal activities are invisible, covert, ad hoc and opportunistic and, hence, difficult to control. Parties engaging in them, as well as their specific practices and tactics, *cannot be known a priori*, unless they are so regular that they are tacitly known to the authorities and the broader public. Therefore, informal outcomes are unpredictable – in quantity and in quality –, uncertain, risky and, usually, they are revealed after the fact. At times, informal activities may mix with state/formal activities, if this is beneficial to informal actors. Usually, informal activities are not *visibly* coordinated among themselves. Because most informal actors are small in size and scope, they can adapt quickly to changing circumstances and, hence, secure their survival and satisfaction of their needs.

3. THE INFORMAL SECTOR AND SUSTAINABLE DEVELOPMENT IN SOUTHERN EUROPE – EXPLORING THE RELATIONSHIP

Socio-economic and spatial development results from the intricate interplay of formal and informal sector activities (Figure 1). The particular socio-economic, political, cultural and institutional conditions determine, to a considerable extent, the magnitude of contribution and the significance of impacts of each sector on observed socio-economic and spatial patterns. The sustainability of a given development pattern can be assessed and evaluated using the three criteria of economic efficiency and welfare, environmental protection, and social equity and justice. A development pattern gets a rating between 'yes' (the criterion is fully satisfied) and 'no' (the criterion is not satisfied at all) on each criterion. It is judged sustainable if all three criteria are satisfied to the maximum possible degree and unsustainable if none of them is satisfied. Intermediate situations exist in between the two extremes depending on the relative degree to which each separate criterion is satisfied. Thus far, only the *direct* impacts of the formal sector on sustainable development have been examined and analysed (arrow 1 in Figure 1). However, improper functioning of, or changes in the formal system, as discussed in the previous section, give rise to informal activities which, in their turn, impact on sustainability. Hence, a comprehensive and holistic analysis should include accounts of the genesis of informal activities and of their impacts on the sustainability of development (i.e., arrows 2, 3, 4, 5, 6). This is attempted in the next section using the three criteria of sustainability. The criteria are examined separately for purposes of analysis and exposition only, since all types of impacts – economic, social and environmental – are interrelated and draw from common causes. The analysis focuses on Southern Europe but parallels to other countries are indicated.

3.1 The economic criterion - economic efficiency and welfare

Sustainable development requires the elimination of poverty. Economic prosperity and vitality, however, are guaranteed when material and immaterial resources are used rationally and efficiently over time. Hick's (1946) definition of sustainable income as "the maximum value which (a man) can consume during a week and still expect to be as well off at the end of the week as he was at the beginning" gives perhaps the exact operational meaning of economic efficiency in the context of sustainable development (properly transferred to socio-spatial entities – cities, regions, nations).

The neo-classical notion of efficiency, commonly used to define and analyse economic sustainability, reflects the preoccupation with the formal sector only as it assumes a centralized model of *formal* economic organization. As complete information as possible is required about: (a) the identity and goals of all present and future parties concerned (resource owners and users); (b) the amount of resources they command, control and use; (c) the constraints on current and future resource use. Optimization of a social welfare function subject to constraints generates efficient solutions to resource allocation problems. In other words, based on these solutions, policies are designed that require planning and coordination within and among the relevant resource markets in order to materialize. Because of present and future uncertainty, the 'precautionary principle' calls for conservation of resources and avoidance of waste (O'Riordan 1993; Perrings 1991).

Things, however, are not as ideal and perfect in the real world as in the abstract neo-classical economic paradise. Even in the case of institutionalized and regulated activities, incomplete knowledge of the present and the future, policy implementation difficulties, problematic and costly coordination of the many parties involved, among others, hamper the economically efficient use of resources and render questionable the neo-classical, economic, and centralized view of reality. More importantly, however, the imperfections of the formal economic system and of the policies and measures devised to control it – regulations, taxation, technological improvements, etc. – generate a parallel, informal economic system. How this system influences economic efficiency and economic welfare and, ultimately, economic sustainability is examined next.

3.1.1 The bad news – Negative influences of the informal sector on economic efficiency

The main negative influences of the informal sector on economic efficiency and welfare can be grouped into six categories: (a) refutation of complete knowledge, (b) distortion of factors markets (labor, capital and resources), (c) distortion of products and services markets, (d) distortion of competition, (e) distortion of the official measures of economic performance, (f) non-durability.

(a) *Refutation of complete knowledge.* Perhaps the most fundamental problem with informal activities is that they nullify the complete knowledge requirement of the neo-classical notion of economic efficiency. Even when informal actors are known, or when formal actors act informally, their goals are unclear and changing, their actions are covert and opportunistic, the outcomes of their activities are unpredictable and uncertain. Things worsen when unknown actors are revealed gradually through their actions. The situation is serious when informal activities participate considerably in local affairs as in several agricultural, industrial and tourist areas in Italy, Portugal, Spain and Greece. The development of squatter housing in Southern European, but also in other first and third world cities, illegal land subdivision, haphazard urban expansion, diffuse industrialization, illegal work, induced development in tourist areas, illegal dumping of hazardous wastes, illegal water abstraction for drinking and irrigation purposes, are all manifestations of resource-consuming, 'invisible' actors acting opportunistically outside the official system (Sassen-Koob 1989; Vaiou and Hadjimichalis 1997; Williams and Windebank 1995).

Overall, the informal sector creates a decentralized model of economic organization that implies that it is impossible or very difficult: (a) to specify completely and adequately the formal social welfare optimization problem, (b) to generate realistic and applicable solutions and policy proposals and (c) to intervene formally, coordinate, plan and regulate economic activities for the purpose of efficient resource use. Informals may coordinate among themselves (Romanos and Chifos 1996; Vaiou and Hadjimichalis 1997) – although still opportunistically and without a visible plan – but not with the formal, official sector where responsibility and authority for efficient resource use generally lies.

(b) *Distortion of factors markets.* Informal labor is the most extensively studied aspect of the informal sector in Southern Europe. Pull and push factors have created an oversupply of labor (cheap, unskilled or, deskilled) with a parallel drop in the remuneration of labor, thus distorting the normal, competitive functioning of labor markets. Chief *pull* factors include the high (but, frequently, seasonal) demand for labor from small to medium enterprises (SMEs) in particular sectors (agriculture, manufacturing and services) and places (e.g. fruit groves in Central Macedonia, Andalucia, Valencia and rice fields in the Pado valley), technological changes (especially in agriculture and services such as tourism), and the intense international competition. *Push* factors include international and internal migration, taxation, welfare contributions and the rigidity of institutions that govern formal labor. The 'Third Italy', Braga and the axis along the river Duro in Portugal, Madrid and the coast of Alicante up to Barcelona, Attica and Macedonia in Greece exhibit high concentrations of informal labor (Vaiou and Hadjimichalis 1997).

A set of interdependent factors explains the uneven temporal and geographical distribution of informal employment: (a) the nature of the enterprises involved and the opportunities offered for pluri-activity – specialization in food products, drinks, textiles, garments and tourism combined with purely agricultural jobs, (b) the role of family and kinship – large pools of unemployed women and dense, secure and stable networks of relatives, (c) the geographical proximity of SMEs and their greater than usual 'affinity' with particular places which function – especially for smaller enterprises – as security against market uncertainties and the daily problems of production, information, distribution, specialized labor availability, (d) the particular location of the labor markets which makes them accessible to immigrants (Roberts 1994; Leontidou 1990; Vaiou and Hadjimichalis 1997).

Informal activities usually secure funding from relatives and friends or from illegal lending money markets (demanding high interest rates). In this latter case, the cost of capital increases, thus distorting capital markets in general. The impact of monetary policies could also be upset if (unexplained) growth in the currency is due to illegal activities (Smith 1981). The same is true for the cost of land and other resources when these are exchanged in informal markets. Little research exists on these issues, however, and their occurrence is gauged indirectly from other characteristics of informal activities.

As regards technological innovation, a dual situation exists. According to one point of view, informalization may reduce or eliminate incentives for technological progress, innovation and investment as entrepreneurial efforts center on better strategies to conceal clandestine activities (Ybarra 1989). Small enterprises in remote, depopulated areas of Southern Europe specializing in traditional sectors (family agriculture, petty trade and services) or in marginal areas of cities (Mezzogiorno, Ipiros, South Peloponissos, some islands, Galicia and regions in the interior of Spain and Portugal) fall in this category. They employ so-called 'defensive flexibility' strategies; these are characterized by restricted flexibility in production, numerical flexibility (labor intensity), low technology for low quality products, dependence on third parties, low remuneration, and frequently unacceptable working conditions. However, firms that develop international linkages exhibit 'aggressive flexibility'; i.e., combination of flexibility in the organization of production with functional flexibility, use of high technology for the design and production of quality products, relative independence of choice and satisfactory wages and working conditions (Lyberaki 1991; Vaiou and Hadjimichalis 1997).

(c) *Distortion of products and services markets.* Informal production usually employs informal channels of distribution. Oversupply of products normally leads to lower prices in informal markets compared to the prices of equivalent products in formal markets. However, for certain products and services, consumers may face higher (e.g. dwellings, rents) or lower to zero (e.g. the non-remunerated labor of women) prices if these products are not provided by the formal system (e.g., due to public spending cuts). Consequently, product prices are not comparable and they do not serve their main function as signals for over- or under-production. In addition, the absence of quality controls in informal markets does not guarantee product safety and reliability. Product markets are, thus, distorted exhibiting extreme fluctuations in supply (over- or under-), prices (usually, inflation) and quality. In Greece, the magnitude of informal activity could be indirectly assessed by the disproportionate magnitude of private compared to total consumption (Kanellopoulos 1992). In addition, by comparing private consumption to production in the corresponding economic branches, the magnitude of underground sectoral production in 1982 was assessed. In manufacturing, the underground was found to be 78%, in transport and communications 31.4%, and in 'other services' 28% of GDP. In 1988, the figure for manufacturing rose to 90% while in transport and communications to 46.2% (Kanellopoulos 1992).

(d) *Distortion of competition.* Distorted products and factors markets lead naturally to distortion of competition that, in its turn, may distort these markets further. Big, mostly formal businesses operate in parallel with small, informal ones. A highly efficient, high-wage sector of the economy functions alongside a low productivity, low-labor cost sector (Smith 1981). Secure wage employment contrasts sharply with insecure, informal employment. SMEs integrated in international markets compete with others that simply survive. Buoyant, thriving cities and regions exist together with stagnating, "pockets of crisis and unemployment" (Vaiou and Hadjimichalis 1997). This formal-informal divide is not conducive to 'healthy' competition and, presumably, to efficiency in market operations. Over- or under-supply of products and factors send signals that are not always reliable and representative of the real magnitudes of the respective supply and demand. In general, tax and rules evasion render competition between firms that pay taxes and those that do not unequal (Kanellopoulos 1992), while at the same time indirectly subsidizing – however temporarily – certain sectors and businesses (Hadjimichalis 1995) and, thus, distorting the conditions of fair competition. Lack of competition makes the development of local regions more dependent now than in the past on international markets (more specifically, on the locational decisions of multinational firms). Regions that base their economic dynamism on international links thrive while those, in the same country, that do not, become marginalized from the new economic trends.

(e) *Distortion of the official measures of assessing economic performance.* As informal labor, informal exchanges of money, production, consumption, income, profits, employer contributions, among others, are not declared and recorded in official statistics, several formal economic indicators commonly used cannot be assessed accurately and cannot serve as reliable indicators of economic performance. For example, the level of unemployment can be overestimated when informal work may range from a conservative 5% to as high as 20% of the level of declared work (CEC 1991). The real magnitude of the national product and income are usually underestimated. In Greece, Kanellopoulos (1992) estimated that in 1982 the GDP was underestimated by 27.6% and in 1988 by 31.2%. He mentions also Italy's GDP underestimated by 20% (1975-77), Portugal's by 22% (1986), Spain's by 15-25% (1987). Under such circumstances, with inflation overstated and employment understated, Smith (1981: 45) observes that "the European economies are not as depressed as they are made out to be".

These unreliable estimates of measures of macroeconomic performance make the formulation and implementation of efficient and effective economic (especially, stabilization) policies questionable as "if, for example, the underground economy is growing while the official data show stagnation, measures to reinforce effective demand will cause unexpected inflation because the slack productive capacity is lower than what the official data show" (Kanellopoulos 1992: 40). The same seems to apply to regional or local level policies as these, too, require reliable estimates of economic performance (regional product, income, unemployment) in order to provide the right assistance measures. Hadjimichalis (1995) cites the example of Kastoria, Greece, where the formal and informal labor markets totaled 18-20.000 workers in the late 80s representing 38% of the total population and 90% of the economically active population. However, official statistics offered an image of crisis purposively conveyed by the locals in order to obtain subsidies, tax exemptions and other privileges from the state.

(f) *Non-durability*. The opportunistic character, close bonds to the formal sector and the imperatives of big capital do not ensure the durability, long-term efficiency and viability of informal activities when the conditions that favor their proliferation change over time. Successful areas of the Third Italy have been showing signs of dysfunction since the late 1980s. In Emiglia-Romana, the most powerful motor manufacturing firms get centralized and, in turn, are being bought out by larger firms in Milan, Turin, Frankfurt and Munich. Similar developments occur in the electronics and food industries (Vaiou and Hadjimichalis 1997). Over-production and competition for (sensitive) agricultural products, the seasonality of demand for labor, the opening of new, even cheaper, labor pools in neighboring countries (e.g. Bulgaria in the case of Greece), the slow pace of technological innovation in SMEs, new technologies, immigration, the unwillingness of young locals to accept problematic working conditions and low wages, overpopulation in tourist regions and environmental pollution, all undermine the viability of the most vulnerable informal activities as well as of the regional economies that depend on them. The structure of the family undergoes deep changes causing crisis in informal activities (Vaiou and Hadjimichalis 1997; Vinay 1987). Heavily developed tourist areas are loosing their dynamism recently as competitors emerge in other countries and existing facilities become congested. For these reasons, questions have arisen as to whether the success and dynamic effectiveness of the informal, 'endogenous' development model of the previous decade can be generalized (Hadjimichalis 1995). From the viewpoint of sustainability, the main issue is that informal activities may not secure long-term economic efficiency and welfare.

3.1.2 The good news – Positive influences of the informal sector on economic efficiency

The literature reveals certain positive economic influences of informal activities. At the individual and family level, particular groups find a job and make a living, secure housing or obtain other forms of support from the informal sector that, frequently, is the only alternative open to them. The 'beneficiaries', however temporary, include immigrants to Southern Europe, 'marginal' groups such as unskilled workers and women, inhabitants of depopulated and underdeveloped agricultural regions (Alicante in Spain, Mezzogiorno in Italy, the Provence-Alpes-Côte d'Azur area in France).

At the local and regional level, well known cases reveal the contribution of the informal sector to economic welfare. The most celebrated is that of the Third Italy in the 1970s and 1980s (Cappecchi 1989; Vaiou and Hadjimichalis 1997), but also areas around Madrid and Lisbon, urban, agricultural and tourist regions in Greece and other Southern European countries. As mentioned before, flexible specialization and geographical proximity of similar, small units, opportunities for pluri-activity, subcontracting and abundant, cheap labor have helped businesses in these regions to adapt to changing economic situations and become integrated in international markets (Vaiou and Hadjimichalis 1997).

The aggregate effects of informal activities on economic efficiency and welfare are, however, not easy to assess, both in general and for Southern Europe in particular. But the evidence presented previously suggests that, in most cases, the main prerequisites for the sound operation of competitive markets to produce a 'strictly' efficient allocation of resources are rarely satisfied. On the one hand, the formal system does not function properly and, on the other, it generates informal activities that aggravate the problems. The positive economic impacts of informal activities are mainly localized, place-specific and short- to medium-term and they do not appear to outweigh the negative impacts. In general, the economic efficiency criterion for sustainable development seems to be violated most of the time in Southern Europe.

3.2 The environmental criterion - Environmental protection and conservation

Environmental protection, the cornerstone of sustainable development, is promoted through a variety of physical, legal, institutional, fiscal and economic policies and means, all of which target formal economic activities at all spatial levels. However, the formal systems of environmental control and protection are not always efficient and effective, as several environmental problems persist, not to mention those countries where these systems are inadequate, loosely implemented or totally absent. In addition to the incomplete formal environmental protection regime, informal activities also put pressure on the environment, although, as discussed below, certain of them may work in the opposite direction.

3.2.1 The bad news – Negative influences of the informal sector on the environment

The adverse physical and environmental impacts of informal activities are well documented world-wide and in Southern Europe (Hardy and Ward

1984; Leontidou 1990; Marquette and Pichon 1997; Pichon 1996; De Soto 1989). The notorious development of squatter settlements, widespread in Southern Europe as in many third world countries, has been extensively analysed but this is not the sole manifestation of informal activities in exurban space. The phenomenon of exurbanization– first and second homes, tourist, industrial and recreational facilities – is mostly the result of informal processes utilizing space in haphazard and formally unplanned ways. Numerous institutional, legal, socio-economic and political factors, the discussion of which is beyond the scope of this paper, supports an *extensive* model of exurban land allocation and facility siting (Economou 1995). Its most important adverse impacts are discussed below. Contrary to those caused by formal developments, these are unpredictable in quantity, intensity, space and time.

Exurban land is consumed indiscriminately without any consideration of its intrinsic quality and suitability. The loss of prime agricultural land in the periphery of large conurbations to non-agricultural uses in the process of suburbanization and exurbanization is well known (Economou 1995; Vaiou and Hadjimichalis 1997). Moreover, since these 'consumers' of land rarely follow a formal plan, land use conflicts are frequent and lead to economically inefficient use of exurban space.

Forest ecosystems and ecologically fragile or aesthetically sensitive regions frequently fall 'victims' of unplanned, exurban development. Forests are cleared 'violently' by setting fires to create space for housing (Briassoulis 1992), wetlands are dredged, filled illegally and then built, structures are erected within environmentally protected areas, garbage is disposed of carelessly, untreated sewage from illegal facilities is discharged into water bodies (as in the case of Kastoria, Greece cited in Vaiou and Hadjimichalis 1997). The result is the loss of precious ecosystem resources, irreversible damage and destruction of the habitats of endangered species (such as the monk seal and the sea turtle *caretta-caretta* in the Mediterranean), and aesthetic pollution of unique landscapes.

Other environmental and physical resources, such as water and infrastructure (roads, water supply and sewage systems) are used frequently beyond their carrying capacity as informal activities generate additional loads and demand. For example, illegal tourist facilities attract extra tourists and, thus, generate higher demand for water, more traffic and waste loads in tourist areas than those normally expected from formal developments. The same applies to illegal first and second home developments that utilize resources without contributing to their creation and maintenance (through, e.g., taxes and user fees) as is the case with formal developments.

Informal activities cause adverse physical and environmental impacts in urban areas of Southern Europe, as it happens in areas similar to those of the developed and the developing world where they abound (Sassen-Koob 1989; Vaiou and Hadjimichalis 1997). Dense concentrations of small, informal enterprises within the urban fabric form manufacturing districts that mix with other urban uses (usually high density residential and recreational), produce haphazard housing development violating various building codes and cause frequent land use conflicts. Moreover, they generate additional traffic and contribute to already congested transportation networks, destroy old, traditional buildings, produce industrial wastes, utilize natural and manmade resources (air, water, infrastructure) and detract from the visual quality of urban neighborhoods. Informal enterprises operate in complete violation of prevailing health, safety and environmental standards causing frequent accidents (explosions and fires) and other environmental problems (e.g., groundwater contamination, ambient and indoor air pollution, etc.). They therefore place humans and the urban environment at risk. Vaiou and Hadjimichalis (1997) cite the unhealthy working conditions in informal firms in Thessaloniki and Kastoria operating in homes (women working piece-work). Similar examples can be reasonably expected in other Southern European countries. In both the urban and exurban cases, although the impacts of a single activity may not be serious, their *cumulative* environmental impacts may be significant, as too is the case with formal activities.

The adverse environmental impacts of informal activities raise several concerns. These invisible (to the eye of the law) activities cannot be brought under the authority of the formal environmental protection regime of regulation and control. The current formulation and implementation of environmental policies are, thus, seriously called into question as they do not account for a significant group of users of environmental resources and services. Compliance by the formal sector only is doomed to ineffectiveness as most environmental resources and sinks are shared in common by all activities and the informal ones do not contribute their share of environmental use and protection. Rational environmental planning and management are hampered as the magnitude and timing of the demands of the informal sector for environmental resources and services are unknown and because formal activities compete for scarce resources with the informal ones, the latter not being subjected to plan and control.

3.2.2 The good news – Positive influences of the informal sector on the environment

Certain informal activities make, however small, positive contributions to environmental protection. Informal (voluntary) environmental organizations and groups defend the environment in many countries all over the world including those of Southern Europe. Although the effectiveness of their actions is not known, it may not be negligible, at least at the local level. The protection of endangered species, fire protection of suburban forests, protests against siting of hazardous facilities are frequently aided by informal groups dedicated to the cause of environmental protection. However, their impacts are localized as they are small in size, possess limited resources, power and authority and usually lack coordination with official environmental protection agencies.

A variety of other informal, income-earning activities contribute positively to environmental management whenever states fail, wittingly or unwittingly, to do so. Romanos and Chifos (1996), among others, cite examples from third world countries of informal garbage collection, disposal and recycling, water provision, light vehicle transport services and urban agriculture which spare the respective cities unhealthy living conditions as well as the state its management burden. Certain governments either encourage or participate illegally in these activities (Desa 1990; Sanyal 1987; Sinha and Amin 1995), while others make formal arrangements with informal activities as in the case of refuse collection in Cairo (Assaad 1996). Comparable Southern Europe research results are not available yet, at least to this author's knowledge, although evidence of such activities exists. In Greece, for example, gypsies tour in the islands during the summer and collect aluminum cans that they sell to recycling companies contributing partly, in this way, to a cleaner environment and to resource conservation. Similarly, informal agriculture seems to be practiced but no relevant studies exist.

An aggregate assessment of the environmental impacts of the informal sector is not possible as critical information is missing, at least for Southern Europe. In addition, environmentally benign informal activities rarely mix with environmentally detrimental ones such that positive influences cancel out negative ones. But the existing evidence points to the negative side. Both in the short- and in the longer run, informal activities render the achievement of environmental protection questionable and the sustainability of development shaky. Informal actors consume unknown and unpredictable quantities of environmental resources and services. They render formal planning and management efforts problematic as they are not (and cannot be) controlled by effective environmental legislation, contributing, thus, to economically inefficient use of environmental resources and services.

3.3 The social criterion – Social equity and justice

Intra- and intergenerational justice in the distribution of the costs and benefits of development is the third basic tenet but the most thorny aspect of sustainable development. It is difficult to define both conceptually and operationally. It is associated with poverty alleviation, equality of opportunities, enhancement of socio-cultural diversity, empowerment and citizen participation in local affairs. It implies redistribution of social and economic benefits within and between generations. The means by which the multifaceted demand for social equity can be met as well as the relative ease of this task vary with the spatial scale as well as the prevailing social, economic, and institutional conditions. Currently, most formal socio-economic systems, regardless of their spatial scale, do not function efficiently and fairly and they are blamed for the unsustainability of development. Moreover, they spur informal activities which cause further inequalities. The evaluation of their impacts is neither easy nor straightforward because, the more the informal economy expands, work situations and social conditions become more heterogeneous and the class structure of a given society becomes blurred (Castells and Portes 1989). Additionally, information is limited for Southern Europe on all aspects of this criterion. Nevertheless, a first attempt is made below.

There is no more good news. Informal activities may contribute to employment and income generation for poor families and individuals (especially women) world-wide and in Southern Europe, especially when the alternatives offered by the formal sector are limited (Castells and Portes 1989; Lubell 1991; Roberts 1994). They may benefit greatly both small and big businesses contributing significant profits and saving them taxes and welfare contributions. As the successful examples of micro-capitalist development in Southern Europe demonstrate, they may mobilize local resources and revive urban economies. A host of informal activities (e.g. mutual-aid societies, cooperatives, etc.) provide a broad variety of tangible and intangible social services (e.g. day and elderly care, credit provision), thus dispensing with the need for state provision or intervention (Putnam 1993). But, as many analysts point out, the short- and long-term social cost of these direct, 'positive' impacts is considerable.

At the individual level and in the short run, informality simply serves to avert temporarily acute situations of poverty. Many male but especially female workers, persons under 14 and, recently, foreign immigrants to Southern Europe receive very low remuneration and work without any social security or insurance coverage, supporting, in this way, the high incomes of informal entrepreneurs (Vaiou and Hadjimichalis 1997). These pecuniary, tangible dis-benefits are accompanied by intangible dis-benefits which may be more serious and insidious. Informal workers face physically and psychologically adverse working conditions and labor relations. They are assigned unhealthy and hard tasks, they work without any elementary safety precautions and frequently under polluted conditions (e.g. fur coat workers in Kastoria, Greece). When international competition becomes keen, employers pass the pressures on workers demanding more output without additional pay. Finding a job in the informal market relies on word of mouth and certain social networks whose *spatial fix* is important for their operation and success. The family life of women piece-working at home is disrupted as work interferes with family functions, causes fatigue, stress, accidents, interior pollution and isolates the workers from their social context (Vaiou and Hadjimichalis 1997).

At a collective level, the prevalence of informal activities has serious consequences. Informal forms of labor are not recognized as ‘regular’ in official contexts (e.g. in the European Union). Consequently, all those who engage in them, and especially women, are not considered as “social partners” (Hadjimichalis 1995: 214). Labor unionism is low especially in geographical areas of high informal labor concentrations – e.g. in Kastoria, in the regions of Third Italy – as needy workers are willing to compromise longer term benefits for shorter term survival (Vaiou and Hadjimichalis 1997). Naturally, such areas attract more informal businesses and there are indications that they may be associated with conservative societies (e.g. Kastoria, Greece cited in Vaiou and Hadjimichalis 1997). Furthermore, the powerful groups in control of informal activities (businessmen, realtors, money lenders, etc.) form *informal power structures* that operate in parallel (or, in co-operation…) with the formal ones. These inhibit variously the functioning of the formal system, maintain and reproduce the informal relationships of the past and nurture more inequalities. As an example, building associations and local interest groups (land owners, entrepreneurs, engineers) oppose formal urban and regional planning, block the passage of related legislation and perpetuate conditions favorable to small, individual interests but detrimental to the broader society and the environment (Economou 1995).

In the longer run, informal work reproduces and perpetuates the inferior position of workers (especially women working at home) and their dependence on wage labor and reduces their opportunities to find a job in the formal sector (Vaiou and Hadjimichalis 1997). Similarly, informal first and second home and tourist developments are retroactively legalized, thereby encouraging the continuation of the social practices that generate them. It is interesting to note that settlers associations (in first and second home illegal developments) are formed to promote the interests of their members and of their places (!). More importantly, however, changes in the conditions that support informal activities may destabilize and create crises in the informal sector. Such changes include: capital movements to places where the accumulation terms are more favorable, changes in legislation (e.g. building codes, urban legislation), real estate and rental price increases (affecting the locational advantages of informal businesses), dissolution of family ties, changes in value systems and life styles (e.g. young men and women rejecting traditional activities), policy changes (e.g. CAP, GATT). As most of these activities are heavily localized, at least in Southern Europe, the long term consequences are grave especially for individuals and social groups with low geographical mobility (women and minorities) as well as for regions with weak intra- and inter-regional industrial linkages.

Overall, then, informal activities foster social and spatial segregation instead of integration – a division between powerful 'haves' and powerless 'have nots', between wage and informal workers, between affluent and marginal regions; in other words, conditions fostering social injustice within and between generations. Their invisible and clandestine nature does not permit true integration of informals in the social structure, does not allow their participation in formal decision making and, naturally, does not induce real redistribution of income, wealth and power. Hence, the quest for empowerment – a critical precondition of sustainable development – becomes questionable (though not impossible). At least in the Southern European context, it has to be re-examined in the light of the particularities of this region's informality. In general, then, the informal sector may not serve well the purpose of social equity and, consequently, the achievement of social sustainability.

3.4 Informality and sustainability – The connection and its implications

The preceding discussion leads to two main conclusions. Firstly, in contemporary times, real or perceived deficiencies and structural flaws of the formal socio-economic, political and institutional system encourage the development (or, the continuation) of informal activities. These same flaws create unsustainable socio-economic and environmental conditions. In fact, the literature on both subjects documents that informality and unsustainability share several common causes although differences exist in certain respects. If the formal system was functioning homeostatically – within the limits of its total carrying capacity and with mechanisms that would ensure its resilience and stability (e.g. flexibility and adaptation) – then, it would sustain itself on all fronts. It would satisfy social needs and, consequently, there would be no *real* need for alternative, informal ways to develop (excluding criminal activities).

Secondly, the achievement of sustainable development, considered as the simultaneous satisfaction of economic, environmental and social objectives, depends on the impacts of both the formal and the informal system. But there is a vicious circle of unsustainability and informality where the former generates the latter, which in turn aggravates, in general, already unsustainable conditions, and so on. Even the potential benefits of certain informal activities are negated by the many negative, aggregate and longer term impacts which the majority of them have on the environment, the economy and society. Sustainable development policy-making and planning, then, are bound to be ineffective, at best, if they only target the formal sector or if they assumes that this sector is currently functioning efficiently or that it will function appropriately to deliver desirable results. The informal sector should be considered explicitly in its relationship both to the formal sector as well as to the sustainability of development in particular socio-economic and political settings.

In Southern Europe, this relationship assumes a particular form. Informality is a psychological and cultural trait of the European South, closely intertwined with the social, economic, political and other aspects of life in this region. In historical terms, informal activities have long existed, in part because of the absence of (or the perceived need for) formal arrangements for a multitude of social needs. The reasons they have developed and are developing are specific to given countries or regions within countries, since Southern Europe is a geographic region of great socio-cultural, political and environmental diversity. Until the 1970s, it seems they raised no major concerns about the imbalances they may have been causing to the economy, society and the environment of the region (although serious objections to this statement can be raised on several grounds). Since that time, however, both the external and internal environment of informal activities has undergone dramatic changes. Globalization and restructuring of the world markets, technological advances, in and out-migration, demographic growth, socio-economic and political changes in the European South as well as an increase in the number of informal activities spurred by these changes figure prominently among them. Crises are developing in several, once thriving,

Southern European regions. As a consequence, both numerically and qualitatively, the adverse impacts due mostly to informal activities have become serious, raised concerns and been recognized.

The natural implication of the situation outlined above is that some form of intervention is needed to restore the current imbalances, manage the informal sector and its relationship to the formal sector and, hopefully, facilitate the transition to sustainability in the European South. More specifically, the real issue is, on the one hand, to seize control of the conditions and forces that foster detrimental informal activities and, on the other, to facilitate and exploit beneficial informal activities.

4. CONCLUDING REMARKS AND FUTURE RESEARCH NEEDS

Sustainable development policies, in general, should target *simultaneously* all three objectives – environmental protection, economic efficiency and social equity. From this perspective, there exists no 'pure' sustainable development policies to date (O'Riordan and Voisey 1998). Most policies focus on environmental protection and rarely consider the integration of all three objectives. More importantly, all policies, including those of the European Union and its member-states, are directed towards formal, regulated activities. The importance of informal activities was, until recently, underestimated and, usually, they faced a neutral, at best, or a negative policy environment. Hadjimichalis (1995) observes that European Union policies have a socio-economic preoccupation with the center and the North which can neither capture the particularities and recent changes of the South nor address them properly.

The purpose of this chapter was to open up discussion about the role of the informal sector in policy-making and planning for sustainable development in the particular context of Southern Europe, where informality has a long tradition compared to Northern Europe. The analysis suggests that, *overall*, the informal sector makes more of a negative than a positive contribution to the sustainability of development and that choosing some form of policy intervention is necessary. Such intervention should be complemented by more thorough analyses in specific settings and for specific issues, the aim being to develop concrete and implementable schemata. In any event, it will take a long time for these schemata to deliver the desired results, since they will necessitate institutional changes that depend critically on deeper cultural adaptation to new socio-economic realities. Certain, broad guiding principles for policy and plan design to address the particular relationship between sustainable development and the informal sector in Southern Europe include:

- Policy and planning frameworks should be flexible, anticipatory and adaptive to accommodate the informal sector (if unavoidable and socially desirable) and to coordinate formal and informal spatial and socio-economic development activities.
- Timely satisfaction of social wants through formal arrangements will probably deflect potential conflicts and discourage alternative, informal ways from gaining the lead.
- Policies should have systems of checks and balances to encourage desirable (sustainable) forms of development and avert undesirable ones.
- It is preferable to identify and try to remove the causes of unsustainability which may overlap with the causes of informality, rather than to employ 'troubleshooting' and 'out of sight, out of mind' strategies. In several cases, institutional reform may be unavoidable as institutions of the past cannot address modern needs in a sustainable way.
- Quasi-informal arrangements may avert complete informality and mitigate its economically inefficient, environmentally destructive and socially undesirable impacts.
- Integrated, mutually reinforcing policies and properly coordinated measures – for both the formal and the informal sectors – are preferable to single-sector policies.
- Policies for the informal sector should be sensitive and adapted to the socio-economic, cultural and political particularities of Southern Europe but, from the perspective of sustainable development, they should be coordinated with macro-level (European Union, global) policies.
- Locals should be involved in decision making by means that overcome the Southern European distrust to the government; otherwise, lack of empowerment will threaten the achievement of sustainable development.

The preliminary examination of the role of the informal sector in sustainable development in Southern Europe uncovers a rich empirical, theoretical and policy-related research agenda. Empirical research needs to be undertaken in a variety of socio-economic and cultural settings on various spatial levels in Southern Europe. Coupled with the extant rich literature on the 'Southern question', it should be directed towards identifying the micro- and macro- conditions and factors that favor the emergence of informal activities and evaluating their impacts on the sustainability of development. It will seek to identify and analyse cases where informality has caused (or, is causing) unsustainability and cases where it has contributed (or, is contributing) to sustainable development. Theoretical notions already proposed to explain the particularities of Southern Europe, such as Putnam's (1993) "social capital", deserve further empirical exploration in the context of the informal sector's contribution to sustainable development.

Very importantly, case studies may shed light on the influence of current European and national policies on the development of informal activities. They may also serve to explore approaches to policy design that prevent similar situations from developing in the future. Comparative analyses will further indicate and explain differences of policy impacts among the regions of Southern Europe. From an even broader perspective, comparative case studies of informal activities in Northern and Southern European settings will reveal similarities and differences in their emergence, determinants and impacts. These studies should inform policy-making for sustainable development in different geographical settings.

Theoretical research may identify various forms and manifestations of informality and their patterns of relationships with the sustainability of development, some of which may be particular to Southern Europe. The extent to and conditions under which the causes of informality overlap with the causes of unsustainability is a subject worth examining since common policies can be designed to address both. *Ex ante* analyses of alternative policy approaches are also necessary to assess their effectiveness in managing the informal sector and contributing to sustainable development. Based on both empirical and theoretical research, development of sustainability indicators may incorporate operational expressions of informality to capture its impacts on the sustainability of development.

In general, considerations of the informal sector's emergence and impacts should penetrate all policy-related research (e.g. Strategic Environmental Assessment, evaluation, institutional reform) so as to support enlightened sustainable development policy-making, suited to local particularities and resistant to changes in their broader context. For the countries of Southern Europe, finding ways to make the transition to sustainability is particularly critical since they "do not possess, for the moment, the political power to react to the coming marginalization as they are heirs of a weak society of citizens where there are no strong institutions and central authorities operate on the basis of client relationships" (Hadjimichalis 1995: 81).

REFERENCES

Assaad, R. (1996) Formalizing the informal? The transformation of Cairo's refuse collection system, *Journal of Planning Education and Research* **16**, 115-126.

Beckerman, W. (1994) Sustainable development: Is it a useful concept? *Environmental Values* **3**, 191-209.

Briassoulis, H. (1992) The planning uses of fire: Reflections on the Greek experience, *Journal of Environmental Planning and Management* **35**, 161-173.

Briassoulis, H. (1999) Sustainable development and the informal sector: An uneasy relationship? *The Journal of Environment and Development* **8**, 213-247.

Cappecchi, V. (1989) The informal economy and the development of flexible specialization in Emiglia-Romana, in A. Portes, M. Castells and L.A. Benton, *The Informal Economy: Studies in Advanced and Less Developed Countries*, The John Hopkins University Press, Baltimore and London, pp.189-215.

Castells, M., and Portes, A. (1989) World underneath: The origins, dynamics and effects of the informal economy, in A. Portes, M. Castells and L.A. Benton, *The Informal Economy: Studies in Advanced and Less Developed Countries*, The John Hopkins University Press, Baltimore and London, pp.11-40.

CEC (Commission of the European Community) (1991) *Employment in Europe*, Office for Official Publications of the European Community, Brussels.

De Soto, F. (1989) *The Other Path: The Invisible Revolution in the Third World*, Harper and Row, New York.

Demertzis, N. (1997) Greece, in R. Eatwell, *European Political Cultures: Conflict or Convergence?* Routledge, London, pp.107-121.

Desa, Y.D. (1990) *Monitoring and Evaluation of Public Hydrants and Water Terminals in North Jakarta*, Final Report. Joint Cooperation of National Development Planning Board of Indonesia, the United Nations' Children's Fund and Y.D. Desa.

Economou, D. (1995) Land Use and Unauthorized Development: The Greek Version of Sustainability, Proceedings of the Scientific Conference on "Regional Development, Regional Planning and the Environment in the Context of the United Europe", Pantion University, Athens, Greece, 15-16 December.

Georgakopoulou, B.N. (1988) Informal formations and informal economy, *Review of Social Sciences* **69**, 3-29.

Griffin, R. (1997) Italy, in R. Eatwell, *European Political Cultures: Conflict or Convergence?* Routledge, London, pp.139-156.

Hadjimichalis, C. (1995) The Southern Ends of Europe and European Integration, Proceedings of the Scientific Conference on "Regional Development, Regional Planning and the Environment in the Context of the United Europe", Pantion University, Athens, Greece, 15-16 December.

Hardy, D., and Ward, C. (1984) *Arcadia for All: The Legacy of Makeshift Landscape*, Mansell, London.

Hicks, J.R. (1946) *Value and Capital*, Oxford University Press, Oxford.

Kanellopoulos, C. (1992) The Underground Economy in Greece: What the Official Data Show, Discussion Paper No. 4 Athens, Centre for Planning and Economic Research.

Lele, S. (1991) Sustainable development: A critical review, *World Development* **19**, 607-621.

Leontidou, L. (1990) *The Mediterranean City in Transition: Social Change and Urban Development*, Cambridge University Press, Cambridge, MA.

Lobo, F.M. (1990) *Irregular work in Spain*, Office for Official Publications of the European Community, Brussels.

Lubell, H. (1991) *The Informal Sector in the 1980s and the 1990s*, Development Centre of the Organisation for Economic Cooperation and Development, Paris.

Lyberaki, A. (1991) *Flexible Specialization: Crisis and Restructuring in Small Industry*, Gutenberg, Athens.

Marquette, C., and Pichon, F. (1997) *Population Dynamics, Labour Patterns, Land Use and Deforestation in Ecuadorian Amazon*, Paper presented at the Opening Meeting of the International Global Environmental Change Research Community, Laxenburg, Austria, 12-14 June.

Mingione, E. (1990) *The case of Greece*, Office for Official Publications of the European Community, Brussels.

O'Riordan, T. (1993) Interpreting the Precautionary Principle, CSERGE Working Paper PA 93-03 London and Norwich, Centre for Social and Economic Research on the Global Environment.

O'Riordan, T., and Voisey, H. (1998) The politics of Agenda 21, in T. O'Riordan and H. Voisey, *The Transition to Sustainability: The Politics of Agenda 21 in Europe*, Earthscan, London, pp.31-56.

Perrings, C. (1991) Reserved rationality and the precautionary principle, in R. Constanza, *Ecological Economics*, Columbia University Press, New York, pp.93-108.

Pichon, F. (1992) Agricultural settlement and ecological crisis in the Ecuadorian Amazon frontier: A discussion of the policy environment, *Policy Studies Journal* **20**, 662-678.

Pichon, F. (1996) The forest conversion process: A discussion of the sustainability of predominant land uses associated with frontier expansion in the Amazon, *Agriculture and Human Values* **13**, 32-51.

Pinto, A.C., and Nunez, X.M. (1997) Portugal and Spain, in R. Eatwell, *European Political Cultures: Conflict or Convergence?* Routledge, London, pp.172-192.

Piore, M., and Sabel, C. (1984) *The Second Industrial Divide*, Basic Books, New York.

Portes, A., Castells, M., and Benton, L.A. (eds.) (1989) *The Informal Economy: Studies in Advanced and Less Developed Countries*, The John Hopkins University Press, Baltimore and London.

Putnam, R. (1993) *Making Democracy Work: Civic Traditions in Modern Italy*, Princeton University Press, Princeton, NJ.

Roberts, B. (1994) Informal economy and family strategies, *International Journal of Urban and Regional Research* **18**, 6-23.

Romanos, M., and Chifos, C. (1996) Contributions of the urban informal sector to environmental management, *Regional Development Dialogue* **17**, 122-155.

Sanyal, B. (1987) Urban cultivation amidst modernization: How should we interpret it? *Journal of Planning Education and Research* **6**, 22-35.

Sassen-Koob, S. (1989) New York City's informal economy, in A. Portes, M. Castells and L.A. Benton, *The Informal Economy: Studies in Advanced and Less Developed Countries*, The John Hopkins University Press, Baltimore and London, pp.60-77.

Sinha, A., and Amin, A. (1995) Dhaka's waste recycling economy: Focus on informal sector labour groups industrial districts, *Regional Development Dialogue* **16**, 173-195.

Smith, A. (1981) The Informal Economy, *Lloyds Bank Review* **141**, 45-61.

Toman, M.A. (1992) The difficulty in defining sustainability, *Resources* **106**, 3-6.

Vaiou, D., and Hadjimichalis, C. (1997) *With the Sewing Machine in the Kitchen and the Polish in the Fields: Cities, Regions and Informal Labour*, Exantas, Athens.

Vinay, P. (1987) *Women, Family and Work: Symptoms of Crisis in the Informal Economy of Italy*, Proceedings of the Samos Third International Seminar, University of Thessaloniki, Thessaloniki.

Williams, C.C., and Windebank, J. (1995) Black market in the European Community: Peripheral work for peripheral communities, *International Journal of Urban and Regional Research* **19**, 23-39.

Ybarra, J.A. (1989) Informalization in the Valencian economy: A model for underdevelopment, in A. Portes, M. Castells and L.A. Benton, *The Informal Economy: Studies in Advanced and Less Developed Countries*, The John Hopkins University Press, Baltimore and London, pp.216-227.

Chapter 4

Sustainability and Tourism

EU Environmental Policy in Northern and Southern Europe

Carlo Ruzza

1. INTRODUCTION

The importance of tourism world-wide can hardly be over-estimated and is constantly growing[1], particularly in Europe[2]. Economic activities such as tourism have direct consequences for delicate ecosystems. In a fragile and densely populated environment such as the Southern European coastal one, neglect of the consequences can be catastrophic. Short-term concern for pressing economic imperatives, and, often, limited environmental awareness, result in continuing environmental deterioration and hinder tourism in Southern Europe.

This happens despite a concerted EU-level effort to protect and improve the environment. The Greening of Southern Europe is a difficult, but necessary task that European policy-makers have pursued for several years and across a variety of sectors. Over the years, EU-level environmental policy has acquired relevance in the treaties, and European institutions have produced a large amount of environmental legislation that now permeates a wide and increasing number of policy sectors. However, studies of the environmental conditions of key sectors such as tourism do not allow for an

[1] Some recent data from the World Tourism Organization encapsulate it: "Tourism is the world's largest growth industry with no signs of slowing down in the 21st century. Receipts from international tourism have increased by an average of 9 per cent annually for the past 16 years to reach US$423 billion in 1996. During the same period, international arrivals rose by a yearly average of 4.6 per cent to reach 594 million in 1996. WTO forecasts that international arrivals will top 700 million by the year 2000 and one billion by 2010. Likewise, earnings are predicted to grow to US$621 billion by the year 2000 and US$ 1,550 billion by 2010." (WTTO 1998)

[2] Europe is the international leader in terms of arrivals, and the industry is growing at a sustained rate. See Environmental Law Review, May 1996: 150.

K. Eder and M. Kousis (eds.), Environmental Politics in Southern Europe, 101–126.

optimistic assessment of the situation. Rather, environmental regulations are neglected, distorted and delayed.

To explain this, I will argue that, at the European level, specific path-dependent and resource-dependent factors have produced a policy process which is inspired by environmental principles. It proceeds through the inclusion of environmentally concerned policy actors, of a career structure in environmental policy that frequently rewards environmentalism, and of an overall institutional legitimacy of environmentalism, which is approved from the top as a reason to justify a specific EU role. Conversely, there is a general agreement that, for cultural and institutional reasons, Southern European environmental policy suffers from delayed and ritualistic incorporation of European policy. I will examine how European policy-making provides an opportunity structure for such delayed and ritualistic incorporation.

There are two sections to this chapter. The first section will examine how policy is produced at the EU level in the interaction among interested actors, mainly political and bureaucratic actors, businesses and NGOs. The second section will analyse the process of Europeanization of environmental and tourist policy, and will identify what is distinctive to them in terms of policy-making style and the sector's distinctive institutional histories and peculiarities. In the conclusion, I will reconsider the dynamics of Southern European incorporation of such policies.

In approaching environmental decision-making it is useful to conceive of the environmental policy community as lodged in a multi-organizational field, where different types of organizations tend to espouse certain views and, on the basis of typical stances, approach the policy process and negotiate with others to produce policy decisions. Successful environmental regulation only results when actors, their motives, and typical approaches, converge on high environmental standards. The main actors are, first, state actors, who are the main gatekeepers, producers, and implementers of environmental policy. Second, they are the business organizations, whose role is very important.. Increasingly in recent years, a global ethos favoring deregulation has, to an extent, reduced the power of state actors and increased the power of business. Equally, a change of ethos away from centralist bureaucratic structures and towards the inclusion of a variety of social partners and a variety of territorial levels has undermined state dominance in environmental policy. Third, there are public advocacy groups, social movements' representatives and NGOs, all of whom play a significant role in environmental policy.

A variety of collegial and corporatist fora have been included in the European policy process and an ethos of subsidiarity has emerged, which empowers regional and sub-regional territorial levels. This de-centralized system of extensively debating organizations is encouraged by policy principles recently emerging from EU institutions, such as the principles of civic dialogue, social partnership, and subsidiarity.

2. POLICY-MAKING IN TOURISM AND ENVIRONMENT AT EU LEVEL

2.1 The historical background

A brief history of environmental regulation in the EU will provide the background for considering current dynamics. The Treaty of Rome made no particular reference to the Environment (since it was a product of the 1950's and at that time environmental problems were not yet high on the public opinion agenda). However, after the UN conference on the Human Environment held in Stockholm in 1972, the EEC Heads of State and Governments met in Paris and recognized that "economic expansion is not an end in itself". This sentence implied that in order to improve the quality of life in Europe one also needed greater protection of the environment. As a direct consequence, the First Action Programme on the Environment was adopted by the Council of Ministers on November 1973. This Programme explained in large detail the actions to be taken to deal with what were increasingly perceived as urgent pollution problems. It set out goals and objectives and defined principles for the development of an environmental policy. For example, it established the Polluter Pays Principle, that is the costs of preventing and eliminating nuisances must be borne by the polluter. This Programme adopted a comprehensive approach and sought to address water and air pollution problems, to improve the management of waste, and to protect wildlife and habitats. Above all, it called for urgent work by the Commission services and provided tight deadlines. The 1973 Action Programme on the Environment provided the framework for environmental policy in the EC.

In the seventies, over a hundred legislative texts of major importance to the quality of life in the Community, had already been adopted.[3] However,

[3] They related mainly to water, air pollution, chemicals, waste treatment, noise abatement, the protection of species and natural resources and international actions. In March 1984 the Commission of the European Communities published the report *Ten Years of Community Environment Policy*. In it, Karl-Heinz Narjes, evaluating the outcomes of the EC Environment Policy, proudly stated that a lot had been achieved even if not all the deadlines had been met.

one of the major limitations of the Environmental Action Programmes of the seventies was the fact that the Programmes were mainly directed at problems with the various media, i.e., with air, land, water, or wildlife. To a great extent, the environment was considered in terms of these different media and most actions adopted by the Community related specifically to them. Over the years, the limitations of this approach have become more and more visible. Many of the problems which have emerged relate not to individual media but to the transfer of pollutants or other substances from one part of the environment to another (for example cadmium, acid deposition, nuclear wastes). Consequently, action to limit environmental damage is most effective when it is taken at source, rather than separately in each sector of the environment. Frequently, therefore, there is a need to develop policies related not to the media themselves, but to the flows which link them. Out of this has come an awareness of the need for a preventive, cross media approach to the environment. For example, the control of chemicals in the global environment.

The action programs of the seventies (1973 and 1977), as opposed to the policies followed in the eighties, were essentially corrective, that is, oriented to solving existing problems. The major shift in environmental policy occurred in the action program of 1983. It concentrated more on preventive action. The Community's third Environmental Action Programme was adopted in February 1983. It was a major turning point and marked a considerable evolution over the first program. The prevention of environmental problems became the main emphasis and the crucial principle for the further development of the national environment policy envisaged under the program. The third program committed the community to the progressive and preventive integration of environmental requirements into the planning and execution of all actions, that is, on all economic sectors that can have significant effects on the environment.

The program adopted a progressive approach and recognized that "the resources of the environment are the basis of – but also constitute the limits to – further economic and social development and the improvement of living conditions". Consequently, work progressively concentrated on the development of new instruments aimed at the prevention of environmental problems and also at better management of resources, that is, environmental impact assessment, the establishment of an environmental information system and the promotion of clean technologies. The Community's environmental impact assessment directive – which came into force on 3 July 1988 – is a weapon which has given force to this important principle. It integrates ecological awareness into the planning and decision-making process in all sectors, notably agriculture, the oil industry, energy, transport, tourism and regional development.

From then on certain categories of projects – e.g., oil refineries, thermal power stations, chemical installations and motorway construction – must be subjected to an impact assessment. This assessment has to identify the effects of a project on human beings, fauna and flora, on soil, water, air, climate and the landscape, on the interaction of all these factors, and on material assets and the cultural heritage. According to this directive, the competent planning authority must take into account information and opinions received in the environmental study before making its decision. The public must be consulted and can propose alternatives. In summary, over the years a sizeable body of environmental laws and regulations has emerged at the EU level.

Many of these laws are differentially implemented in member-states. Thus, although EU institutions are a powerful source of environmental knowledge and laws across a wide, but by no means comprehensive sets of policy areas, their practical impact is mediated by differing levels of commitment to the environment and by selective attention to different areas.

Although limited by a de-facto and in most areas de-jure lack of control over implementation processes, the power of EU institutions is increased when financial allocations presuppose the respect of environmental preconditions and when the assessment of environmental compliance is moved at EU level. This has typically happened with the distribution of structural funds.

Since 1988 ERDF Framework Regulation introduced environmental protection as a criteria in the allocation of structural funds, assistance from the EIB and from other existing financial instruments. It also required member-states to supply information to enable the Commission to assess the environmental impact of the funds of Objective 1 regions. These are 'less developed regions' and include several Southern European areas of tourist relevance. One can generalize by pointing to research that indicates that the influence of the Commission is related to its role as a financial provider. The greater the EU contribution to member-states' structural spending, the more significant is the Commission's role (Bache 1998: 100).

2.2 Europe, tourism, and the environment

At the EU level, whilst acknowledging the subsidiarity issues that it impacts, the role of tourism in the Union was recognized for the first time in the Maastricht treaty. This recognition was based on a growing awareness of tourism's importance in economic terms, its impact on labor markets, and its connections to consumer protection as well as cultural and natural heritage. The first visible initiatives included the declaration of 1990 as the European year of tourism and then the First Plan of Action in Favor of Tourism[4], which promoted competition among public and private bodies for pilot projects on rural, environmental, social and cultural tourism.[5]

At the EU level during 1989-1993 over 3,000 MECU were provided to co-financing investment and infrastructure projects for tourism and to upgrade cultural and historical resources under the Structural Funds. In particular, these funds were intended to support tourism in under-developed and declining regions. Tourism policy has also been elaborated in the context of education initiatives financed through the European Social Fund.[6]

Given the scale of activities in the tourist sector, its territorial dispersion, and its use of environmentally sensitive areas, it is clear that it can constitute a major environmental threat. However, since both the environment and tourism are potentially extremely broad policy areas, what kind of sector is considered appropriate for environmental initiatives is a political decision that has emerged out of a complex process of agenda setting. It is difficult to define and examine the environmental sector, but, however defined, it is substantial in terms of size of industry and labor markets, and is expanding to encompass new areas.[7] In the tourist sector, the environmental dimension is crucial, and comes readily to the mind of consumers. With varying levels of concern it can be argued that, in their search for pristine nature, tourists are frequently buying 'environment'. Environmental considerations were already stressed in the First Plan of Action in Favor of Tourism. The Commission has proposed to include environmentally insensitive tourist developments in the Environment Impact Assessment Directive, (Annex II), and has sponsored relevant demonstration projects such as the Integrated Coastal Zone Management Demonstration Programme.[8] A decisive intervention on the environmental impact of tourism emerged with *Towards Sustainability*, which expressly targeted the environmental impact of tourism. Taking as its point of reference the influential Fifth EC Environmental Action Programme

4 Decision 92/421/EEC of 13.07.92 in accordance with article 235 of the Treaty.

5 See ELR96: 150.

6 Also relevant are the various programs drawn up in response to the objectives defined in Articles 126 and 127 of the Treaty.

7 See Sprenger (1997), summarized in CEC 97 XXIII291/97: 43-57.

8 This is a joint initiative of Directorate Generals XI (Environment), XIV (Fisheries) and XVI (Regional Policy and Cohesion). This program is based on a series of demonstration projects located in a wide variety of settings. However the tourist theme is not the main focus.

approved by the Council in 1993, there clearly emerges from the document the intent to incorporate environmental awareness into a wide number of sectors, including tourism.[9] There are particular themes that have been identified as being in need of special attention and which are connected more or less directly with one or the other sector. For instance, the document singles out urban environments and coastal zones, which appear particularly relevant for tourism. And it lists a number of policy instruments ranging from research to public education and financial support.

The effectiveness of EU institutions in 'greening' tourism has varied and so have the reasons for success and failure. A 1995 review of the 5th Environmental Action Program pointed out that the least success was achieved in policy areas such as tourism – an area that is very fragmented –, because it is difficult to reach widely dispersed policy actors, and conversely in agriculture – which is highly integrated and has well organized vested interests –, because it is difficult to change the culture and operating procedures of such an area.[10]

However, some results have been achieved. As the 5th action program review indicates, the Commission's *Action Plan to Assist Tourism* supported a number of sustainable tourism pilot projects (18 MECU budget until 1995) and established the 'European Prize for Tourism and the Environment' to encourage States' environmental awareness in tourism (including the cultural dimension). In April 1995, the Commission produced a Green Paper on *The Role of the Union in the Field of Tourism* to launch wide consultation on the EU's role in tourism. The Green Paper focusing on sustainable tourism, reviewed the actions carried out by the EU in the tourism sector and the related policy instruments, stressing the need to link the three interconnected areas of tourism, consumer affairs, and natural and cultural heritage.

9 The sectors particularly singled out are: Industry, the Energy Sector, Transport, Agriculture, and Tourism.

10 This review notes: "Integration of environmental considerations into the different target sectors has made progress, but at varying speeds. If integration of environmental considerations is generally most advanced in the manufacturing sector, where legislation has existed for twenty years and where the economic advantages were quickly seen, it is least apparent in agriculture and in tourism, for diametrically opposed reasons. On the one hand, the Common Agricultural Policy is a system which was put in place at a time when food security needs. They were paramount in Europe, into which it is difficult to introduce fundamentally new directions, and from which nature conservation has been largely isolated. On the other hand, tourism is a highly fragmented and diverse sector where a very large number of economic and other interests need to be reached before the effects of change can be felt." COM(95) 624, 10.1.1996 – Progress Report on implementation of the European Community Programme of Policy and Action in relation to the environment and sustainable development 'towards sustainability'.

However, progress is slow. Administratively, there is a unit in DGXXIII in charge of tourism. Its small size and budget have limited its influence. It is clear that tourism as a specific sector lacks, at the EU level, the political power to make an impact. The major initiative proposed in recent years has yet to be approved, despite an already very long gestation.[11] Yet funds and influence trickle down to member-states through a variety of channels. It has been estimated that tourism indirectly receives up to ECU 1 bn a year from a variety of unrelated programs administered by a host of Commission services, but not necessarily in coordination with the small tourist unit.[12]

Although it works in close collaboration with sectoral organizations such as the WTO, it does not address a broader environmental constituency of, for instance, environmental NGOs. Tourism policy-making has taken and continues to takes place as much, if not more, in DG III, DG IV, DGV, DG VII, DG XI, DG XVI, DG XXI, DG XXII.[13] In addition, compared with other areas, the tourism industry is rather small and fragmented, limiting its ability to lobby. In summary, with tourism we have a policy area that is deeply enmeshed in the EU policy environment, but in a weak and dispersed way.

[11] A December 1997 issue of Euro-Info encapsulates the history of the initiative and the attitudes prevailing among many policy-makers in the tourism sector: The 'Tourism Council', in its meeting of 26 November 1997, did not come to a political agreement on the proposal for the first multi-annual program in favor of European tourism *Philoxenia* (1997-2000 – COM(96)168 final) as two delegations were unable to adhere to the compromise proposed by the Luxembourg Presidency. Presented by the Commission in April 1996, the Commission proposal had been examined under the Italian and Irish Presidencies. The dossier had not advanced further in the Council due to the opposition of several delegations who invoked the principle of subsidiarity in the field of tourism. The Luxembourg Presidency thus presented a compromise text which, while taking account of the principle of subsidiarity, provided for a five-year program (1998-2002). The actions foreseen were of a horizontal nature and focused on improving coordination and strengthening co-operation between the actors involved in tourism at the European level. At the final press conference, Christos Papoutsis, Commissioner responsible for tourism, expressed his "profound disappointment" at the Council's inability to adopt the *Philoxenia* Programme (the legal base for the proposal, article 235 of the Treaty, requires unanimity). He also underscored that it is "extremely disappointing that only five days after the Luxembourg Summit for Employment, the Council was unable to adopt a program in favor of a sector so important in the creation of jobs, a fact which has been unanimously recognized and emphasized by the Ministers in the Council conclusions under the heading of 'Tourism and Employment'. (Euro-Info n. 106).

[12] See Paola Buonadonna 'Going places with tourism' The European 15-22 May 1996.

[13] Respectively Internal Market and Consumer Affairs, Competition, Employment (ESF, Free movement of labor), Social Affairs and Education; Transport; Environment; Science, Research and Development; Regional Policy (ERDF); Customs Union; Coordination of Structural Instruments (Mediterranean Programmes) (Akehurst 1995: 220).

Yet certain common traits emerge in much new legislation on tourism. One is, at least at the level of declarations, a concern with the environment, with the sustainability of tourist initiatives. Why is this concern emerging in a variety of disjointed initiatives? What are the implications for the future of European tourism? On the basis of interviews with actors from various policy areas connected to tourism, I am going to examine in the rest of this paper the implications of the emergence of environmental ideas in tourist policy. I will attempt to explain their origin and diffusion in the Brussel's policy environment.

2.3 Institutional activists and EU institutions

A stress on environmentalist policy principles and high environmental standards is prevalent at the EU level for a variety of reasons. First, when environmental policy emerged, a large number of environmentalist regulators from Northern countries were selected as civil servants and put in a cohesive bureaucratic unit – now DGXI – which over the years acquired influence and established an effective environmental regulatory system. Because of their background, rooted in the environmental movements of the eighties, these civil servants incorporated in their bureaucratic processes elements of principled policy-making. As the stronger partners of policy communities, they were able to instill as taken-for-granted an ethos of concern for the idea of environmental sustainability.

Their ideal commitment was and is made easier by the EU institutional structure. In addition to the issue of selective recruitment of institutional activists, one has to consider which interests are served and what are now taken-for-granted assumptions of the EU environmental policy process. Environmentalism emerged at a time when the chronic need for legitimacy of the process of European construction could be addressed by presenting Europe as a powerful partner in the struggle for ecological modernization. In creating a pro-environmental bureaucratic unit, institutional architects created a reward structure for environmentalist civil servants.[14] In the EU institutional structure, while following their principles, at least for a time, institutional activists were also furthering their careers.

A separate issue concerns the role of universalist collegiality in the relationship between environmental and tourism policy. At both the EU and

[14] Through environmentalism, in a specialized policy unit, institutional activists can easily pursue both budget maximizing and bureau-shaping strategies (Dunleavy 1991). In addition, most DG's have distinctive ideologies, and at least discursive compliance to them is generally expected from officials Michelmann 1978).

member-state levels, the closer one is to locations where institutional activists are present and powerful, the more collegial decision-making is present and the inclusion of NGOs is encouraged. In weakly environmentally identified policy areas, such as tourism, however, the dominant policy discourse is one of economic growth. Environmental considerations find a role only in this context. Nevertheless, to a certain extent the diffusion of environmental legislation offsets at the EU level resistance from weak policy sectors. Environmental objectives become less relevant in areas that are not central to the environmental regulating community. Environmental policy principles are more central in the Directorate General 11 than in other directorates, such as the tourism policy unit. However, because of the sizeable institutional activism represented in Directorate General 11, and the consequent environmentalist ethos, there is still substantial attention to the connection between the two sectors. Environmental neglect does, however, result from a lack of policy coherence. It sometimes results from support in the context of structural funds for initiatives that are environmentally damaging. And this can happen at the same time as other pro-environmental initiatives are supported.

Furthermore, of the variety of possible environmental considerations, only those that do not depress tourist demand tend to be given prominence. More generally, all types of environmental concerns are filtered by a corporatist ethos that tends to exclude environmental behavior which is appropriate, but does not clearly benefit any of the partners of a policy community. Environmental solutions are sought that benefit economic growth and political interests, and even attract support for pressure groups. These are not necessarily the only or even the best solutions. Thus, for instance, one sees a large number of environmental solutions that stress a technological environmentalism where, instead of limiting consumption, economic interests are included so as to provide commercial technical solutions, or where an aesthetic environmentalism is encouraged which instead of tackling threats to the ecosystem, stress is placed on beautifying areas that can attract tourists. NIMBY environmental expenditure is sometimes encouraged if there are powerful political interests at stake, even if it means less resources for more pressing problems. Considerations of Environmental Social Justice are neglected as socially excluded populations are underrepresented in decision-making fora. Anti-modernist environmental solutions of the kind that affect economic interests are marginalized in favor of expanding tourist demand, even if that necessitates additional environmental remedies.

The marginalization of certain types of environmentalist ideas that takes place in corporatist environmental policy and in collegial groups that only include a limited number of social actors, appears further accentuated when one considers that policy debates taking place in the media and in public political discourse have only an indirect connection with detailed policy-making. Whilst at the level of public discourse environmentalist ideas may be aired and supported by a variety of actors, at the level of concrete policy decisions these ideas are frequently seen as scarcely relevant, of unclear applicability, and difficult to implement at the level of decision-making on detailed issues.

Focusing on the level of a Brussels-centered multi-organizational field, it would be possible to identify a broader level where policy cultures are formed and action is legitimated. The more public the process of decision-making is, the more the legitimacy-granting aspect of public discourse becomes relevant. A distinction needs to be made between this public level of policy discourse on the environment and detailed policy-making. It is at this public level that environmental sustainability is a key concept. It emerges repeatedly in key Commission documents, conferences, study groups, and advisory fora. However, at the level of detailed policy deliberation, this concept only finds indirect resonance. In the Brussels environment, institutional activists exercise more relevant control on the connection between the two levels than do other agencies. The virtual absence of a civil society, the weak representation of social movements (Mazey and Richardson 1993), and the limited media control on policy process, make institutional activists the only guarantors of environmental principles.

2.4 EU movements and public-interest lobbies

European institutions collaborate actively with numerous NGOs in a variety of policy issues. The Commission estimates that annually over 800 MECUs of EU development assistance (of which 196 involve co-financing) is channeled through NGOs and is seeking to further increase their role (CEC 1997: 7-8). Environmental NGOs are substantial beneficiaries of several programs. The boundaries between political action and work on specific projects (such as green tourism demonstration projects, or studies) is often flexible, and funding benefits both aspects to the point that it is unnecessary to draw a line between NGOs and social movements in EU lobbying.[15]

[15] My interviews of environmental NGOs in Brussels revealed that in general they prefer funding for the organization rather than for specific projects that typically come 'with strings attached', but there is in EU institutions a tendency to increase accountability by restricting the use of resources, and funding benefits both the daily running of organizations and its broad political goals, and specific projects. In general, the channeling of resources from NGOs to social movements is well documented. See for instance Biekart (1995: 65) and his contribution in Edwards and Hulme (1995).

In Brussels there are seven NGOs active in the environmental field.[16] Some have been operating in close collaboration with EU institutions for a long time: the European Environmental Bureau – an umbrella group representing 1500 European environmental groups – has been active for 20 years. Others are much less established. They all suffer from a substantial lack of resources which hampers their effectiveness.

The role of environmental activists in promoting environmental legislation is difficult to investigate, since quite conflicting opinions are expressed about them. According to several civil servants, environmental activists have been rather ineffective. Environmental groups, whose main approach is to collaborate with institutions, tend to present their contribution as substantial, whilst outsiders minimize institutional availability. There are cases where environmentalist pressure has clearly been heeded, as in the Commission's involvement in the beginning stages of the Twyford Down case (Edwards and Spence 1994: 111). Despite the number of NGOs, overall in Brussels there are fewer than 15 people who specifically follow and monitor Community activities from the point of view of environmental non-governmental organizations (Krämer 1992: 124). Yet the role of environmentalists is potentially important for the commission as a consequence of the perceived need for democratic legitimacy. In a Communication from the Commission (by DG XXIII) the importance of voluntary organizations is clearly stressed and the sector is extensively reviewed with the intent of increasing their resources and influence.[17] The awareness has also permeated sectors of the bureaucracy. However, the attention of institutional actors is proportional to the NGOs ability to collaborate effectively. That is, to represent a real constituency, to provide valuable information to counter-balance industry's studies, and to take a 'constructive approach' in consultation meetings.

This is particularly difficult for the entire NGO sector as hierarchical rules of representation are often contested, knowledge is expensive to acquire and, in parts of the movement, is distrusted or subordinated to a moral stance. Yet over the years the environmental movement has undergone a process of institutionalization, and by and large has accepted the language of science. In Butter's and Taylor's words, "the rising persuasiveness of environmental and global change data has contributed to shifting the essential

[16] EEB, Greenpeace, Transports and the Environment, Friends of the Earth, WWF, Climate Network and Birdslife.

[17] Communication from The Commission on Promoting The Role of Voluntary Organizations And Foundations In Europe – DG XXIII, 1995 (en/23/95/00891100.W00(EN). DG VIII also has a document entitled 'Digest of Community Resources Available for Financing Activities of NGOs and other Decentralized Bodies Representing Civil Society in the Fields of Development, Co-Operation and Humanitarian Aid'.

thrust of modern environmentalism towards an increasingly thoroughly 'scientized' Weltanschauung and mode of movement's strategy" (Benton and Redclift 1994: 233). This trend tends to be even more magnified by the institutional selection of interlocutors in Brussels, partially marginalizing the anti-technology, anti-market and anti-science voices in the environmental movement. Groups such as Green Peace however, are still keeping a presence in Brussels even if their attitude is often more critical than other groups.[18]

A quite different situation, however, occurs with the rest of the movement. As Smith (1993: 48) points out, what is crucial in the relation between pressure groups and the state is the type of the relationship, and not so much the amount of resources a group has. If the state's actors have an interest they can build up groups. This is in part what happens to environmentalism in Brussels. The EEB (European Environmental Bureau) in particular receives a substantial portion of its resources from DGXI and is in frequent consultation with civil servants and Euro-parliamentarians.

In summary, the environmental movement has undergone an extensive process of institutionalization; yet it has failed to significantly increase the power of traditional environmental groups in the EU (Mazey and Richardson 1993, Rucht 1997). In the terms of Giugni (1995), their acceptance by the system is real but still ad hoc. They have a sensitizing impact on small areas of decision-making. Nonetheless, environmentalist themes have affected EU decision-making through environmentally-sensitive bureaucrats and a global cultural change in environmental discourse. In this sense, the influence of the environmental movement should be re-assessed as crucial at the stage of discourse formation in the policy domain, if limited at most stages of the policy process.

In the tourism field not all groups have the same competence. It is an area that is more central to the WWF and Friends of the Earth, who have been involved in Commission-sponsored projects, than to the EEB. Attitudes characteristically vary, with Greenpeace more pessimist about the tourist industry's ability to green itself or it being forced to do so.[19]

[18] For instance a volunteer commented: "I think one of the main obstacles is this theology of liberalism and open markets. This is a catastrophe for an environmentalist. Because if the main target is free market then it is very difficult because everything you would try to think, every new idea that is coming through the political channel is not good."

[19] See for instance 'The Mediterranean New Order Ignores Ecology' Greenpeace Mediterranean Report – Amman, Jordan, 29-31 October 1995, which expresses skepticism on the possibility of change and warns that "an uncontrolled tourism boom would lead to the destruction of coastline areas, to sea pollution and to wasting water needed in the vital agricultural sector".

2.5 EU institutions and business

The European system of representation has from the start taken a pro-business stance – an ethos that is still entrenched in EU institutions. For instance, a civil servant typically argued that 'it is very important that the community industry is the first in the industrial game'. Civil servants often believe that business influence and competence should be enlisted to help the state run the economy. This is advocated not through the integration of business organizations into the state, but at the meso-level (Smith 1993) through a system of extensive and diffused consultations.[20] Business organizations generally welcome this role. In the field of the environment, multinational corporations, associations of industrialists, and even smaller companies generally have the resources and technical expertise to attempt to argue their case and to sell their ideas, at least in part, to the relevant committees. But they know that to regulators in charge of protecting the environment, a purely resource-maximizing approach would be unacceptable. Conversely, merging environmental and wealth-creation frames by advocating sustainable development appears legitimate.

Extensive lobbying and the display of more environmentally supportive attitudes were often associated. In general one notes, for instance, some chemical industries spend substantial sums in environmental advertisements and are among the best sponsors of environmental charities. Concerns about the environment are exhibited by all the highly regulated corporate actors, or by state contractors, such as chemicals manufacturers, biotechnology firms, and water treatment companies. However, it is not the case that business simply acts strategically. Rather, those actors that are more involved in consultations come to develop a common culture with those who regulate them. While some business lobbyists see the concept of sustainable development as a difficult issue which cannot be ignored, many others view this concept as an issue of central concern to the future growth of their industries.

Obviously, not all business actors are supportive of the concept of sustainable development. Some actors are skeptical about the intelligibility of sustainable development, but they still recognize its importance. Even in

[20] For instance, Lehmbruch and Schmitter (1982) examine the interactions between corporatist elements and pluralist societies – a society which, in Schmitter's terms, is based on a system of interest representation in which the constituent units are organized into an unspecified number of multiple, voluntary, competitive non-hierarchically ordered and self-determined categories which are not specifically licensed, recognized, subsidized, created or otherwise controlled in leadership selection or interest articulation by the state (Schmitter 1974: 85). See also the contributions to Cawson (1985).

cases where consultants do not directly endorse the policies debated by the Commission, some nonetheless demonstrate that they have adopted both the language of the EU environmental networks as well as the general goals behind the environmental policies.[21]

This emphasis on sustainability is becoming a common cognitive scheme for some industries. Studies of corporate cultures have often pointed to the binding role of organizational cultures. They provide corporate identity as they provide differences between organizations, but they also provide similarity by importing into business organizations the cultural markers of each historical period. If environmentalism is to a certain extent an accepted value in society, for those who operate in an environmental field it is difficult to deny its relevance. Clearly, those firms who produce environmentally sustainable goods and services (including green tourism) will promote sustainability more than those whose main concern is to resist regulation. But, as previously pointed out, roles are not infrequently reversed, and regulation is welcomed for a variety of reasons, such as first mover advantages. Hence there is no clear correspondence between environmental attitudes and market position.

Whether supportive of, or merely concerned about sustainable development, business lobbyists are in Brussels because they are concerned about environmental regulation. However, this concern is not necessarily a defensive stance. In the environmental field, the relationship between business and the regulators can at times be very strained, but can also be a symbiotic one. Civil servants encourage industry's lobbying, because they need the information and desire the influence on member-states' politics that business can acquire. But there is at times some dismay at the persistence and frequency of attempts to exert influence.

Relations between business lobbies and environmentalists are at times strained and infrequent in the absence of a bureaucratic or political organizer, but they can be constructive and are considered useful by both actors. Business realizes the potential threat of a direct appeal to public opinion and needs information. As both industries and NGOs become more global, political regulation loses some of its power, and direct relations between the two become necessary. Events such as the Brent Spar fiasco have clearly con-

[21] This is indicated for instance by a representative of the chemical industry who said: "For example, they (DGII) speak about what they call the PER, polluting emissions register. But I immediately say that here at this organization we have the impression that they have taken the wrong direction for implementing the objective of providing the various stakeholders of environmental information. A register is not a good instrument. But the objective is right in the concept of *sustainable development* and of partnership."

veyed to industry the message that expert opinions or political influence are not sufficient. As for environmentalists, they are often convinced that in many instances there are affordable environmental solutions to business problems that business simply must be informed of, and there is of course a different willingness to accept environmentalist suggestions in different sectors of business.

Among industry representatives in the tourist sector the awareness of the necessity of a 'green turn' is widespread. The tourist industry, acting, like other industries, with a short term perspective, finds it difficult to abandon successful strategies for the radical re-conversion that an environmental perspective would require. Yet in several national and supra-national arenas the industry discourse stresses the importance of better environmental practice. There are abundant declarations of principle. For instance, the World Travel and Tourist Council (WTTC) has environmental guidelines prepared by the International Chamber of Commerce Business Charter for Sustainable Development. There are several industry-organized environmental schemes that give certificates of good environmental practice to thousands small operators. For instance, Green Globe, with over 500 members in over 100 countries, is an environmental management program for travel and tourism companies. A subscription fee grants member the right to use the schemes' logo in the literature and commits them to the priorities specified in the 'Agenda 21 for Travel and Tourism industry: Towards Environmentally Sustainable Development' produced by the WTTC, World Tourism Organization and the Earth Council. Members are enticed by stressing that good environmental standards reduce overhead and operating costs, give competitive advantages positioning the company "at the leading edge of the sustainable tourism movement" (Green Globe web page, 1998), improve relations and satisfaction, and achieve global recognition through the well publicized green globe scheme. Such schemes are not merely driven by commercial considerations. They reflect the efforts of organizations on a continuum that, at some level, merges with social movements and NGOs.

Although tourism can bring economic prosperity, it also has great potential to reinforce social inequality and economic dependency. Over the past decade, governments, NGOs, communities, and environmental groups have become aware of the downside of tourism and of the need for tourism to be developed 'sustainably'. As with any sustainable development, finding a model of sustainable tourism development is a complex endeavor. Implementing it is an even more complex problem as the following analysis will show.

3. THE EUROPEANIZATION OF ENVIRONMENTAL AND TOURIST POLICY – THE CASE OF SOUTHERN EUROPE

During the last thirty years a large number of environmental laws and regulations have come to Southern Europe from the Community level. Southern European resource dependency on EU sources and it late and weak environmental legislation give EU institutions a key policy-shaping role (Hanf and Jansen 1998). EU laws and regulations express a conception of the centrality of environmental issues that has emerged in Europe-wide debates and has resulted in a global attempt at ecological modernization. Since tourist policy is clearly connected to environmental policy, much regulatory change has also come from EU level environmental policy. In addition to specific policy contents, the EU also increasingly exports policy styles. For instance, there is in many important policy areas, such as in regional policy, a formal stress on partnership and subsidiarity. It amounts to pressure to develop a corporatist decision-making structure of the type prevalent in the Brussels institutional environment. After reviewing specific institutional structures which have an impact on environmental policy, I will examine the consequences of this policy style on policy outcomes.

3.1 Three modes of policy-making

For analytical purposes it is useful to make a distinction between three fundamentally different approaches that a policy process can take. First, it can take a bureaucratic approach, in which a taken-for-granted stance towards the policy process prevails. The values, operating procedures and goals are organizationally sanctioned and go unquestioned by state actors. Weber's words when he describes officials as acting dispassionately and professionally well encapsulates this ethos.

Secondly, there is a policy-making approach in which actors pursue their interests in corporatist negotiations. This is the kind of policy resulting from interest intermediation where interests are pre-constituted and negotiations strive to find a common denominator between policy actors. Rules are not fixed and are re-defined in the process of arriving at acceptable compromises. Thirdly, there is a policy-making approach based on a moralization of policy positions, driven by the affirmation of principles, where controversies are solved by discursive attempts at consensus formation. The way these three policy approaches settle disputes is different. The bureaucratic approach relies on procedures, and, when these are uncertain, on authoritative interpretations passed along chains of command. Coercion by authoritative sources is accepted and taken for granted by bureaucrats and imposed on non-state actors. The corporatist approach relies on intermediation based on actors' relative power. The principled approach relies on collegiality (Waters 1989), where even weak policy advocates may participate in a collective search for universalist solutions. The self-understanding and organizational discourse of the main actors of environmental policy – civil servants, business and political actors – is closer to, respectively, a taken-for-granted approach, a corporatist resource-dependency approach and a principled approach. However, there are in effect frequent and substantial deviations from this model, which I believe help to explain the differences between Northern and Southern Europe that result from environmental policy at the EU level.

In real policy-making the three modes will often be present simultaneously, though in different measures, and the same actors will at times utilize all approaches. Institutional and cultural dynamics will determine their prevalence. That said, it is important to keep them conceptually distinct. Distinguishing between them will allow one to understand why, at times, public-pressure groups have an impact on policy processes even when their negotiating power is limited – for instance, when their potential threat of disruption is limited. It also allows one to understand how their ideas came to have an impact on European environmental policy and why they are included in policy-making even when their inclusion is not mandated. The merits and feasibility of these three approaches are also the subject of an EU-wide debate, which refers more broadly to the appropriate role of the state and of interest groups in shaping social change. Using these policy-style ideal-types I will be able to address this debate in terms of the particularity of Southern Europe.

These three approaches operate differently in different policy sectors, as bureaucratic behavior, interest representation and advocacy function in different ways in different policy contexts and within different sectors. The Southern European tradition of large but often ineffective bureaucratic states has typically approached policy problems in a regulatory fashion. States with a strong interventionist tradition in economic policy, large state sectors, and a recent history of authoritarianism demonstrate a reluctance to delegate policy to economic actors and even more to social actors. At the EU level, weak decision-making mechanisms which force policy based on consensus have promoted a fragmentation and technicization of political decisions and a constant search for consensus-based decision-making. This has promoted a participatory and corporatist policy style. However, the fundamental orientation of Southern Europe for bureaucratic policy-making and at the EU level for corporatist style has been tempered by recent events and the impact of new policy actors. In Brussels' institutions the strong environmental movement that has emerged in Northern European countries has found a voice, which has resulted in elements of collegiality in policy-making. In Southern Europe, environmental movements have been weaker, but a recent societal emphasis on business and the impact of the EU level have introduced elements of the corporatist style. Thus, there is a substantial difference in policy styles in the two contexts, the transnational EU style and national styles in Southern Europe.

This difference is not reflected equally in the different sectors. For instance, the sector of tourism has a different institutional structure and policy principles from the core environmental sector. I will now proceed to identify the distinctive features of the two sectors and their interrelations in order then to illuminate the prospects of the three policy-making approaches – the corporatist, collegial and bureaucratic.

3.2 Environmental exceptionalism

The difference between corporatist and collegial decision making – that is a difference between mediating represented interests and searching for universalist solutions, which in principle include the entire society – is particularly relevant when, in the policy process, there are social movements activists who are professionally concerned with representing civil society. While it is debatable whether activists do so effectively, it is clear that they make contributions of time and energy to represent views that are left out from other policy processes. This is typical of environmental groups connected to the new social movements of the eighties, which are by now well entrenched in several countries. Among the early popularizers of environmental concerns, a key role was played by environmentalist social movements, together with sectors of the media and certain epistemic communities. These three influences were often indistinct, either in terms of ideas or often even in terms of actors, since, particularly at the onset of environmental activism, there was an overlap of roles. As is often the case, social movement groups with different degrees of institutionalization unwittingly support each other, as the extreme ones attract media attention and are instrumental in agenda setting, and the institutionalized ones offer credible alternatives to the public and policy makers (Gamson and Modigliani 1989). Unlike epistemic communities, their range of views is wide enough to accommodate different causal beliefs and knowledge bases (Haas 1992: 18), and yet their moralization sustains commitment and occasionally high risk activities. They have been a notable source of ideal innovation in environmental policy. Their presence distinguishes environmental policy from several other policy sectors.

Approved policy principles are difficult to translate into specific action. The presence of activists matters when policy is so broad and complex that a variety of actors can easily re-interpret principles in their own terms, even at the cost of neglecting the public good. This is typically the case in new policy areas where there is still little consensus on the intersubjective validity of policy principles and their application. It is the case in highly technical fields where information is unequally distributed and can be used instrumentally. It is the case in areas where scientific knowledge is limited and there is high uncertainty. All these conditions pertain to environmental policy. The uncertainty of environmental risk and of policy remedies make it very versatile in the use that different organized interests can put it to, but it also obscures where interests lie. However, the presence of activists can ensure a measure of control in the way policy principles are re-defined by interested partners. They represent an element of universalism that in other policy areas only occasionally emerges from controls from the media and the political system. They also help actors to re-evaluate their interests and to question certain previously taken-for-granted assumptions. As they often point out, they endeavor to show business that a pro-environmental stance is not necessarily in conflict with economic interests.

Questioning assumptions has important consequences. This becomes clear if one considers that environmental policy is also distinctive because interests in the policy community are sometimes reversible, and change over time. In many older policy areas interests are more stable. Questioning assumptions can then lead actors to re-define their positions in the light of a different evaluation of their interests. For instance, business is often reluctant to accept environmental regulation, but at times it welcomes it because it offers first mover advantages to parts of regulated sectors. It creates niches for consumer behavior that allow lifestyle witnessing – consumer behavior that expresses and reinforces environmentalist social identities. It provides market certainty that comes from mandated homogeneity of standards. New technologies can offset environmental costs, which in any event can often be passed on. Relative market positions are also not changed after the introduction of new operating rules if an entire sector is modified. For these reasons, sectors of business can at times be advocating either stringent or lax regulation. Similarly, political actors can frame environmental issues in various ways, making them compatible with both left and right perspectives. Under the banner of environmentalism they can stress environmental justice, aesthetic values or NIMBY concerns, making it politically versatile. In this versatility lies the reason for the success of environmental regulation and also the fact that it can be utilized to cater to vested interests.

If ambiguity is the rule of the game, there are, however, mechanisms that limit it, but they work better in some contexts than in others. First, the more cohesive a policy community is, the more it needs agreed-upon concepts, which it utilizes as tools in debating. Typically, across the different types of policy actors, the concept of *sustainable development* acts as a catch-phrase that guides regulatory discourse (Ruzza 1996). It gives identity and professionalism to the environmental regulation issue network. At the same time, the ambiguity of the concept allows for conflicting framings of the state of the environment, of regulatory priorities, and conflicting interpretations of scientific evidence, and creates the basis for a debate. Furthermore, as I will argue, at EU level it serves several different and even conflicting interests of the environmental policy community. Where the environmental policy community is less cohesive this tool is less relevant. Second, it is a tool that is more relevant to core environmental policy areas than to peripheral ones. It will only trickle to peripheral ones if the environmental community is able to impose it across the board. As we move from core environmental policy issues to issues that are also central to other policy communities, such as tourism, a different set of principles comes to play a role. These principles interact, merge and re-define environmental principles. It is important to analyse the ways in which different institutional settings and decision-making mechanisms influence how environmental issues are framed in separate policy communities, such as the one regulating tourism.

3.3 Tourism and environmental policy

Having emerged in predominantly poor areas which were then offered the opportunity of rapid growth, there is in the tourist sector great stress laid on unrestrained economic development. However, in recent years, a heightened sensitivity to environmental issues is emerging at the global level. Nonetheless, this sensitivity, though well represented within EU institutions, does not easily develop or trickle down to small tourist operators in Southern Europe. The situation is somewhat different within policy-making circles. As a new, fragmented and often neglected policy sector, tourism has incorporated personnel, policy values and operating rules of connected and more central sectors such as economic policy and also environmental policy. That is, where the environment is an important sector, at the EU level and in much of Northern Europe. Where the environment is also a weak area, as in

Southern Europe, a more distant relation develops, with some connections resulting from European-level regulations.

In the interaction between environmental policy and tourism policy there also emerges the same variability of positions and the same uncertainty that characterize other aspects of environmental policy. In the tourism sector, actors with an interest in sustainability can change rapidly. Pro-environmental approaches can be used to protect or create monopolies of tourist operators or to open new markets such as agri-tourism. They can be used instrumentally, as when, for example, they only involve a superficial re-branding of locations. Frequently, however, short-term economic concerns dominate and calls for environmental protection are ignored. As in the environmental field, emerging tourism policy principles, such as those of subsidiarity and of partnership that are institutionalized in EU regional policies – an area with strong connections to tourism –, can be utilized to represent wide sectors of civil society or to exclude them when decision-making is reduced to interest intermediation between prominent local economic interests.

To ensure environmental sustainability it is therefore necessary that the policy value of sustainability is institutionalized in the tourist sector, both in collegial and in corporatist decision-making fora, in the same way that it is institutionalized in core environmental policy areas. This means making sure that in corporatist fora environmental sustainability becomes a taken-for-granted policy value and that in collegial fora the wider environmental interests of civil society are represented. In addition, it is necessary that there is awareness of the instrumental re-definitions that key environmental principles tend to undergo when re-interpreted by vested interests. Two questions arise, one factual and the other programmatic: Who are the actors who presently best represent environmental values and controls? How can they best be empowered? These questions need separate answers at the EU and Southern European levels. In the following section I will argue that while at the EU level environmentally universalist viewpoints have been culturally available and institutionally expressed in the comitology, this has not happened in Southern Europe.

3.4 Multi-level governance, policy actors and institutional activism

To understand the opportunity structure for Southern European policy-making, we need to consider the particular relations between the EU and Southern European policy contexts. Among European Studies specialists there has for years been a debate about whether EU politics is the reversible outcome of intergovernmental relations or whether it is a distinctive and cohesive political system to be studied without a main disciplinary reference to international relations. Scholars favoring the second approach have begun to observe the various nested political games that policy actors play while attempting to influence the same events at different levels of governance, and have described this focus as multi-level governance. They point out that with the extension of qualified majority voting, which occurred after the Single European Act and the Maastricht and Amsterdam Treaties, most crucial policy areas, including some with powerful environmental implications, such as transport, trade, agriculture and core environmental policy, are now largely decided at supranational level. Some analysts argue that even a multi-level governance approach does not fully take into account the extensive integration of the various levels which is now occurring, and they propose a more radical integration of EU and member-state political analysis.

It is unquestionable that a pervasive EU integration is occurring in environmental and tourist policy due to factors such as integrating regional policy, the resource-dependency of Southern European member-states, as well as to issues of cross-border pollution, that have stimulated mutual controls of different territorial levels on environmental policy. This does not imply a quick demise of the distinctive policy traditions of the different levels, it implies only that they have to reckon with each other's characteristics. Briefly, this 'reckoning' means that they have to take into account on the one hand specific regulations and on the other general policy principles.

In terms of specific regulations, a significant policy event has been the 1988 ERDF Framework Regulation, which addressed the environmental impact of economic development by requiring that member-states receiving assistance in the form of Structural Funds give evidence of the environmental soundness of their expenditures. Objective 1 regions (the 'less developed' regions) are also required to supply appropriate information to the Commission so that it can evaluate the environmental impact of funds received. Another important step forward in the Europeanization of environmental and tourist policy has been the Maastricht Treaty adoption of the *Fifth Action Programme* and the recognition of the Community role in regulating the environment (Bache 1998: 76, 87). These and the indirect impact of other integration measures has effectively constrained environmental policy in Southern European states.

In terms of principles, one indirect but important outcome is the production and dissemination of policy ideas from the center to the periphery. This happens because agenda-setting is the specific task of EU institutions, and particularly the Commission. The recent emphasis on subsidiarity has reinforced member-states' freedom to implement principles differently, but that has meant that EU policy has gradually moved to a higher level of formulation of policy ideas.

Whilst the importance of ideas has frequently been stressed in the field of policy analysis[22], it is less clear how they matter – who brings new ideas to the policy process and how. Above I argued that social movements, in this case environmental movements, have been powerful channels for new ideas in the policy process through their impact on activists. However, it must be noted that environmental actors are not only movement activists co-opted in decision-making fora but also other institutional actors who accepted movement ideas not in order to avoid disruption coming from protest actions, but because of their principled stance, their ideological and sometimes personal roots in environmental movements and their notions. I would like to call them *institutional activists* (Santoro and McGuire 1997). Environmentalism is a stance that they are likely to express in collegial fora and to institutionalize as taken-for-granted in corporatist fora.

If institutional activism is a powerful shaping force in the policy process of the two contexts, the EU multi-organizational field and national policy contexts in Southern Europe, it is evident from the section above that its influence is not only greater at the EU level, but also takes place along different channels – more political and less bureaucratic. Southern Europe lacks, in terms of this analysis, the power of institutional activism and the shaping force of non-institutional actors. This is the theoretical hypothesis proposed to explain the specific problem of the Greening of Southern Europe.

4. CONJECTURES ON SOUTHERN EUROPE

There is a particular prominence of environmental themes at the level of global political discourse, a level that encompasses the discourse of several transnational non-state actors and supra-national levels of governance, such as the EU. A consensus this widespread cannot easily be disregarded. It provides normative constraints on national actors in the member-states. Such normative constraints become effective only in specific institutional environments which provide an opportunity structure for either compliance, instrumentalization, or sheer neglect. Southern Europe seems to be a combination of all three modes of processing the normative constraints of EU policy-making.

At present, the normative effect is only occurring to a small extent, and this explains environmental policy failure. Norms are insufficiently chan-

[22] See Radaelli (1995) and Yee (1996) for reviews.

neled from the EU to Southern Europe due to the limited weight of institutional activism and social movements, although a variety of environmental protests have taken place in the tourist sector (Kousis 1999). No selective recruitment has helped institutional activists, few collegial fora help social movements. Corporative intermediation instills mimetic behavior only to a limited extent in an industry internally divided and unable to express coherent interests. Small-scale industry, which might be influential in local decision-making, is unfamiliar with the taken-for-granted assumptions apparent in EU-level fora. Bureaucrats are ineffective in coercing operators to uphold environmental standards because of generalized non-compliance rooted in insufficient resources to effect controls, in political corruption, and ultimately in cultural resistance to environmental values.

In this situation, while EU-level influence on Southern European policy continues to increase the slack between legal incorporation and behaviors of ritualistic adoption, misinterpretation, and distortion of EU policy principles does not decrease. As one goes from core environmental policy to peripheral policies such as tourism, this slack means that, at best, environmental considerations come to function as the glue that cements the otherwise unconnected policy problems and preferred solutions of a host of vested interests. The idea of sustainable tourism creates new connections between institutions and individuals. It is a sort of garbage can (March 1988) where problems and solutions are tossed in and taken out by interested actors, but where on a large scale tourism continues to be environmentally unsustainable.

REFERENCES

Akehurst, G. (1995) European Community tourist policy, in P. Johnson and B. Thomas, *Perspectives on Tourist Policy*, Mansell, London, pp.215-231.

Bache, I. (1998) *The Politics of EU Regional Policy*, Sheffield Academic Press, Sheffield.

Benton, T., and Redclift, M. (eds.) (1994) *Social Theory and Global Environment*, Routledge, London.

Biekart, K. (1995) European NGOs and performance in South America, in M. Edwards and D. Hulme, *Non Governmental Organisations: Performance and Accountability*, Earthscan, London, pp.63-72.

Cawson, A. (ed.) (1985) *Organized Interests and the State: Studies in Meso-Corporatism*, Sage, London.

CEC (Commission of the European Community) (1997) Examples of Initiatives in Favour of Employment in Tourism Supported by the European Community, Paper No XXXIII/291/97 Luxembourg, Author.

Dunleavy, P. (1991) *Democracy Bureaucracy and Public Choice*, Harvester, New York, NY.

Edwards, G., and Spence, D. (1994) *The European Commission*, Longman, Harlow - Essex.

Edwards, M., and Hulme, D. (eds.) (1995) *Non Governmental Organisations: Performance and Accountability*, Earthscan, London.
Gamson, W.A., and Modigliani, A. (1989) Media discourse and public opinion on nuclear power: a constructionist approach, *American Journal of Sociology* **95**, 1-38.
Giugni, M. (1995) Outcomes of New Social Movements, in H.-P. Kriesi, R. Koopmans, J.W. Dyvendak and M. Giugni, *New Social Movements in Western Europe*, Minnesota University Press, Minneapolis, MI, pp.207-237.
Haas, P.M. (1992) Epistemic communities and international policy coordination, *International Organization* **46**, 1-35.
Hanf, K., and Jansen, A.-I. (eds.) (1998) *Governance and Environment in Western Europe*, Longman, Harlow - Essex.
Kousis, M. (1999) Tourism and the environment: A social movements perspective, *Annals of Tourism Research* **27**, 1-22.
Krämer, L. (1992) *Focus on European Environmental Law*, Sweet & Maxwell, London.
Lehmbruch, G., and Schmitter, P.C. (eds.) (1982) *Patterns of Corporatist Policy-Making*, Sage, London.
March, J. (1988) *Decisions and Organizations*, Oxford University Press, Oxford.
Mazey, S., and Richardson, J. (1993) Environmental groups and the EC: Challenges and opportunities, in D. Judge, *A Green Dimension for the European Community*, Frank Cass, Essex, pp.109-128.
Michelmann., H.J. (1978) *Organizational Effectiveness in a Multinational Bureaucracy*, Saxon House, Farnborough, UK.
Radaelli, C.M. (1995) The role of knowledge in the policy process, *Journal of European Public Policy* **2**, 159-183.
Rucht, D. (1997) Limits to mobilization: Environmental policy for the European Union, in J. Smith, C. Chatfield and R. Pagnucco, *Transnational Social Movements and Global Politics: Solidarity Beyond the State*, Syracuse University Press, Syracuse, NY, pp.195-213.
Ruzza, C. (1996) Inter-organizational negotiation in political decision-making: EC bureaucrats and the environment, in N. South and C. Samson, *Policy Processes and Outcomes*, Macmillan, London, pp.210-223.
Santoro, W.A., and McGuire, G.M. (1997) Social Movement insiders: The impact of institutional activists on affirmative action and comparable worth policies, *Social Problems* **44**, 503-519.
Schmitter, P.C. (1974) Still the century of corporatism? *Review of Politics* **36**, 85-131.
Smith, M.J. (1993) *Pressure, Power and Policy*, Harvester, New York, NY.
Sprenger, R.-U. (1997) *Employment in the Environmental Sector*, Paper presented at the European Conference on 'Environment and Employment', 26-27 May, Brussels.
Waters, M. (1989) Collegiality: Bureaucratization and professionalization: A Weberian Analysis, *American Journal of Sociology* **94**, 945-972.
Yee, A.S. (1996) The causal effect of ideas on policies, *International Organization* **50**, 69-108.

II

SUSTAINABILITY FROM ABOVE AND FROM BELOW IN SOUTHERN EUROPE. COMPARATIVE STUDIES AND COUNTRY STUDIES ON COLLECTIVE ACTION IN SOUTHERN EUROPE

Chapter 5

Competing Claims in Local Environmental Conflicts in Southern Europe[1]

Maria Kousis

1. INTRODUCTION

Within the study of environment-related social conflict, unlike environmentalist claims which have been a subject of more frequent research, counter-claims of the challenged groups constitute an issue that has only recently become a research topic (McAdam, McCarthy and Zald 1996). The counter-claims of challenged groups, in particular the responses to the demands and actions of environmental mobilizations, remain to a large extent unexplored. In addition, the types of agencies or bodies which the mobilizers approach, directly or indirectly, in order to seek assistance, intervention, or simply to be heard, have not been examined systematically.

Social conflict arises when one interest group makes negative claims about another (Tilly 1987). The major actors of environmental politics and conflict in contemporary Western societies, i.e., market, state, and civil society groups, compete for the use of natural resources (Eder 1996b; Schnaiberg 1994). Driven by profit motives, producers apply all forms of control capacity to seize exchange values in markets and to affect the related state positions. By contrast, environmental movement organizations and

[1] The data for this chapter originate from project EV5V-CT94-0393 funded by DG XII of the European Commission. I am more than grateful for the invaluable collaboration and support of my partners, Susana Aguilar Fernández and Teresa Fidélis, as well as all the research assistants, especially Ilse Borchard. Many thanks to Allan Schnaiberg and Klaus Eder for their constructive comments and their support. Any remaining errors are my own.

K. Eder and M. Kousis (eds.), Environmental Politics in Southern Europe, 129–150.

participants usually focus on a wide array of use-value interests in ecosystems. The state assumes a dual role, since on the one hand, as a facilitator of economic growth, it must turn to environmental resources for their exchange values, while on the other, as a social legitimator, it must secure the capacities of ecosystems to produce the use-values of its constituents (Schnaiberg 1994).

While the institutionalization of the environmental movement, especially its associational wing and its related neo-institutional discourse as a whole, is now widely recognized (Eder 1996a, 1999), scholars do point to another of its wings, grassroots environmental activism, as more radical in both demands as well as action forms (Gould, Schnaiberg and Weinberg 1996; Rucht and Roose 1999; Kousis 1999b). Mainly a product of persistent environmental activism, discourse in this local context avoids the institutional route.

The major protagonists of local environmental conflicts are community mobilizing groups, the users, owners or interest groups of sources or (in)activities responsible for ecosystem intervention, and the bodies approached by the mobilizers for assistance in environmental protection (Kousis 1999a). Each of these groups is characterized by distinct environmental discourses. Schnaiberg's (1994: 39-42) thesis of competing claims delineates the major issues around which environmentalist-oriented challengers and their antagonists (the state and producers) make their demands. According to Schnaiberg, within a framework of "managed scarcity", the counterclaims of challenged group typically respond to environmentalist claims by referring to the problem's severity, causal issues, and cost-benefit issues. Tilly's (1978) proposition to classify action repertoires is extended to cover claims and counter-claims.

This chapter aims to bring to the surface evidence on environmental claims and counterclaims between local contenders, the groups they approach for assistance, as well as the groups they challenge in relation to ecosystem damages in Greece, Spain and Portugal, from the end of their authoritarian regimes to 1994. What are the claims of these groups regarding the impacts produced by ecosystem intervention and which types of resolutions do they propose? Have the types of approached groups changed through time in Southern Europe? What does the evidence reveal about such groups approached as EU agencies and large environmental organizations? Which are the major challenged groups and how do they tend to respond to the demands of the mobilizers? The implications of the findings will be discussed for environmental policies and practices within wider, contemporary Southern European political culture.

2. LITERATURE REVIEW: CONTENDERS, HELPERS, AND ANTAGONISTS

2.1 Community-based environmental contenders

In contrast to the continuing belief that 'civic culture' in Southern European countries sanctions non-cooperative and non-compliant [i.e., contentious] behavior as regards the environment (La Spina and Sciortino 1993), during the last three decades, Southern European societies witnessed the growth of a new form of popular/issue public (Eder 1996b) involved in significant environmental protest (Gil Nave 2000; Jiménez 1999; Barcena, Ibarra and Zubiaga 1995; Louloudis 1986; Diani 1995; Aguilar Fernández 1997; Aguilar Fernández, Fidélis and Kousis 1995; Dede 1993; Close 1999). This 'issue public' has usually appeared either later, in the form of environmental associations, or earlier, in that of more informal and flexible groups making different types of claims about the protection of the environment. Environmental associations have increased, especially since the mid-1980s (Holliday 1997; Botetzagias 2000; Gil Nave 2000), while grassroots environmental groups have flourished since the end of the military dictatorships in the 1970s (Kousis 1999b). Local resistance against environmentally damaging, government-promoted, industrial projects or urban pollution occurred, albeit rarely[2], even under authoritarian regimes in Southern Europe (Castells 1983; Louloudis 1986; Spanou 1995).

Who are these local environmental contenders? The international literature has called them community, eco-populist, or grassroots groups (Szasz 1994: 72-74; Weinberg 1997), local citizen-worker movements (Gould, Schnaiberg and Weinberg 1996), working class activists (Cable and Cable 1995), and new middle-class movements (Eder 1995). Qualitative and quantitative studies distinguish a wide variety of local groups participating in environmental mobilizations. Among the more prevalent are residents[3], and/or their committees, local government, local environmental groups, labor and trade unions, and local activity clubs (Kousis 1999a).

Grassroots groups differ from environmental organizations in the way they deal with the environmental problems they are faced with (Gould, Weinberg and Schnaiberg 1993; Carmin 1999). Whereas, upon institutionalization, the associations have softened their tactics in addressing powerful

[2] Especially for Greece and Portugal.

[3] Including neighbors (especially in Spain).

actors in control of sources and activities that enable them to intervene in the ecosystem (Eder 1996a), local groups remain more confrontational, directly challenging those actors (Brand 1999; Rucht and Roose 1999).

Community-based environmental mobilizers have been especially important within a given geographical area and seem to be increasing, especially since the 1980s. Unlike professionally organized mobilization, these mobilizations are – given community dynamics and needs – simultaneously more temporary and richer in nature (Tilly 1994) and are not usually supported by larger environmental organizations (Szasz 1994; Kousis 1999b). The participants of such environmental movements usually come into conflict with and challenge political and economic power-holders over issues of ecosystem use, thus compelling the latter into non-institutionalized discourse.

Although all of these environmental activists make claims concerning health, economic, and ecosystem impacts, different importance may be attributed to these by each group, given their different sets of priorities. Simultaneously, for the same reasons, the resolutions they propose are likely to differ (Schnaiberg 1994). While residential groups will attempt to secure health and the related ecosystem protection for their families via more radical propositions, such as halting the disturbing activity, environmental groups will tend to place higher values on nature conservation, via a less radical and more institutionalized route, i.e., via the application of environmental rules and regulations. Finally, worker's-groups (such as labor and trade unions) are economically threatened and therefore tend to be concerned about economic issues as well. Therefore, by comparison, they are more likely to highlight economic impacts and to propose solutions that will be more beneficial in economic terms.

In general, movement participants, regardless of whether they take direct or indirect action, call for power holders to take the crucial measures to solve the problem (Tilly 1994), either by soliciting their assistance or by challenging them.

2.2 Groups approached

When faced with a serious environmental problem, local environmental activists normally approach a variety of bodies in their attempt to resolve it. Activists may decide on how to proceed with resolving the problem depending on how these bodies respond to their calls (Freudenberg and Steinsapir 1992; Cable and Cable 1995). Although these bodies are dealt with in qualitative case studies of new social movements (Gould, Schnaiberg and Weinberg 1996), in quantitative ones they remain unexplored (Kriesi et al. 1995).

Grassroots environmental groups seeking ecosystem protection have turned to local governments, state agencies, and the courts. State or local governments face the dilemma of either helping their citizenry or accommodating economic growth (Schnaiberg 1994). Thus, their responses to the activists are not likely to yield the desired solution. At the same time, partial support is to be expected, most likely in the form of promises or procedural types of assistance. Lying outside state interests and roles, the courts are likely to be firmer in their stance and more effective in terms of outcomes.

In addition, local environmental activists are more typically known to resort to large environmental organizations for assistance (Gould and Weinberg 1991; Kousis 1997, 1999b). During the past two decades these organizations have become not only large on the national level, but also branches of internationally operating organizations as well (Kousis 1999b; Rootes 1999). They usually offer crucial external resources to the local activists, in the form of scientific and technical information, tactical planning, in addition to support for their position. They may also synchronize movement strategies cross-nationally (Kousis 1999b). The empirical question therefore is: How often are these organizations approached and during which period have they been more visible in Greece, Portugal and Spain?

EU agencies have also been approached as legitimators of public demands. Environmental activists have pursued their goals through the European Court and the European Parliament, taking advantage of the new opportunities provided by the emerging EU structure (Marks and McAdam 1996). Has the entry into the EU shifted power away from the nation-state as regards the strategies of environmental activists in the Southern European states? What are the implications for the environmental movements of these countries?

2.3 Groups challenged

Power holders who are challenged by local environmental contenders are usually producers, states, or international agencies. In control of some form of organization, they are all directly or indirectly involved in the decisions and policies that lead to or intensify ecological marginalization (Schnaiberg 1994; Gould, Schnaiberg and Weinberg 1996; Kousis 1998a; Perrow 1997). The elimination of ecosystem 'use values' is achieved by corporations/economic producers, which are permitted to externalize the environmentally damaging costs of production and limit corporate liability (Cable and Cable

1995). Driven by profit motives, producers apply all forms of control capacity to seize the exchange values in markets and to affect the related state positions. As far as causal issues are concerned, they attempt either to reject or downplay the existence or severity of the problem, or to attribute it to the general developmental trends of the time. As regards cost-benefit matters, they may offer job opportunities or compensations to outweigh the negative costs of environmental degradation. Given their motives, they would be the least likely of the other challenged groups to open negotiations, promise ameliorating measures, or proceed with correcting the problem (Schnaiberg 1994: 39-42).

The state is challenged – usually directly and less often indirectly – more often than the producers in most environmental social mobilizations (Kousis 1999b; Close 1999). First, the state assists the exploitation of environmental resources by producers for their 'exchange values'; secondly, it must also preserve the ecosystems' capacities so that they can produce the 'use values' for the citizenry (Gould, Weinberg and Schnaiberg 1993). There is some evidence that governmental agencies seek to achieve minimally acceptable levels of environmental remedy at the least possible economic cost, usually via non-structural technological fixes (Gould 1991). Given their conflictual roles, it may be expected that state and sub-state groups will take similar – yet not as strict – positions in defense of ecosystem exploitation, while simultaneously consoling the mobilizers by opening dialogues or promising to take ameliorative measures. Thus, state and local government groups can be expected to be less negative than producers in the counter-claims towards the activists.

Studies have only begun to shed light on the responses or counter-claims of challenged groups to environmentalist demands (Brand 1999). Preliminary evidence related to challenged groups involved in tourism activities shows very limited positive responses on the part of entrepreneurs, local government, and the state (Kousis 2000; Close 1999). Compared to an earlier period it nevertheless becomes evident that the state is responding more to claims from different types of environmentalist groups in Southern European states (Kousis, Aguilar Fernández and Fidélis 1996; Aguilar Fernández 1997; Jiménez 1999; Gil Nave 2000; Kousis 1997; Close 1999). The European Union (EU) has been viewed as a further resource of collective action (Imig and Tarrow 1999; Close 1998). EU's development policies have been described as environment-threatening, especially for its less well-off Southern European member states (Kousis 1993; Close 1998).

The sections that follow use evidence on Southern European local environmental conflicts first, to test Schnaiberg's thesis of competing claims.[4] Are groups more interested in the ecosystem for its 'exchange values' (such as producers, the state) more likely to make claims favoring economic growth and less likely to make strong claims for environmental protection than those groups (e.g. residents and local environmental groups) interested in the ecosystem's 'use values'? Are there groups (such as local governments) which make claims for both – environmental protection and growth – simultaneously? Even among the mobilized, do differences in claims exist, depending on the type of mobilizing group (e.g. workers' groups versus environmental groups)? Secondly, the data is applied to test the hypothesis of Marks and McAdam regarding the effects which shifting power from the nation-state to the European Union had on the strategies of environmental activists. After entry into the EU, have these activists pursued their goals through EU agencies? What types of responses have they received? Finally, what is the overall character of grassroots environmentalism and of the competing claims in Southern Europe, and what have the implications been for the environmental movement in Southern Europe from the mid-70s to the early 1990s? Is it closer to a citizen-workers' or a middle class movement?

3. COMPETING CLAIMS ON THE ENVIRONMENT IN SOUTHERN EUROPE

The evidence presented here has come out of 'protest case analysis', a content analysis method that compiles information from newspapers on five series of data: location, events, groups, time, and issue-claim linkages.[5] Protest case analysis borrows from protest event analysis on the movement dimension, but also comes close to public discourse analysis (Eder 1996b) since it codes the claim and counter-claim repertoires for all actors involved in a given environmental conflict. This analysis focuses on the discursive character of local environmental conflict.[6]

4 The code categories for the counter-claims began with Schnaiberg's (1994), but were adjusted to the information drawn from the variety of protest cases. Thus, the final specific categories appearing in this work reflect the influence of the data at hand.

5 For more information on the data see Kousis (1999a, 1999b).

6 This work does not adopt a comparative approach. Instead, it examines competing claims within local environmental activism in all three countries as a whole, because the evidence (Kousis, Aguilar Fernández and Fidélis 1996) suggests that, overall, differences in competing claims are not significant across Greece, Spain, and Portugal. Although differences in the patterns and magnitude of local environmental mobilizations across the three coun-

Each case represents collective incident/s in which five[7] or more persons from a specific geographic area, excluding members of the national government, express criticism, protest, or resist and make a visible claim about their health, physical environment or economic status, which, if realized, would affect the interests of some other person(s) or group(s) during a given time period. Each case also contains information on the types of challenged and approached groups and on their responses to the demands of the mobilizers. The data relates to 2,613 local environmental protest cases[8] in Greece, Portugal and Spain from the end of their dictatorial periods through 1994 and was drawn from articles located in the major national newspapers *Eleftherotypia* (GR), *El Pais* (ES), *Jornal de Noticias* (PT), and *Publico* (PT)[9], as well as from ecology magazines *Oikologia and Perivallon* (GR), *Nea Oikologia* (GR), *Integral* (ES), *Quercus* (ES), and *AAVV-Forum Ambiente* (PT)[10] for the same period.

3.1 Grassroots environmental activism

Grassroots environmental mobilizations in Greece, Spain and Portugal are for the most part single, community-based event, more urban than rural (except in Portugal, where the reverse occurs) and usually involving less than 500 participants. Residents are the leaders of these mobilizations, usually followed by local government, local environmental groups, and other groups such as labor and trade unions, cultural activity clubs, and local political party representatives. They protest in the form of mostly non-violent

---

tries do exist (Kousis 1999a, 1999b), the patterns revealed by competing claims in each country showed more similarities than other dimensions of local environmental conflict. The only exception to this appears in the Portuguese producers' responses. Even though the overall tendency among challenged group responses is the same in the three countries, albeit for a small number of cases, private producers appear to respond more positively in Portugal than do their counterparts in Greece and Spain. Future research efforts may draw on multiple sources of information in an effort to provide a more detailed comparative analysis of competing claims related to specific groups of protest cases for each of the three countries.

7 If the number was not mentioned explicitly in the articles, the coders applied rules estimating the number of participants.

8 Approximately thirty percent of these cases were located for Greece, fifteen percent for Portugal, and fifty-five percent for Spain.

9 The Portuguese team, headed by Teresa Fidélis, chose 'Jornal de Noticias', the oldest, independent and reliable newspaper covering such events. But given its main focus on the Northern and Central regions of Portugal, it was supplemented by 'Publico', first appearing in 1990, which covers both national and regional events in two regional dossiers.

10 The Portuguese environmental magazine first appeared in 1994.

actions, which are normally characterized by general claims-making (such as press conferences), courts action, petitions, demonstrations, strikes, and source/road blockades. Violent episodes are highest in Spain (13.5%), followed by Greece (7.3%), and Portugal (1.8%). For the great majority of the cases in the three countries, non-local supporters are absent. The most important of these were non-local environmental organizations and political parties, present in less than one tenth of the cases. About one third of these mobilizations are sustained for periods beyond one year and are therefore considered as community-based movements (Tilly 1994). These are characterized by network extension, intensive protest activity, exposure to more pervasive impacts, and deeper concerns about the natural environment (Kousis 1999b).

Grassroots environmental contenders identify source/s or in/activities intervening in their environment which are more evident in construction and infrastructural activities, waste disposal, and lack or failure of environmental protection policies. For Spain, nuclear power plants and nuclear waste disposal were also more important – given the dynamic of nuclear energy production in the country. The majority of sources already exist (i.e., are not being planned) and are in operation, producing negative effects for the people in the immediate vicinity. They tend to be concentrated in the more developed and urbanized regions of the countries, although in more recent years they seem to be growing in rural areas, along with the expansion of economic growth activities (Kousis 1999a).

A systematic comparison of the regional distribution of grassroots environmental mobilizations shows that rural activists are more homogeneous, less likely to be tied to political parties, and least numerous when compared to the number of urban and cross-regional mobilizations. In general, they mobilize against toxic and non-toxic waste disposal, as well as construction- and extraction-related activities in undeveloped areas. Compared to urban mobilizations, rural activists refer more frequently to economic impacts resulting from their dependence on resource-based economies. Urban activists are more representative of resource-rich groups such as political parties or other formal organizations. They are more heterogeneous and their claims focus more on construction activities in already developed areas, land transport, traffic, military installations, tourism activities, and environmental policies. In contrast to both urban or rural activists, it is coalitions of both that (albeit for a minority of cases) undertake the most intense and violent actions. Mobilizers coming from both urban and rural communities are more political, heterogeneous and extensive in terms of networks, as well as more numerous in their actions (Kousis 1999a).

3.2 Competing environmental claims

3.2.1 Contenders' claims on impacts and proposed resolutions

Amongst a variety of about twenty different types of local groups[11], five are met more frequently in local environmental protest cases. Residents are present in the great majority of cases (73%), while local government and local environmental groups appear in about one third of all cases (30% and 28% respectively). Labor & trade unions as well as social/cultural activity clubs[12] are evident in less than one fifth of the cases (about 17% and 14% respectively). In general, participants come from a variety of groups. The five most important in terms of frequency have been selected in order to examine the kinds of impacts each associates with ecosystem intervention and the types of resolutions they propose for the amelioration of the environmental problem. The related data presented in Table 1 reflect mostly converging, although at times also diverging views. The mobilizers focus mainly on combinations of ecosystem, health, and economic impacts.[13]

Concerning the impacts which stem from specific ecosystem intervention, the data indicate that for all five groups ecosystem impacts are most important, followed by health and economic impacts. The only exception to this involves local labor and trade unions, for whom economic impacts come second – hardly unexpected, given their interest in economic matters. Although there is an overall agreement on the relative importance of the three major impacts, the weight of their importance is not consistent across the five groups. Notable differences appear when comparing identified impacts for local environmental groups with those for local residents. While for the latter there is more of a balance between the three impacts, for the for-

[11] Other local groups which appear less frequently include representatives of political parties, parent-teachers associations, students, hunters, employers, local physicians, religious groups, women's groups, judges, artists, local development agencies, and local agencies affiliated with the state.

[12] Local Political Party representatives come in fifth place (13.3%) immediately after Cultural Activity Clubs (13.6%). They are in themselves worthy of a separate and detailed future study, and thus have not been included here.

[13] A wide range of impacts also included aesthetic, cultural-historical, built-environment, political, psychological, recreational and life-threatening. Health impacts contain realized, suspected, or expected incidents. Economic impacts include those related to property values, decreasing incomes, threat to economic subsistence, as well as plant, animal, and crop destruction. The range of ecosystem impacts extends from the local ecosystem to forests and contains a total of 12 categories of specific ecosystems.

mer, ecosystem impacts are the top priority. The groups most concerned with the economic impacts are local labor and trade unions as well as local government; those least concerned with these impacts are cultural activity clubs and environmental groups. Health impacts are especially important for residents and local government.

Table 1. Impacts identified and resolutions proposed by type of local environmental activists (%)

	Environmental Activist Type				
	Local Residents	Local Government	Labor and Trade Unions	Local Activity Clubs	Local Environmental Groups
IMPACTS					
Health	44.3	45.0	43.0	39.0	32.0
Economic	30.7	40.1	60.1	21.3	24.5
Ecosystem	88.6	90.3	88.8	92.1	97.4
RESOLUTIONS					
Ecosystem preservation	28.6	32.7	29.8	44.4	41.1
Compensation/jobs	6.2	8.1	16.4	3.4	4.2
Env.regulations/policy	41.6	42.4	46.5	50.3	60.1
Ecolog.modernization	15.0	18.1	21.7	13.2	17.2
Halt disruptive source	74.9	75.8	68.0	73.3	73.6
equal say in decisions	6.9	10.4	10.1	7.0	7.2
Total no. of cases	1891	785	456	356	734

Note: Percentages were calculated within type of local group for each category of impact or resolution, which was coded as a dichotomous yes/no variable. Thus percentages do not add up to 100. The total number of cases is 2,613.

The second half of Table 1 depicts the resolutions[14] proposed by local environmental activists. Once more, there is a clear overall convergence

[14] The resolutions presented in the table have been re-coded from a total of 19 categories. Ecosystem preservation consists of 'preservation/conservation/environmental management of a wild life/rural area', 'preservation of a cultural heritage area', and 'preservation of green zones/spots in already developed areas'. Compensation and jobs/employment have been merged into one category. Environmental regulations includes 'the creation of environmental protection rules and regulations, laws/policy, or protected area', 'the implementation of regulatory rules, existing laws', as well as 'proper regional and urban planning'. Ecological modernization contains less polluting technologies: 'main process related', and 'nonmain process ones' such as end of pipe), and 'Environmental Impact Assessment studies'. Halt of Disruptive Source includes 'removal/relocation', 'removal and restoration of area', 'plans not actualizing/annulment of challenged group's plans', 'decrease in production activities', 'permanent shut down of source'; 'permanent stop of

between the groups regarding the measures needed to resolve the environmental problem. Halting the source or activity disrupting the ecosystem is the most predominant resolution proposed, followed respectively by the creation or implementation of environmental regulations, the preservation of the local ecosystem, and the application of modern ecological technologies. The remaining two types of resolution are about the same in importance, both ranking last in terms of frequency. Surveying the different types of mobilizing groups, some differences, albeit small, are evident. Two groups appear to differ, although to a limited extent, from the rest. While local environmental groups tend to favor ecosystem preservation and environmental regulations more so than others do, labor and trade unions are more supportive for resolutions proposing ecological technologies as well as compensations and employment opportunities. Simultaneously, the latter are also least supportive when it comes to halting the operation of the source of ecosystem disruption. This alignment is indicative of the interests associated with each of the mobilizing groups. Nevertheless, although these small diversions do, to a certain degree, reflect variations in the views of the five most salient environmental activist groups, there is a very clear priority given by all to resolutions favoring more environmental protection. It should be noted here that, in the great majority of cases, more than one resolution is proposed by the mobilizers.[15]

3.2.2 Approached bodies and their responses[16]

Before proceeding with contentious tactics, local environmental activists may approach one or more bodies from a broad range of organizations, agencies, and actors[17] in order to request assistance in solving the problem. This occurred in more than one half of all cases (57%). State bodies are more frequently approached (26%) than any other ones, followed by substate bodies (16%), the courts (8%), the EU (4%) and large environmental organizations (3.2%). Based on the information provided in the national

---

ongoing projects or eco-disturbing activities', and 'total restoration of affected area'. Equal say or participation in decision-making comprises a category of its own.

[15] Given the plethora of combinations, each of the twenty categories has been coded as a dichotomous yes/no variable.

[16] It should be noted that, although information on mobilizers' claims was abundant, the same did not hold for counterclaims. Information on how the approached or challenged groups responded to the mobilizers' claims was limited, as shown in the tables which follow. This is an inherent problem in using media accounts, which tend to define 'news' as 'front-stage' activities far more so than 'backstage' ones..

[17] 22 categories have been coded in total.

newspapers, the responses of the approached bodies to the requests of the mobilizers vary extensively, from very negative to very positive.[18]

Table 2. Responses* of bodies approached for assistance

Response	Sub-state	State	Courts	EU	ENGOs
No specific information	22.8	41.1	39.2	51.9	16.7
Indifference; support to polluter	7.7	10.2	6.1	5.8	1.2
Recognition of mobilizers	19.2	12.6	14.0	8.7	33.3
Jobs/compensation	1.0	0.1	-	-	-
Promised solution	10.8	5.3	0.5	2.9	-
Temporary halt of activity	3.2	5.3	7.0	-	-
Aid +	27.6	19.1	22.0	24.0	46.4
Resolution	4.7	3.8	9.3	3.8	1.2
Other	3.0	2.5	1.9	2.9	1.2
Response total (N)	100(404)	100(678)	100(214)	100(104	100(84)

* The major/final response
+ Types of Aid: implementation, procedural, technical, legal, economic, organizational, publicity

Note:
Sub-state contains regional government (SP:70 cases), freguesias (PT: 67 cases), local government (257 cases), regional parliament (SP: 10 cases).
State includes government/central state, ministries, governing party, (352 cases), central state representative at local level (294 cases), national parliament (32 cases).
Courts consists of regional, and supreme courts.
EU includes the European Commission, the European Court, and the European Parliament.
ENGOs *are large environmental organizations, usually non-local, and possibly international in origin (see Kousis, 1999a).*

Table 2 presents the five most important approached agencies and their respective major[19] reply. Information on the responses of these important actors in environmental conflict remains limited. This information was not available in about half of the cases that approached EU bodies, less than half for state and legal bodies, about one fourth for sub-state agencies and less

[18] For each of the 22 categories of approached bodies, 24 different types of possible responses were coded.

[19] Each body can have multiple responses throughout the conflict. The major one has been used in this analysis.

than one fifth for large environmental organizations. In the majority of cases these bodies did offer assistance to the mobilizers. The most prevalent type of reply across all five groups, appearing in about one fourth of the cases, comes in the form of procedural, technical, legal, economic, and organizational aid as well as aid in implementation and publicity. This response is more frequent in environmental non-governmental organizations (ENGOs) and least offered by state bodies. The second most important response across all groups (in 16% of cases) appears in the form of recognition of the mobilizers, it being highest for ENGOs and lowest for the EU. This form of moral support may also include the recommendation to be patient with the situation. The third most important response (9% of cases) is a temporary halt of the ecosystem-disturbing activities, or total/partial resolution of the problem. The courts play the most crucial role here, followed by sub-state, state, and EU bodies. Although in most of the cases these five types of bodies have responded positively to the mobilizers' call, in another one tenth of the cases they did not; instead, they expressed indifference, or supported the source of ecosystem disturbance.

Local environmental activists have not approached the same bodies for assistance since the 1970s, as illustrated in Figure 1. The number of all bodies approached appears to be increasing, especially since the late 1980s. To an extent, this reflects the number of environmental action forms. Although a thorough examination is required to explain this, based on previous work it may be said that both structural and cultural factors are behind this trend (Kousis 1999b). The first decade after the fall of the authoritarian regimes in the three countries witnessed the rise of many social mobilizations (for Spain see Koopmans 1996). As regards local environmental activism, qualitative changes are evident in the type and quantity of environmental problems claimed before and after the mid-80s. Increases in environmental awareness may also affect the apparent increases. It is difficult at this point to predict whether these increases will continue, or at what point a decrease will occur in the protest-case wave of the early 90s.

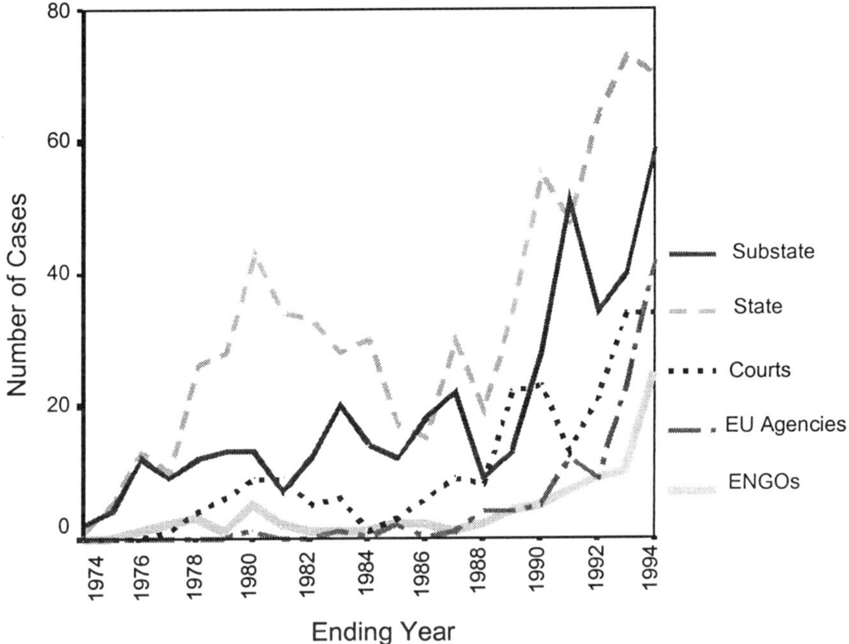

Figure 1. Bodies approached in local environmental cases for Greece, Spain and Portugal 1974-1994

Grassroots environmental activists are not influenced by the same factors as environmental associations are. For the latter, changes in the political opportunity structure may be more important, while the former may be influenced more by combinations of other cultural, ecosystem, or economic factors. What is certainly obvious is that local environmental activists approach the state more often, followed by sub-state agencies, the courts, EU agencies, and large environmental NGOs. More drastic increases are evident for the two groups which, although least important in terms of frequency, have gained special significance during the last decade in Southern Europe. The figure depicts different tendencies before and after 1988. Since the late 1980s a noticeable shift has occurred. Although the state is still the most prevalent body, in the later period mobilizers resort to it less often than they did earlier (leading to a decrease from 65 to 52%). Simultaneously, they approach EU bodies and ENGOs much more often than they did in the earlier period (from 3 to 14% for the former, and from 5 to 10% for the latter). After the mid-1980s, criticisms of EU entry had subsided in all of the three countries and activists sought assistance from an authority with a more 'developed' environmental policy. At the same time, ENGOs rapidly increased during this period, given the expanded domestic and international political opportunities (Gil Nave 2000; Jiménez 1999; Kousis 1999b).

3.2.3 Challenged groups and their responses

In local environmental protest cases, the ecosystem-disturbing source/s or (in)activity/ies are usually associated with the social user, whom mobilizers challenge as being responsible for the problem.[20] In most cases they challenge more than one user, usually the state or local government, as well as a producer.[21] The state is challenged more often than any other group (58%), followed by producers (53%) and sub-state (41%) groups. As elsewhere in social conflict, challenged groups react to the demands of the mobilizers in a variety of ways.[22] The main responses of the three most important challenged groups are illustrated in Table 3.

For the groups that were challenged, specified information was available in approximately half of the cases for state and producers alike, and in about 64% of cases for challenged sub-state groups. In general, surveying all three challenged groups, the most outstanding specific reply has been a negative one, i.e., that of continuing operations or showing indifference to the mobilizers' demands. When the replies shown in Table 3 are grouped in negative (1-3), neutral (4-6) and positive (7) ones, then the following succinct pattern emerges. The challenged state and producers rarely respond positively. They do so in 5.6% and 5.3% of cases respectively, while the challenged sub-state does the same in 10% of cases. By contrast, all three usually respond either negatively or neutrally. Producers are characterized by the most negative replies, in one third of the cases, while state and sub-state reply negatively in about one fourth of the cases. Producers are also more likely than the other two groups to stop mobilizers from protesting using all available means, such as force, legal means, threats of job loss, pleas or refusal to open talks, in about one tenth of the cases. Sub-state groups respond more neutrally (in one third of cases), followed by state (in about one fifth of cases), and producers (less than one tenth of cases).

[20] For a detailed presentation on how environmental claims were coded see Kousis (1999b).

[21] Challenged groups have been coded using 18 specific categories (Kousis 1998b)

[22] The replies of the challenged groups were coded using twenty-two different categories (Kousis 1998b); some of these have been merged in more general ones for the purposes of Table 3.

Table 3. Responses/ counter claims* of the challenged groups (%)

Responses	Sub-state	State	Producers
No specified information	36.1	49.4	49.9
Stopped mobilization**	6.2	5.3	8.9
Indifference; continued operation; no response	14.9	15.5	17.3
Jobs/compensation/compliance too costly***	5.8	3.9	6.8
Temporary halt of activity	1.5	2.5	2.0
Recognition/Negotiations	5.8	4.9	2.5
Measures/alterations promised	15.2	8.8	4.0
Measures taken+	10.2	5.6	5.3
Other	4.3	4.1	3.3
Total++	100(1074)	100(1521)	100(1374)

* The major/final response
** by force, by law, by job loss threat, by refusal to talk, by plead
***as well as considering problem as insignificant
\+ some, most, all; better alternative chosen.
++ Since more than one group may be challenged totals exceed the total number of cases (N=2,613)
Note:
Sub-state includes regional government (266 cases), and local government (808 cases).
State consists of central state (1106 cases), and central state representatives at the local level (415 cases).
Producers contains private entrepreneurs (974), state producers (236) semiprivate/semistate producers (96), farmers (48) and fishermen(20).

The major antagonists of the mobilizers are the state[23], the producers, and local government groups. When information is available, the data show that whereas the state and local government respond more positively than negatively, the producers stand out as the group which responds very negatively. It is extremely rare that they give positive responses; they usually counterclaim that insignificant/no offences are produced. This pattern is more evident in Greece and Spain. In Portugal, the producers seem to respond more positively than in the other two countries, albeit for a small number of cases (Kousis, Aguilar Fernández and Fidélis 1996).

Related research and evidence suggests that missing data on counterclaims are likely to be more negative than positive towards the mobilizers,

[23] The state is normally also challenged in cases where the project is funded/operated by a European Union user. Usually these projects are large but small in number. Thus, a separate EU coding category was not applied. These were coded in the 'other' category and will be analysed in a later phase.

thus reinforcing the documented pattern presented above. Many local environmental conflicts tend to pass through periods of dormancy or waiting which do not receive media attention. After initial promises of measures and temporary halting of operations, and once intense mobilizations subside, powerful interest groups return, insisting on the continuation of usually costly operations. This process/stage of return, which is also more underreported in the media than the protest stage, normally takes a considerable length of time - sometimes years – to conclude and thus is more difficult to trace in the printed sources.

4. CONCLUSION

This examination of competing claims in local environmental conflict illustrates the way in which networks of community groups contribute to a non-institutional discursive context. These contenders, though for the most part urban, can also be rural, community-rooted activists who may be citizen-workers (Gould, Schnaiberg and Weinberg 1996) or of working class (Cable and Cable 1995) or new middle-class (Eder 1995) background. Although to a very limited extent, there exist, even among mobilized groups, differences in claims, depending on the type of the mobilizing group (e.g. workers' groups versus environmental groups). This supports Schnaiberg's view that worker's-groups are economically threatened and therefore tend to be concerned about economic issues as well. Thus, even when mobilizing for environmental protection, labor and trade unions are by comparison more likely to highlight economic impacts and to propose solutions that will be more beneficial in economic terms.

In contrast to the continuing beliefs (La Spina and Sciortino 1993) that Southern European countries sanction non-cooperative and non-compliant behavior as regards the environment, the data at hand show that grassroots environmental mobilizations made a relatively strong appearance in the 1970s and 1980s, increasing especially from the mid eighties through the early nineties. These are mostly autonomous acts of resistance not tied to political parties or any form of authority. At the same time, most are not related to the larger environmental movement organizations. Grassroots environmental activists claim that they confront significant ecosystem, public health, and economic impacts. Therefore, they usually demand the discontinuation of the ecosystem-destructive activities, the creation or implementation of environmental regulations, the preservation of the local ecosystem, and the application of ecological modernization technologies.

The evidence also supports those qualitative studies which point to the importance of approached bodies. In Greece, Spain, and Portugal these are mainly the state, sub-state bodies, the courts, European Union agencies and large environmental organizations. As Schnaiberg (1994) posits, although state and sub-state groups may also be challenged by the mobilizers, prior to this, they may also be approached for assistance. Most of these bodies do provide assistance, the nature of which differs from one body to another.

Is this a Europeanizing society? The data also support the related hypothesis of Marks and McAdam. A shift of power from the nation-state to the European Union has occurred, as is reflected in the strategies of environmental activists. Since the entry into the EU, these activists have also pursued their goals through EU agencies, as well as through large environmental organizations, many of which are branches of international ones, such as the WWF or Greenpeace.

Finally, the findings provide support for Schnaiberg's (1994) thesis. Groups interested in the use value of local ecosystems, such as residents and environmental groups, come into conflict with groups more interested in the exchange values of ecosystems, such as producers and the state. While the former pressure the latter to recognize and take measures to alleviate the negative damages and impacts of their activities, the latter rarely respond positively. The available evidence thus indicates that the outstanding reply from challenged groups has been a negative one – they either continue operations or show indifference to the mobilizers' demands. Furthermore, in this data set, counter-claims focus less on causal issues than on problem severity or cost-benefit issues. Indifference or no response on the part of the challenged groups imply that no problem exists, or that it is only minor. At the same time, the temporary cessation of operations, or promises to solve the problem suggest difficulties in accepting the economic costs that must be taken to correct the problem. Thus, on the basis of the interests involved and their related responses, challenged groups in Southern Europe try to promote the continuation of their activities. Some groups (such as local government) are caught in the middle (between challengers and challenged) and subsequently make simultaneous claims for both environmental protection and growth.

Finally, further research on competing claims, especially counterclaims, will garner the detailed data required for further study as long as the focus is put on specific groups (not the population) of environmental conflict cases, and use is made of multiple sources of information.

REFERENCES

Aguilar Fernández, S. (1997) El Reto del Medio Ambiente: Conflictos e Intereses en la política medioambiental Europea, Alianza Editorial, Madrid.

Aguilar Fernández, S., Fidélis, T., and Kousis, M. (1995) Encounters between social movements and the state: Examples from waste facility siting in Greece, Portugal and Spain, "Alternative Futures and Popular Protest", Conference Papers, April 4-6, Faculty of Humanities and Social Sciences Manchester, The Manchester Metropolitan University.

Barcena, I., Ibarra, P., and Zubiaga, M. (1995) Nacionalismo y ecología. Conflicto e institucionalización del movimiento ecologista vasco, Los Libros de la Catarata, Madrid.

Botetzagias, I. (2000) Network Analysis of Greek Environmental Movement Organizations, Paper presented at the Workshop "Environmental Organizations: A Comparative Assessment", ECPR Joint Sessions, April 14-19, Copenhagen.

Brand, K.-W. (1999) Dialectics of institutionalization: The transformation of the environmental movement in Germany, Environmental Politics **8**, 35-58.

Cable, S., and Cable, C. (1995) Environmental Problems, Grassroots Solutions: The Politics of Grassroots Environmental Conflict, St. Martin's Press, New York, NY.

Carmin, J. (1999) Voluntary associations, professional organizations and the environmental movement in the United States, Environmental Politics **8**, 101-121.

Castells, M. (1983) The City and the Grassroots: A Cross-Cultural Theory of Urban Social Movements, Edward Arnold, London.

Close, D.H. (1998) Environmental NGOs in Greece: The Acheloos campaign as a case study of their influence, Environmental Politics **7**, 55-77.

Close, D.H. (1999) Environmental crisis in Greece and recent challenges to centralized state authority, Journal of Modern Greek Studies **17**, 325-352.

Dede, I. (1993) Ökologiebewegung in Griechenland und in der Bundesrepublik Deutschland, Peter Lang, Frankfurt.

Diani, M. (1995) Green Networks. A Structural Analysis of the Italian Environmental Movement, Edinburgh University Press, Edinburgh.

Eder, K. (1995) Does social class matter in the study of social movements? A theory of middle-class radicalism, in L. Maheu, Social Movements and Social Classes: The Future of Collective Action, Sage, London, pp.21-54.

Eder, K. (1996a) The institutionalization of environmentalism: Ecological discourse and the second transformation of the public sphere, in S. Lash, B. Szerszynski and B. Wynne, Risk, Modernity and the Environment: Towards a New Ecology, Sage, London, pp.203-223.

Eder, K. (1996b) The Social Construction of Nature. A Sociology of Ecological Enlightenment, Sage, London.

Eder, K. (1999) Taming Risks through Dialogues: the Rationality and Functionality of Discursive Institutions in Risk Society, in M.J. Cohen, Risk in the Modern Age. Science and the Environment, Macmillan, London, pp.225-248.

Freudenberg, N., and Steinsapir, C. (1992) Not in our backyards: The grassroots environmental movement, in R.E. Dunlap and A. Mertig, American Environmentalism, Taylor & Francis, Washington, pp.27-35.

Gil Nave, J. (2000) Environmental Politics in Portugal? Ph.D. Thesis, European University Institute, Florence.

Gould, K.A. (1991) The sweet smell of money: Economic dependency and local environmental political mobilization, Society and Natural Resources **4**, 133-150.

Gould. K. A., Schnaiberg, A., Weinberg, and A. S., (1996) Local Environmental Struggles: Citizen Activism in the Treadmill of Production, Cambridge University Press, Cambridge, MA.

Gould, K.A., and Weinberg, A.S. (1991) Who Mobilizes Whom? The Role of National and Regional Social Movement Organizations in Local Environmental Political Mobilization, Paper presented at the meetings of ASA, August, Cincinnati, OH.

Gould, K.A., Weinberg, A.S., and Schnaiberg, A. (1993) Legitimating impotence: Pyrrhic victories of the modern environmental movement, Qualitative Sociology **16**, 207-246.

Holliday, I. (1997) Living on the edge: Spanish greens in the mid 1990s, Environmental Politics **6**, 168-175.

Imig, D., and Tarrow, S. (1999) The Europeanization of Movements? First Results from a Time-Series Analysis of European Collective Action, 1985-1993, Paper prepared for presentation at the Conference on Cross-national Influences and Social Movements Research, Mont Pélerin, Switzerland.

Jiménez, M. (1999) Consolidation through institutionalisation? Dilemmas of the Spanish environmental movement in the 1990s, Environmental Politics **8**, 149-171.

Koopmans, R. (1996) New social movements and changes in political participation in Western Europe, West European Politics **19**, 28-50.

Kousis, M. (1993) Collective resistance and sustainable development in rural Greece: The case of geothermal energy on the island of Milos, Sociologia Ruralis **33**, 3-24.

Kousis, M. (1997) Grassroots environmental movements in rural Greece: effectiveness, success and the quest for sustainable development, in S. Baker, M. Kousis, D. Richardson and S. Young, The Politics of Sustainable Development: Theory, Policy and Practice within the European Union, Routledge, London, pp.237-258.

Kousis, M. (1998a) Ecological marginalization in rural areas: Actors, impacts, responses, Sociologia Ruralis **38**, 86-108.

Kousis, M. (1998b) Protest Case Analysis: A Methodological Approach to the Study of Local Environmental Mobilizations, Working Paper No. 570, Center for Research on Social Organization, Department of Sociology Ann Arbor, MI, University of Michigan.

Kousis, M. (1999a) Environmental protest cases: The city, the countryside and the grassroots in Southern Europe, Mobilization **4**, 223-238.

Kousis, M. (1999b) Sustaining local environmental mobilisations: Groups, actions and claims in Southern Europe, in C. Rootes, Environmental Politics, Special issue: Environmental Movements: Local, National and Global, pp.172-198.

Kousis, M. (2000) Tourism and the environment. A social movements perspective, Annals of Tourism Research **27**, 486-489.

Kousis, M., Aguilar Fernández, S., and Fidélis, T. (1996) Grassroots Environmental Action and Sustainable Development in Southern European Union, Final Report, European Commission, DG XII, Contract No. EV5V-CT94-0393.

Kriesi, H.-P., Duyvendak, J.W., Giugni, M., and Koopmans, R. (1995) New Social Movements in Western Europe. A Comparative Analysis, University of Minnesota Press, Minneapolis, MN.

La Spina, A., and Sciortino, G. (1993) Common agenda, Southern rules: European integration and environmental change in the Mediterranean states, in J.D. Liefferink, P.D. Lowe and A.P.J. Moll, European Integration and Environmental Policy, Belhaven Press, London, pp.216-234.

Louloudis, L. (1986) Social demands: From environmental protection to political ecology, in C. Orfanidis, The Ecological Movement in Greece after Rain Publications, Athens, pp.8-21.

Marks, G., and McAdam, D. (1996) Social movements and the changing structure of political opportunity in the European Union, West European Politics **19**, 249-278.

McAdam, D., McCarthy, J.D., and Zald, M.N. (eds.) (1996) Comparative Perspectives on Social Movements: Political Opportunities, Mobilizing Structures and Cultural Framings, Cambridge University Press, Cambridge, MA.

Perrow, C. (1997) Organizing for environmental destruction, Organization and Environment **10**, 66-72.

Rootes, C.A. (1999) Acting globally, thinking locally? Prospects for a global environmental movement, Environmental Politics **8**, 290-310.

Rucht, D., and Roose, J. (1999) The German environmental movement at a crossroads? Environmental Politics **8**, 59-80.

Schnaiberg, A. (1994) The political economy of environmental problems and policies: Consciousness, conflict and control capacity, Advances in Human Ecology **3**, 23-64.

Spanou, K. (1995) The beginning of environmental policies in Greece: The dynamics of the political-administrative agenda during the dictatorship, in K. Spanou, Social Demands and State Policies, Sakkoulas, Athens.

Szasz, A. (1994) Ecopopulism: Toxic Waste and the Movement of Environmental Justice, University of Minnesota Press, Minneapolis, MI.

Tilly, C. (1978) From Mobilization to Revolution, Addison-Wesley, Reading, MA.

Tilly, C. (1987) Social Conflict, Working Paper No.43, Center for Studies of Social Change.

Tilly, C. (1994) Social movements as historically specific clusters of political performances, Berkeley Journal of Sociology **38**, 1-30.

Weinberg, A.S. (1997) Local organizing for environmental conflict: Explaining differences between cases of participation and nonparticipation, Organization and Environment **10**, 194-216.

Chapter 6

Greek Rural Society and Sustainable Development

George A. Daoutopoulos, Myrto Pyrovetsi and Eugenia Petropoulou

1. INTRODUCTION

The success of post-war agricultural development in Greece is a well-known story. In recent years, however, it is a story that has come up against its own intrinsic boundaries and contradictions. The main parameters of post-war developments in Greek agriculture included the constant growth of total agricultural output, the securing of new markets, and the increase of agricultural income. Yet, prospects for the future are haunted by several worrying problems. Greek agriculture is in crisis (Maravegias 1989).

Both success and crisis are the result of a deliberate modernization process formulated during the last fifty years. The promotion of modern agricultural practices and the increasing dependence on costly external inputs in order to maintain productivity, had significant socio-cultural and environmental impacts. For example, knowledge and practices of management methods which have sustained production in traditional agricultural systems in Greece for thousands of years under the auspices of the so-called Community Mode of Production system (CoMP) have declined or vanished, leading to accelerated soil erosion, accumulation of toxic chemicals and the Greek farmers' reliance on diminishing and unpredictably priced petroleum resources (Kousis and Petropoulou, forthcoming). The dissolution of this mode of production due to the penetration of market forces that brought the Capitalist Mode of Production (CaMP) was detrimental for Greek rural areas, resulting in a loss of jobs, further economic disadvantages for small-scale farmers, and increasing specialization of livelihoods (Pyrovesti and Daoutopoulos 1999). Hence, the need to identify the ecologically sustainable features of traditional agriculture is now being recognized, as part of a strategy which can be integrated into present-day economic pressures and appro-

K. Eder and M. Kousis (eds.), Environmental Politics in Southern Europe, 151–173.

priate agricultural technologies in order to insure long-term productivity (Altieri and Anderson 1986).

The analysis to follow examines the foundations of the modernization paradigm, as shaped by different top actors (e.g. multinationals, EU, Greek state), and the current prospects of development which can be accomplished through the sustainability transition. It can be argued, therefore, that the notion of sustainability is probably the most appropriate concept for utilizing and summarizing the wide variety of new development tendencies at the local level, as defined and practiced by bottom actors (e.g. large and small-scale farmers, local authorities, local communities, etc.). Local, and until recently forgotten and obscured resources, such as the capacity to combine agriculture with the conservation of natural resources and landscape; the possibility of producing high quality products and establishing new interlinkages with consumers; the challenge of responding to the environmental crisis with more adequate plans than the ones proposed by the central state and the 'art' of farming economically, are central to this new approach (van der Ploeg and van der Dijk 1995). This chapter discusses a few of the main trends found in these new initiatives and, above all, explores the potential influences of these new perspectives. It highlights how powerful social actors induce the sustainability transition in Greek rural areas, as well as identifies the way in which local actors are affected in terms of social, economic and cultural means - tracing their positive or negative responses to the recent modernization paradigm.

The following sections: delineate the features as well as the destruction of the Community Mode of Production (CoMP) that secured a sustainable livelihood for Greece's rural population; pinpoint the specific characteristics of rural space which survive under the present Capitalistic mode of production (CaMP) and which favor a sustainable path, and; bring attention to the important impacts of International and European Union policies. Finally, this study aims to shed light to approaches that join traditional management practices with modern technological innovations and question whether these can meet Greece's needs to improve rural living standards and reduce dependence on costly foreign technologies.

2. COMMUNITY MODE OF PRODUCTION AND THE ENVIRONMENT

In Greece and the Mediterranean area in general, people have managed to survive for centuries in a harsh terrain with small fertile lands and little rain, due to a system of cultural adaptation, the so-called 'Community Mode of Production' (CoMP) (Karavidas 1981; Sakiotis 1995). The CoMP differs markedly from the 'Capitalist Mode of Production' that has replaced it over the last five decades. The CoMP was a unique model of social, economic and cultural organization which existed throughout much of rural Greece until the late 1940s, living traces of which can still be found in marginal rural areas of the country. The CoMP was basically a subsistence model of production whereby people produced goods to meet their own needs for food and energy, while resources were maintained in such a way as to provide the long-term survival of the community. Those needs were met by production on the household farm, in the nearby forests, in the household vegetable garden, and in the household barn (meat, milk). Any surplus was traded, only under favorable conditions, for the purchase of raw materials (fabrics, leather, metals, household kitchen items, tools, etc.) to be used in another production cycle, the *'household unit of production'*. This economic organization, that was not based on monetary exchange, was part of a series of relations and functions geared to meet the needs of the people in the community. Needs, of course, were based on an entirely different mentality (thriftiness) than the present lifestyle. The functions of the typical household in the community were so diverse and interrelated that they permitted a balanced relationship with the environment. As a result of balanced exchanges and the wise use of resources through recycling, there were no waste products or garbage to be disposed of.

For example, prior to the sixties livestock manure had traditionally been the key to maintaining agricultural productivity in subsistence farming systems in Greece. The subsistence success of the Greek farming system at that time was crop-livestock integration, which allowed farmers to manage an efficient nutrient recycling system centered around small ruminants, who converted the nitrogen fixed by legumes into manure when legume crop residues were used as fodder (Martinos, Louloudis and Daniel 1988). This, of course, involved high labor inputs by farmers, who needed to keep animals tethered within the compound during the rainy season, collect crop residues and weeds for fodder, and then transport their manure back to the fields.

Consequently, the nutrient dynamics of the Greek traditional rural household had been an important development, not only in terms of the limited use of inorganic fertilizers, polycultures, and the incorporation of animal and crop production in a mixed type of farm, but also in diversifying farmer's income resources. Farmers' engagement both in crop and livestock production allowed them to cope with risks, whether they be environmental (drought) or economic (price fluctuations) (Dimen and Friedl 1976).

Sustainable utilization of rural natural resources, as revealed above, indicate the unconditional dependence of farmers on their surrounding environment. A supportive example of this wise use of rural and natural resources can also be illustrated by the Lake Prespa area, presently a National Park, where natural resources and humans co-existed for hundreds of years, thanks to balanced exchanges. In the past, Prespa's farmers harvested the reeds from the lake shores and used them as animal feed and building material (thatch roofs and walls). The benefits of this practice were twofold. First, by removing every year large quantities of organic matter, they improved the physical properties of the lake water, preventing eutrophication and at the same time providing highly nutritious feed for the livestock (Pyrovetsi 1984). Secondly, buildings (e.g. houses and farm buildings) were constructed with mud and reeds extracted from the lake. Even when the mud-reed constructed buildings were demolished, this same material was recycled in the environment. Water from the lake was used directly by humans and animals until the late 1960's. From 1970 onwards, coinciding with the modernization of Greek agriculture and the construction of an irrigation system (Pyrovetsi and Gerakis 1987), reeds were no longer harvested and nutrients in the water increased due to agricultural runoff, thus causing lake eutrophication and making the water unfit to drink for both humans and animals.

The second major characteristic of traditional agricultural management in Greece was farming diversity, which involved the practices of intercropping, crop rotation, and fallowing. It is, however, impossible to say categorically whether a mixture will result in better yields than the monocropped alternatives, except perhaps for legume - non-legume mixtures (Gliessman 1990). In Greek Macedonia, cultivation of legumes (e.g. Phaseolus spp) together with a cereal crop, such as maize, reduced the need for nitrogen fertilizers. Bacteria present in specialized nodules that develop on the roots of legumes can fix nitrogen directly from the atmosphere making it available to plants. The cultivation of cereals and legume crops together can both improve total yields and stabilize production (Gliessman 1990). Thus, long before fertilizers were invented, indigenous knowledge had discovered natural ways to benefit plants that need a lot of nitrogen, without applying chemical fertilizers. Moreover, diversity in crop species in Greece reflected the needs of subsistence farmers to meet the dietary needs of their families.

Furthermore, pest attack was frequently reduced in intercrops (Thurston and Parker 1995). Host plants were more widely spread and so harder to find; one species might trap a pest; or one species might repel the pest; and/or predators might be attracted. Weeds were also more likely to be suppressed by mixtures. Greek subsistence farmers used to leave their land fallow for at least three years or exchange seeds because they had observed that any particular variety tended to suffer from pest problems if grown continuously on the same land for several years. With farm-to-farm variations in cropping systems, the resulting temporal, spatial, and genetic diversity conferred at least partial resistance to pest attacks, making it less risky and more productive among the few mountainous Greek farmers who used low energy inputs (Altieri and Anderson 1986).

Farmers on the mountainous or hilly villages of Greece used sloping terraces (pezoules) in order to slow the rate of rainfall runoff and thereby soil erosion (Kosmas et al. 1995). Within the field, they could manage runoff by grading the slope of the furrows to minimize rapid runoff and erosion and yet avoid pooling of water around crop roots. Many uplands have been terraced for cultivating cereals, vines, olives and other crops for centuries. In Lesvos, for example, large areas were and still are cultivated with olives in individual terraces, while in several hilly areas of Epirus, Peloponesos and the Aegean islands, terraces have been constructed by stone walls. Soil had been removed from river deposits by animal traction in order to fill in the constructed terraces. Returning to the Prespa case, when the terraced fields in the hills were abandoned, as a result of the construction of the irrigation system in the low lands (Pyrovetsi and Gerakis 1987), hill soil erosion increased to the point that sedimentation in the lake made it shallower and highly eutrophicated.

In Greece as well as in other traditional farming systems throughout the world, the simplest strategy to check the surface flow of water and thereby to conserve soil and retain water for irrigation, is to construct earth banks across the slope that act as a barrier to runoff. In some Greek mountainous communities the earth banks are reinforced with vegetation, such as crop stalks or trees, in order to ensure better stability. Such vegetative bunds are partly permeable, so crops planted in front of the bund also benefit from water runoff (Altieri and Anderson 1986). Moreover, on the Aegean islands, local architecture incorporated water conservation practices. Houses were equipped with a specially constructed roof to collect rain water and store it to irrigate the homestead garden and animals. Both conservation management techniques required high labor costs to construct and maintain sloping terraces and to retain water runoff.

From what has been discussed so far, CoMP, along with traditional management techniques such as soil and water conservation and nutrient cycling methods in Greece, were based on labor intensive means of management. Human labor has historically been more available than other forms of energy or natural resources in rural Greece, so management strategies have evolved to make use of this labor potential.

3. MODERNIZATION: DESTRUCTING COMP, AND ESTABLISHING THE CAPITALIST MODE OF PRODUCTION

Greek agriculture has experienced very rapid changes in the post World War II period (Sanders 1962; Shaw 1969). From the perspective of modernization, it has been argued that the Greek agricultural sector had too many farms and that most of these were too small and inefficient (Sarros 1997; Ziogas 1997). Consequently, structural development was seen as a process of selective growth to be made effective through a range of corresponding interventions and mutual arrangements. Thus, the first phase, not only in Greece but in European policy, was one of consolidating farm structures (i.e., land reform in Greece and Southern Italy, plot consolidation and enlargement programs in Belgium, France, West Germany, Spain and the Netherlands) linked to land improved schemes (including drainage and irrigation) and the development of farm-oriented infrastructure (Hoggart, Buller and Black 1995).

The use of chemical fertilizers, insecticides, pesticides, new high-yielding varieties, more machinery and new farming practices expanded tremendously. For example, in 1984 the quantity of fertilizers used per hectare of land was five times higher than the amount used in 1950. Also, in 1982 the overall power of all machinery used in Greek agriculture was slightly higher than the electric power of all stations in operation in that year (Tsatsarelis 1995). The aim was to establish commercial units able to mechanize and absorb the above technologies and to reduce the agrarian population, particularly through the elimination of small and marginal holdings. Although this strategy was intended to strengthen the economic and social structure of rural areas, the aim was closer integration into regional, national and international markets.

It was obvious, therefore, that the development path adopted by Greek agriculture was similar to the one followed by other Western countries. However, it became apparent that such measures could not stabilize rural economies and rural populations; indeed they seemed to intensify the flow of labor out of agriculture and, often, out of the rural areas. By the late 1970s external models of Greek rural development had fallen into disrepute. The continued intensification and industrialization of agriculture came up against the saturation of domestic markets, ecological limits (with rising problems of agricultural pollution and environmental degradation), and a greatly diminished capacity in the urban sector to absorb the surplus rural population. Moreover, the recession of the 1970s resulted in the closure of many branch plants and a growing sense that rural regions which had attracted a great deal of such inward investment were highly vulnerable to fluctuations in the world economy and distant boardroom decisions (Amin 1993). During this process, Greek agriculture lost most of its natural, human and cultural elements that held the key to its sustainable character.

Thus, in contrast to the traditional management strategies, modern agricultural practices or the Capitalist Mode of Production (CaMP) involve the introduction of fossil-fuel energy and synthetic chemical nutrients. 'Replacement' rather than 'maintenance' characterizes the major approach to continued agricultural production. Open flows of energy and inputs of nutrients involve economic trade on a global scale. Production decisions are based not only on local subsistence needs, but also on world market prices and national development policies. The impact of these changes as they affect the ecological, socio-economic and cultural functioning of traditional Greek farming systems are as follows.

4. IMPACTS OF MODERNIZATION

4.1 Impacts on traditional farming patterns

As agricultural expansion forces utilization of more marginal, hilly lands, farmers are forced to use poor fertility soils or soils exposed by poor management. The Greek government has assisted in this reclamation by providing large earth moving machinery to break up the stone like terraces of Peloponnisos and Aegean Islands and form new terraces (Ministry of Agriculture 1992; Kosmas et al. 1995). The work is partially subsidized and requires a fraction of the time that hand preparation involved (days vs. years). However, due to the poor percolation capacity of the soil, rainfall collects and flows out the ends of the fields. Since the new terraces lack furrows to channel the flow of runoff, erosion quickly results in the loss of large areas of the cropping surface.

In depopulated villages there has been a more subtle change, involving the gradual decline in the use or maintenance of soil and water conservation techniques over the last twenty years (Millington 1991). The farmers attribute this decline to the rising cost of labor, but it also reflects other socio-economic and cultural changes. Many farmers' children have moved to urban areas and the farmers themselves often work part-time in local industry or in the tertiary sector, further limiting time available for agricultural activities (Pyrovetsi and Daoutopoulos 1997). The unavailability of household labor, characterizing the more vulnerable nuclear family unit, also means that rural people can no longer maintain the moisture-retaining terraces and hence, harvests are declining. Consequently, more fertilizer is applied in order to offset the loss of nutrients and increase productivity. Deteriorating land productivity, due to a lack of labor input, is perpetuated by a gradual loss of the indigenous knowledge of soil and water conservation practices (Pyrovetsi and Daoutopoulos 1999).

4.2 Impacts on local resource management institutions

Local institutions and indigenous knowledge have contributed to the complex land use regulations that have evolved in traditional Greek villages. Although they are often overlooked by outsiders, these institutions at least used to play important roles in the complexities of communal land tenure arrangements found in Greek farming systems (Dimen and Friedl 1976).

In Greek marginal villages, migration outflows during the 1960s and 1970s restricted the remaining aging population to small-size landholdings. In the 1970s the Greek Forestry Department compounded the problem with policies that further reduced the mobility of local farmers to use larger holdings for grazing (Dimen and Friedl 1976). Traditional authorities that enforced rules on fallow, water and grazing systems were abolished. This process has undermined the rotational, water, grazing systems of Greek mountainous villages, resulting in the depletion of the land resources and falling productivity, thus increasing insecurity, ecological stress and social conflict between farmers and grazers.

4.3 Impacts on local diversity and reciprocity

Farmers who have enlarged field size by grading adjoining fields together in order to allow for the use of larger tractors, have in the process removed hedges, i.e., natural border habitats which are an important source of wildlife diversity. In addition to this trend, the production of single crops, usually maize, wheat or cotton, is on the rise, having been encouraged by EU agricultural policies through local extension services (Lekakis and Kousis 1994). These programs stipulate that a farmer may not intercrop if he wishes to receive government agricultural assistance. The seeds provided in the assistance 'packet' are usually hybrids developed and tested in ideal growing conditions which correspond neither with natural local conditions nor with management practices. Standardization of crop characteristics makes the plants more vulnerable to climatic extremes and the spread of disease or insects that might wipe out an entire harvest. Traditional genetic variability insured production by at least some varieties or individual plants in a variety. The susceptibility of the less diverse farming system is compensated for by the use of increasing amounts of other agricultural inputs such as pesticides and fertilizers. The result is an ecologically less functional and economically more open system.

Finally, in monocropping, the use of machinery and chemical substances may have short-term effects on farm productivity. But at the same time, it may hinder the seasonal character of farm work which was based mainly on communal labor, diversified crops, small land holdings and complex ecological conditions (Bentley 1987). Reciprocity, which was and still is a mechanism found in all areas of the country – as elsewhere in Europe and the USA (Molnar and Korsching 1983) –, takes the form of helping arrangements and informal assistance norms, in order to avoid hiring labor. Reciprocity is, however, affected by the family life cycle of farmers today. A decline in the total population of marginal areas and the migration of young people force the remaining population either to abandon farming altogether or to intensify it using synthetic fertilizers and hired labor. This trend may undermine traditional community relations based on kinship and mutual support (Poole 1981; Pyrovetsi and Daoutopoulos 1997).

4.4 Impact on gender dynamics

Most Greek farmers today use synthetic chemical fertilizers on their crops as a substitute or addition to animal manure. The significance of this change is that the petroleum-based fertilizer is produced outside of the farming system and therefore can only be obtained through exchange for crop products. This results in a greater flow of energy through the system and a dependence on world petroleum price fluctuations. In the past government agencies have offered subsidized chemical fertilizers and advantageous credit terms (Beopoulos and Skuras 1997).

Implicit in the increase in fertilizer use is deeper involvement in the cash economy in order to pay for this input. The sale of crops for income increases the energy and nutrient outflow from the farming system, which is compensated for by greater fertilizer inputs. Many farmers seek employment in local factories to gain supplemental income. With less time to work on their land, they abandon soil conservation measures and use more fertilizer to maintain crop yields.

This process has important consequences for women who must shoulder more work in addition to their responsibilities as mothers (Lazaridis 1995). Whether the out migration, as a diversification strategy, improves family living standards, depends in great part on the amount of money saved and remitted back to the farm for investment into economic activities. As argued by Lazaridis (1995), male off-farm employment increases women's workloads and erodes their position in society as men consolidate access to cash and other resources. While many women suffer disproportionately, some studies show the importance of support from the extended household or family in cushioning the impact on women of male off-farm activities (SOS Sahel 1993).

General statements about the weakness and vulnerability of women can be problematic, however, given local and regional variations and the relative paucity of available information. Much more needs to be known about gender relations in farming as they relate to soil and water management in Greece.

To conclude, this model of traditional (CoMP) versus modern (CaMP) farming systems in Greece has been used as a measure of change in a world of increasing global integration. It also allows us to locate 'leaks' in the cycling of resources, since net losses make it increasingly expensive to sustain agricultural production.

The decline in traditional management practices involves several factors operating at the state and national levels. Greece's attempts to halt the flow of rural-urban migration have included the encouragement of the secondary and tertiary sectors. The attractive wages offered by these employers have drawn farmers away from intensive agricultural management and raised wages within the farm sector, thus making traditional practices unaffordable. Agricultural extension programs have stipulated monocropping as a condition for receiving economic and technical assistance. Regional agricultural seminars do not promote soil, water or nutrient conservation techniques but, instead, encourage systems maladapted to the local environmental conditions, heavy machinery, and external inputs. Government ceilings on crop prices and subsidized fertilizer, further encouraged the adoption of new technologies.

Furthermore, the modern management strategies reflect the availability and use of resources that originate outside the farm-level system and which are exchanged for products of labor from within the system. Heavy machinery replaces lost soil mass by reclaiming severely eroded regions; fertilizer replaces nutrients lost through soil erosion and the harvest of the crop; and pesticides are introduced to compensate for reduced ecosystem diversity which had traditionally maintained resistancc to infestation. Unless modified and adapted to regional conditions, the adoption of these modern strategies can involve a degradation of local cultural and ecosystem functioning and, thus, pose a threat to the sustainability of agricultural production.

Thus far we have pointed to only some of the deleterious consequences of the modern management strategies, not with the intent of criticizing the government for introducing them, nor the farmers for adopting them, but rather to encourage further investigation of the impact of their long-term use so that the road to sustainability can be assured.

5. THE PRESENT STATE OF GREEK AGRICULTURE

As seen in the previous section, the support of product and input prices, infrastructure, technical change, and increased farm size have led not only to an increase in production, but also to detrimental impacts on the local rural environment (Lekakis and Kousis 1994). Many of the modernization policies appear to have worked effectively in changing farm size and increasing productivity for some 'preferred crops' such as cotton, cereals, or tobacco. But products such as wine and beef have seen a decrease which, according to Maravegias (1989), means that the CAP not only increased Greek agricultural production but also altered both its volume and composition.

More specifically, it can be argued that Greek agriculture is caught up in the consequences of successive CAP reforms which have brought about a steady decrease in the total volume of agricultural production. Arable farming is currently in the throes of an on-going crisis (Maravegias 1989) and conventional agriculture is increasingly the target of consumer criticism, as the public begin to voice their serious doubts about food quality (Fotopoulos and Chrisohoidis 1998). Taken together, this means that the Greek agricultural sector cannot be expected to grow in the coming decades: either it will remain at the same level or, perhaps, it will decrease even further. It is not only marketing perspectives that are gloomy (Maravegias 1989), environmental regulation and the increasing attention paid to nature conservation and the maintenance of the landscape, practically exclude any further physical expansion of production either at the macro or micro (farm) level (Beopoulos and Skuras 1997).

Moreover, the successive reforms of CAP, along with Greek state's decreasing income and input support subsidies, have brought severe financial crises for family farmers. They have been squeezed by debt and low product prices. Small family farms have been especially vulnerable since they rely on diverse sources of off-farm income and so are dependent upon the wider success of the rural economy, whereas entrepreneurial farmers have seen this rural transformation as desirable (Kassimis and Papadopoulos 1994). Finally, the decline of the traditional system of crop rotation (maize, beans or other leguminous crops) replaced predators with pesticides, cattle and traditional land management with fertilizers. Government officials now make decisions made previously by local institutions, and local labor groups are replaced with specialist workers and tractors (Altieri and Anderson 1986).

In addition, the outcome of the agricultural modernization paradigm is 'structural involution', elements of which have been detected not only in Dutch agriculture (van der Ploeg and van der Dijk 1995) but also in recent studies of rural Macedonian and Cretan communities (Pyrovetsi and Daoutopoulos 1997, 1999; Kousis and Petropoulou, forthcoming). For example, the farmers of these communities were found to be less satisfied with their incomes and were rather pessimistic about the future. What is interesting to note here is that the once unquestioned solution – structural development at farm and sector level – is no longer self-evident and has lost its legitimacy and persuasiveness. Further structural development is increasingly seen as being part and parcel of the crisis itself. It has been transformed into structural involution (van der Ploeg and van der Dijk 1995). Although the farmers studied implicitly rejected modernization in the context of everyday reality, the search for agricultural intensification was still quite strong. The above paradox implies that the erosive implications of the modernization paradigm and the simultaneous lack of any substantial alternative for rural development can tie farmers to old securities. What is important, however, is that a widespread, still somewhat hidden search for alternatives is also taking place.

This view, however, is further reinforced by a focus group interview of young people from the same villages, both farmers and non-farmers ranging in age from 22-25 years. The majority of them, although less experienced in farming, felt that the intensification of agriculture could no longer be seen as progress unless local farming style and scale changes. They were much more knowledgeable about the impact of agriculture on the environment and had more formal education and training than older farmers (Pyrovetsi and Daoutopoulos 1999; Beopoulos and Louloudis 1997). This new trend may well suggest that while the Greek state, European rural policies, and agricultural sciences remain hopelessly wedded the modernization paradigm, in farming itself a series of new sustainable responses are emerging and together these represent new ways of envisaging the future.

Another way of anticipating Greek farmers' attitudes to this new trend of sustainable development as suggested by the reformed CAP, is to study their views on the relationship between agricultural activities and the environment. But very few studies have been carried out in Greece regarding farmers' sensitivity to agri-environmental issues. A couple of studies (Pyrovetsi and Daoutopoulos 1989; Daoutopoulos and Pyrovetsi 1990) have examined the attitude of a specific group of professional farmers and fishermen towards the wetland ecosystem, and the way in which they use the wetland resources. Others (Kousis 1992; Pyrovetsi and Daoutopoulos 1997, 1999) have examined the impact of farming practices on the environment, and farmers' awareness and willingness to adopt environmentally friendly practices in farming. The crucial role of the state in disseminating information and providing strong financial and technical is usually emphasized (Pyrovetsi and Daoutopoulos 1999; Beopoulos and Louloudis 1997; Kousis 1992).

In addition to the above findings showing Greek farmers' emerging responses to a more sustainable, viable agricultural sector, there are a number of comparative advantages (or, considered from the perspective of the modernization paradigm, disadvantages):

a) The variety of microclimatic and soil conditions and the cultivation of different crops in small parcels contribute substantially to diversity, which is a valuable characteristic of sustainable agriculture (Pretty 1995),
b) Although chemical inputs have increased in Greece, they are still far below the corresponding levels in other European countries. Fertilizer consumption per hectare is 2 times lower than that of the UK and 5 times lower than that of the Netherlands (Tikof 1996),
c) The small scale farming that prevails in the country, along with fragmentation, blocks the prevalence of monocultures over large areas so that one disadvantage of modern agriculture is transformed into an advantage within sustainable agriculture (Sarros 1997),
d) Proximity of farm-land with forests, water bodies and small plains (where intensive agriculture is practiced) offers a variety of ecosystems and habitats for wildlife that is beneficial to agricultural production (crop pollinators, beneficial insects, predators and parasites of pests).

As indicated so far, the pursuit of increased productivity over the course of Greek rural modernization has produced disadvantages in the form of socio-cultural and environmental impacts on the rural population. But the outcome of a few studies (Pyrovetsi and Daoutopoulos 1999; Beopoulos and Louloudis 1997; Kousis 1992) on young, well informed and large-scale farmers' responses to the drawbacks the modernization paradigm, shows a considerable interest in new forms of quality production and environmental control. Moreover, the diversity of the Greek rural environment further supports the transition to a more sustainable agriculture. Despite this potential the development of a more sustainable agriculture in Greece depends not just on the motivations of individual farmers and the favorable physiographic conditions that prevail, but on the action of groups or communities and external institutions as a whole. This makes the task facing Greek agriculture exceptionally challenging.

6. THE CONTRADICTIONS OF THE PRESENT ROUTE TO 'SUSTAINABILITY'

Although it is relatively easy to describe goals of a more sustainable agriculture, things become more problematic when one attempts to define sustainability. Precise and absolute definitions of sustainability, and therefore of sustainable agriculture, are impossible, since it is a complex and contested concept. We could suggest that sustainability involves an equation between environmental requirements and developmental needs. It can be balanced by acting either to reduce stress – ecologist's approach – or to increase the carrying capacity of the environment – economist's approach (UNESCO 1997).

However, in any discussion of sustainability, it is important to clarify what is sustained, for how long, for whose benefit and at whose cost, over what area and measured by what criteria. Answering these questions is difficult, as it means assessing and trading off values and beliefs. Nonetheless, when specific parameters or criteria are selected, it is possible to say whether certain trends are steady, or not. For example, practices causing soil to erode can be considered to be unsustainable compared with those that conserve soil; or forming a local group as a forum for more effective collective action is likely to be more sustainable than individuals trying to act alone (Scoones and Thompson 1994).

At the farm or community level, it is possible for actors to weigh up, trade off, and agree on these criteria for measuring trends in sustainability. But as we move to higher levels of the hierarchy, to districts, regions and countries, it becomes increasingly difficult to do this in any meaningful way. It is essential to recognize that sustainable agriculture does not prescribe a concretely defined set of technologies, practices or policies at these levels. This would only serve to restrict the future options of farmers. Sustainable agriculture is, therefore, not a simple model or package to be imposed. It is more a process for learning and practicing at the local or regional level. Despite increased understanding of the need to transform present day agriculture to embrace sustainable features, there are a number of contradictions that are worth being discussed here. This implicitly means that sustainability is directly or indirectly linked to the roles exercised by powerful actors (Voisey and O'Riordan 1997).

European policy actors for example, increasingly ignore the specific conditions prevailing in each member-state. For example, Aguilar Fernández (1994), with particular reference to the Spanish experience, argues that in Southern countries, and especially in Greece, Portugal and Spain, the most pressing environmental problems attracting public and political concern involve soil erosion, desertification and the risk of forest fires. Yet the EU agenda has been primarily concerned with setting strict limits on air and water pollution. The agenda she argues, is Northern-driven.

Similarly, Regulation 797/85 is directly relevant to more peripheral upland areas of the Northern Union and specifically concerns measures aimed at Less Favorable Areas (LFAs). Within these areas, farmers have access to 'compensatory allowances' in proportion to their land area or the number of 'livestock units'; there are also joint investment schemes which favor fodder production and pasture improvement. Although this policy was specifically designed to promote less intensive and more viable agricultural systems such as animal husbandry, the application of this policy has been detrimental for the islands of Greece and other Southern European states where highly eroded hills and mountains with very limited and irregular rainfall and vegetation are presently being degraded at an alarming rate by increasing sheep and goat populations. Between 1981 to 1991 sheep and goat population in Greece rose by 17.2% and 26.1% respectively reaching an all time record (Tikof 1996).

Moreover, the 1448/88 directive is being complied with and vines are being uprooting. The EU policy to restrict cultivation of vineyards was imposed throughout Europe, uniformly restricting the cultivation in Northern and Southern European regions. Subsidies were given to farmers clearing vineyards and recently olive trees in Greece that are extremely well adapted to the Mediterranean environment. However, the directive provides only a 'one-off' payment to farmers and there are no policy guidelines on post-vineyard land use. When vineyards or olive groves are abandoned, a vegetation succession from agricultural land to fallow land and possibly later marquis will follow (Millington 1991). Consequently, as land abandonment proceeds and vegetation cover increases, soil erosion rates should decline, but in areas of terracing this will be offset initially by increased erosion from terrace breaching. Clearly, the economic impact of land abandonment could mean that people will be forced to leave these marginal areas as the land apparently becomes less productive due to increased soil erosion.

European, environmental policies are not integrated and, frequently, contradict with each other (Baker, Milton and Yearly 1994). Greece, like other Southern European countries did not take full advantage of certain opportunities offered by agricultural (Directive 797/85) and agri-environmental acts (Directive 2078/92) of legislation (Beopoulos and Skuras 1997). Among other factors, this is because the political and institutional 'culture' of Southern states has not, on the whole, taken on board agricultural and environmental prerogatives and is having difficulty adapting to the need of doing so under the pressures of European economic integration. There are several reasons for this: the Southern states tend to be urbanized nations with weakly developed rural administrative capacities, despite having relatively high rural and agricultural populations; there is limited social demand for environmental intervention; and higher priorities tend to be given to economic development (Baker, Milton and Yearly 1994).

An additional economic actor that influence the process of sustainability is the so-called agro-chemical industry which has been mainly encouraged by national state actors (Lekakis and Kousis 1994). First world corporations control seed resources, whose raw materials come from the Third World (Theodoratos 1991). Agricultural policies of the developed countries took the control of agriculture away from local and passed it on to corporate and state groups, thus reducing returns to farmers and local economies. With biotechnology and market power, corporate control has been increasing over agricultural and food production, seeking to protect the markets against cheaper alternatives promoted by sustainable agriculture.

7. THE ROAD TO A 'SUSTAINABLE AGRICULTURE'

There have been an increasing number of European Union and State policy initiatives in recent years, specifically oriented towards improving sustainability of Greek agriculture. Most of these have focused on input reduction strategies, such as IPM[1] and patch spraying[2] (Bougiouris, Zoulias

[1] Integrated Pest Management is the integrated use of a range of pest control strategies, such that pest populations are reduced to satisfactory levels with the minimum use of toxic chemicals.

and Pappos 1993; Gouli and Echaliotis 1994), because of concerns over foreign exchange expenditure or environmental damage. But even these do not fully represent coherent plans and processes that clearly demonstrate the value of integrating policy goals. A thriving and sustainable agricultural sector requires integrated action both by farmers and communities and by policy makers and planners. This implies both horizontal integration, with better linkages between sectors, and vertical integration, with better linkages from the micro to the macro level.

For example, a very common approach, that of providing subsidies to promote the sheep and goat population of an already degraded environment, could be used instead to safeguard the local environment by subsidizing farmers and stockbreeders to lower their animal population, train them to become better stewards and managers of the local environment, and employ them in projects aimed at improving local pastures (Daoutopoulos 1995).

The removal of agricultural input subsidies can, on the other hand, be very costly to Greek farmers in the short term if they are not matched by the development of systems for farmer training, the realignment of exchange rates, and reductions in industrial sector protection. Thus, policy reforms in the sustainable agricultural sector will also have to synchronized with macro, industrial sector and trade reforms.

But will the industrial and trade sector become the facilitators of this reformed (sustainable) agriculture? It is doubtful given their present day practices. The production of new varieties, not as a response to consumer demands but to promote their own products (e.g. genetically altered crops to withstand higher applications of herbicides), runs contradictory to the sustainability principle. In addition, both multinational companies and the accompanying trade sector thrive when they provide uniform solutions for large markets, something that also does not have a sustainable character.

What can be concluded from the above is that national and international institutions have tended to be substituted for local action, thus undermining any existing local initiatives or institutions. Since local groups and institutions have been ignored, many have disappeared entirely. This has led both to increased degradation and to decreased capacity of local people to cope with environmental and economic change.

What is needed therefore is an increasing recognition that 'participation' between external institutions, such as agricultural professionals and rural people, is essential for sustained agriculture (Chambers 1983; Scoones and

---

[2] Patch spraying, used instead of aerial spraying to combat the olive-tree pest dacus olea, involves regular field monitoring and modified spraying systems that allow the application of the chemicals exactly where there are known problems.

Thompson 1994). This means that the wider challenge for agricultural institutions/organizations is to become 'learning organizations' by providing better training to producers, since sustainable agriculture requires more qualified personnel than conventional agriculture. Environmental education, on the other hand, should receive high priority and aim at developing informed and skilled producers who are willing to take action to resolve environmental issues (Hungerford, Peyton and Wilker 1980). The ultimate aim of environmental education is promoting responsible environmental behavior (Lozzi, Lareanet and Marcinkowski 1990). Thus, one of the systematic challenges facing Greek agricultural research and the extension and planning institutions – be they governmental or non-governmental – is to institutionalize approaches and structures that encourage experimentation and a strong participatory approach in planning and action (Daoutopoulos 1980; Daoutopoulos and Pyrovetsi 1995).

According to the Global Biodiversity Strategy (WRT, I.U.C.N., UNEP 1992) local communities should play a fundamental role in stewardship of their natural resources. These conditions cannot be met without community empowerment and organization, which in turn depends on heightening public awareness and concern which can be achieved only by properly informing and educating people in local communities. The case of Aperathou in the island of Naxos is a representative example of how locally planned and implemented projects can achieve impressive results in the management of scarce resources (water) (Glezos 1994).

Although both policy and legal reforms are essential for the transition to a more sustainable agriculture, they are not sufficient conditions. State action is needed to reform both the internal functioning of state agricultural institutions and the way they interact with the interested party, as well as the local and regional farmers associations. The recent establishment of the Council of Agricultural Policy operating at the district, regional and national level is a new policy apparatus that can incorporate local initiatives in the formation of national agricultural policies.

8. CONCLUSION

The traditional model of Community Mode Production (CoMP) versus the modern Capitalist Mode of Production (CaMP) in Greek farming systems has been used in this chapter as a measure of change in a world of increasing global integration. In the past five decades, the traditional systems of management which provided a descent livelihood for the population were based on the conservation of existing resources through intensive maintenance practices. But the pursuit of increasing productivity and consequently agricultural modernization has produced the so-called CaMP which had detrimental effects on Greek rural areas.

External inputs of machines, fossil fuels, pesticides and fertilizers have displaced farmers and Greek rural culture has been put under pressure, as more and more people have been forced to migrate in search of work. Local institutions, once strong, have become co-opted by the state, while farms have been simplified and some resources, once valued on the farm, have become waste products to be disposed of off the farm. Some external inputs are lost to the environment, thus contaminating water and soil, while the overuse of some pesticides causes pest resistance and leads to pest resurgence, encouraging farmers to apply yet more pesticides.

It is obvious therefore, that the Greek government along with international actors, such as the EU, has intervened to transform traditional agricultural systems by encouraging the adoption of modern varieties of crops and modern breeds of livestock, together with associated packages of external inputs (such as fertilizers, pesticides, credit, machinery etc.) necessary to make these productive. In addition, they have supported new infrastructure as well as a range of other policies in support of the modernization paradigm.

But most Greek rural households have different conditions, needs, values and constraints than those of top-down policies. Throughout this discussion we have focused on the possibilities within Greek agriculture – the possibility to move towards a more sustainable sector, an option that can also be seen as a turning point away from external rural development and towards an endogenous form of development.

Although there are many potentially productive and sustainable (e.g. small-scale farms, IPM etc.) resources and technologies available to Greek farmers, a transition to a more sustainable agriculture will not occur without the full participation and collective action of rural people. The development of a more sustainable agriculture depends not just on the motivations of individual farmers, but also on the actions of groups or communities as a whole. Yet, Greek local groups and institutions have been ignored, thus many have disappeared entirely. This has led both to increased degradation and to decreased capacity in local people to cope with environmental and economic change.

Finally, in addition to the valuable local resources/technologies available within the Greek rural context, a third essential prerequisite for Greek agriculture to become sustainable concerns the way external institutions are organized and the way they work with other institutions and farmers. In the process of agricultural modernization, external institutions have tended to ignore and thus suffocate local knowledge and initiative. The complexities involved in achieving diverse and productive sustainable agriculture mean that organizations will have to adopt new ways of working. This implies greater multi-disciplinarity, more structured participation with farming communities in research, extension and development activities, the evolution of learning processes in organizations and the development of a whole new agricultural professionalism itself.

As long as the modernization paradigm dominates policymaking and as long as the Greek state is characterized by restrictive bureaucracy and centralized hierarchical authority, there will be no significant change in the current agricultural situation in Greece. A new, sustainable future for Greek agriculture and the countryside is close at hand, but at the same time very far away.

REFERENCES

Aguilar Fernández, S. (1994) Spanish pollution control and the challenge of the European Union, in S. Baker, K. Milton and S. Yearly, *Protecting the Periphery: Environmental Policy in Peripheral Regions of the European Union*, Frank Cass, Newbury Park, CA, pp.102-117.

Altieri, M.A., and Anderson, C.A. (1986) An ecological basis for the development of alternative agricultural systems for small farmers in the Third World, *American Journal of Alternative Agriculture* **1**, 30-38.

Amin, A. (1993) *The Regional Development Potential of Inward Investment in the Less Favoured Regions Of the European Community*, Paper presented at the Conference on Cohesion and Conflict in the Single Market, Newcastle upon Tyne.

Baker, S., Milton, K., and Yearley, S. (eds.) (1994) *Protecting the Periphery: Environmental Policy in Peripheral Regions of the European Union*, Frank Cass, Newbury Park, CA.

Bentley, J.W. (1987) Economic and ecological approaches to land fragmentation: In defence of a much-maligned phenomenon, *Annual Review of Anthropology* **16**, 31-67.

Beopoulos, N., and Louloudis, L. (1997) 'Farmers' acceptance of agri-environmental policy measures: A survey of Greece, *South European Society & Politics* **2**, 118-137.

Beopoulos, N., and Skuras, D. (1997) Agriculture and the Greek rural environment, *Sociologia Ruralis* **37**, 253-269.

Bougiouris, K., Zoulias, A., and Pappos, G. (1993) Bio-cultivation of Corinthian currants in Aigio Achaias, *Biological Farming* **33**, 21-23.

Chambers, R. (1983) *Rural Development: Putting the Last First*, Longman Scientific and Technical, London and New York.

Daoutopoulos, G. (1995) Humans and natural environment, *Geotechnical Review* **70**, 73-74.

Daoutopoulos, G.A. (1980) The importance of preserving farmland for the Greek economy, *Scientific Publications of the Agricultural Bank (Epistimonikes Ekdoseis ATE)* **14**, 84-113.

Daoutopoulos, G.A., and Pyrovetsi, M. (1990) Comparison of conservation attitudes among fishermen in three protected lakes in Greece, *Journal of Environmental Management* **31**, 83-92.

Daoutopoulos, G.A., and Pyrovetsi, M. (1995) The need of conserving farmland to meet the demand for food, *MEDIT* **3**, 18-25.

Dimen, M., and Friedl, E. (1976) *Regional Variation in Modern Greece and Cyprus: Towards a Perspective on the Ethnography of Greece*, Annals of the New York Academy of Sciences, New York.

Fotopoulos, R., and Chrisohoidis, B. (1998) *Factors that Influence the Market - Decision of Biological Products*, Paper presented at the Fifth Panhellenic Conference on Rural Economy "Restructuring the Rural Space", Athens.

Glezos, M. (1994) Integrated environmental management in Naxos and measures to combat soil erosion, *Geotechnical Review* **59**, 63-69.

Gliessman, S.R. (1990) *Agroecology: Researching the Ecological Basis for Sustainable Development*, Springer-Verlag, New York.

Gouli, G., and Echaliotis, K. (1994) Olive-tree: Olive orchards in the Messinian Mani, *Biological Farming* **36**, 8-13.

Hoggart, K., Buller, H., and Black, R. (1995) *Rural Europe: Identity and Change*, Arnold, London.

Hungerford, H., Peyton, R., and Wilker, R. (1980) Goals for curriculum development in environmental education, *Journal of Environmental Education* **21**, 8-21.

Karavidas, K. (1981 [1936]) *Local Government and Greek Economic Regionalism*, Papazisis Publishing House, Athens.

Kassimis, C., and Papadopoulos, A. (1994) The heterogeneity of Greek family farming: Emerging policy principles, *Sociologia Ruralis* **34**, 206-228.

Kosmas, C.N., Moustakas, N., Danalatos, N.G., and Yassoglou, N. (1995) The effect of land use on soil properties and erosion along a catena, in J. Thornes and J. Brand, *Mediterranean Desertification and Land Use*, John Wiley & Sons, Chichester.

Kousis, M. (1992) Exploring the adoption of renewable energy: The case Biogas plants in Greek agriculture, *Perspectives in Energy* **2**, 99-108.

Kousis, M., and Petropoulou, E. (in press) Local identity and survival in Greece, in T. O'Riordan, *Globalism, Localism and Identity: Perspectives on the Sustainability Transition in Europe*, Earthscan, London.

Lazaridis, G. (1995) Market gardening and women's work in Platanos, Greece, *The European Journal of Women's Studies* **2**, 441-467.

Lozzi, L., Lareanet, D., and Marcinkowski, J. (1990) *Assessment of Learning Outcomes of Environmental Education*, UNESCO, Paris.

Maravegias, N. (1989) *The Accession of Greece to the European Community: The Effects on the Agricultural Sector*, Foundation for Mediterranean Studies, Athens.

Martinos, N., Louloudis, L., and Daniel, N. (1988) *Economic Development and Structural Change in Rural Greece*, Unpublished Report.

Millington, A.C. (1991) Land use transformations and environmental impacts of EC agricultural policy and rural depopulation: Messinia, Greece, in G. Jones and G. Robinson, *Land Use Change and the Environment in the European Community*, Biogeography Research Group and Rural Geography Study Group of the Institute of British Geographers, pp.107-121.

Ministry of Agriculture (1992) Agriculture and the State of Environment in Greece, Internal Report of the Ministry Athens, Author.

Molnar, J.J., and Korsching, Peter F. (1983) Consequences of concentrated ownership and control in the agricultural sector for rural communities, *The Rural Sociologist* **3**, 298-302.

Poole, D.L. (1981) Farm scale, family life, and community participation, *Rural Sociology* **46**, 112-127.

Pretty, N.J. (1995) *Regenerating Agriculture - Policies and Practices for Sustainability and Self-Reliance*, Earthscan, London.

Pyrovetsi, M. (1984) *Ecodevelopment in Prespa National Park, Greece*, Ph.D. Thesis, Michigan State University.

Pyrovetsi, M., and Daoutopoulos, G.A. (1989) Conservation related attitudes of lake fishermen in Greece, *Environmental Conservation* **16**, 245-250.

Pyrovetsi, M., and Daoutopoulos, G.A. (1997) Contrasts in conservation attitudes and agricultural practice between farmers operating in wetlands and a plain in Macedonia, Greece, *Environmental Conservation* **24**, 76-82.

Pyrovetsi, M., and Daoutopoulos, G.A. (1999) Farmers' needs for nature conservation education in Greece, *Journal of Environmental Management* **56**, 147-157.

Pyrovetsi, M., and Gerakis, P.A. (1987) Environmental problems for practicing agriculture in Prespa National Parc, Greece, *The Environmentalist* **7**, 35-41.

Sakiotis, Y. (1995) The Greek community victim of development, *New Ecology* **128**, 55-59.

Sanders, I. (1962) *Rainbow in the Rock: The People of Rural Greece*, Harvard University Press, Cambridge, MA.

Sarros, P. (1997) The problem of succession in the Greek agricultural sector, in C.Z. Karanikas, *Is This the End of Greek Agriculture?* Oikonomikos Tachydromos, 7 February, , pp.12-14.

Scoones, I., and Thompson, J. (1994) *Beyond Farmer First - Rural People's Knowledge, Agricultural Research and Extension Practice*, Intermediate Technology Publications, London.

Shaw, L.H. (1969) Postwar Growth in Greek Agricultural Production, Special Studies Series No. 2 Athens, Center of Planning and Economic Research.

SOS Sahel International (1993) *The Effects Of Male Out-Migration on Women's Management of the Natural Resource Base in Sahel*, Draft from SOS Sahel International.

Theodoratos, T. (1991) Local varieties and the copyright of cultivators, *Nea Oikologia* **86**, 40-41.

Tikof, M. (1996) *Greece: Policy Measures and Practices, and Environmental Benefits from Agriculture*, OECD Seminar on Environmental Benefits from a Sustainable Agriculture: Issues and Policies; Helsinki, 10-13 September 1996.

Tsatsarelis, K. (1995) *Management of Agricultural Machinery*, Aristotle University Press, Thessaloniki.

UNESCO (1997) *Educating for a Sustainable Future. International Conference on Environment and Society: Education and Public Awareness for Sustainability*, UNESCO and the Government of Greece, Thessaloniki.

van der Ploeg, J.D., and van der Dijk, G. (1995) *Beyond Modernisation: The Impact of Endogenous Rural Development*, Van Gorcum, Assen.

Voisey, H., and O'Riordan, T. (1997) *Globality and Locality: Ideas, Issues and Possible Research Questions in Relation to Sustainable Development*, Paper presented at the Opening Meeting of the International Global Environmental Change Research Community, Laxenburg, Austria, 12-14 June.

WRT, I.U.C.N., and UNEP (1992) *Global Biodiversity Strategy: A Policy-Makers Guide*, World Resources Institute Publications, Maryland.

Ziogas, C. (1997) Solutions are awaiting and credits are available, in C.Z. Karanikas, *Is This the End of Greek Agriculture?* Oikonomikos Tachydromos, 9 February, pp.31-33.

Chapter 7

The Ecologist Movement in the Basque Country

From Nationalism to Localism

Iñaki Barcena and Pedro Ibarra

1. INTRODUCTION

A permanent national conflict in the Basque Country, with a violent dimension, has marked the development of its ecology movement. After a brief examination of the environmental movement in Spain as a whole, this analysis focuses on the formation and development of the Basque Ecology Movement (BEM), from the seventies to the present, highlighting its political, organizational as well as discourse-related aspects. The analytical framework that we will employ is influenced by an integrating perspective (Neidhardt and Rucht 1992; McAdam, McCarthy and Zald 1996; Diani 1996) – taking into account the mobilizing structures approach (McCarthy and Zald 1987; Diani 1997), the political opportunity structure approach (Kriesi 1995; della Porta and Rucht 1995) and finally the framing approach (Gamson 1992). This will guide our synthetic definition and analysis of the broad evolutionary tendencies of the movement.

Recent ecologist movements in European and non-European regions use direct action against specific projects such as road-building in Britain or the transportation of radioactive waste in Germany (Taylor 1995; Doherty 1997; Kousis 1998, 1999). Indeed, in some movements the defense of the life world is dominant, whilst in others their environmental focus is much more pragmatic. Some choose ways of mobilizing that are unconventional, whereas others tend to channel the conflict through the courts. Defense of the socially constructed 'locale' is the central axis on which these new ecological movements converge. They are in essence movements that resist the perceived invasion or aggression to their territory by external institutions or elites. According to Preston (1997: 176) a political cultural identity is articu-

K. Eder and M. Kousis (eds.), Environmental Politics in Southern Europe, 175–196.

lated in three dimensions: local, network and memory. In the first, the individual constructs his relationship with the nearby community through a living and informal mixture of narratives, biographies, common sense and the political and cultural traditions of 'his' community. In the second, he extends his identification to other groups situated beyond this local community, and in the third dimension, he bases his identification on the historical memory of broader collective spaces (the nation for example). Such an identity does not necessarily lean towards nationalist positions; in many occasions, it arises against nationalist strategies, or at least keeps a clear distance from them (Smyth 1998). Yet, these campaigns tend to receive the support of nationalist groups.

In examining the evolution of the Basque Ecology Movement, this chapter aims to address the following questions: Since when has the environmental movement in the Basque country showed a strong local orientation? What have been the major internal (organizational and discourse-related) and external (political context) factors influencing its development from the seventies to the present?[1]

2. THE SPANISH CONTEXT

The implementation of public environmental policies in Spain arrived late and with serious shortcomings. Its late entry into the European Community in 1986 marked the beginning of environmental policies by political institutions. Since Spain has always been a receptor (never a transmitter) country of the EU's environmental policy, to a certain extent Spanish environmental policy would be practically non-existent without pressure from the EU. Until 1996, Spain was the only country in the European Community without a Ministerial Department of the Environment.

[1] The following analysis is based both on prior analyses of the ecological movement in the Basque Country (Barcena, Ibarra and Zubiaga 1995, 1996, 1997, 1998; Barcena and Ibarra 1997, 1999; Barcena et al. 1999; Ibarra and Rivas 1996; Tejerina, Fernandez Sobrado and Aierdi 1995; Rivas 1999; Eguzki 1987; Beaumont et al. 1997; Casado da Rocha and Perez 1996) and, above all, on a research work in progress on the transformation of the environmental movement in European Union member-states, including the Basque Country. Protest-event data (1988-97) used in this chapter come from the European Commission funded project on 'Transformations in Environmental Activism' (TEA) (Contract number ENV4-CT97-0514; DG.XII); see Rootes (1997). Additional data, within or outside of this period, derive from different surveys and interviews involving ecological groups as well as books, articles and documents relating to the movement.

New social movements in Spain, including the ecological movement[2], have emerged late and have never managed to form solid movements that extended throughout the whole territory (Pastor 1998). The low level of environmental consciousness compared with other EU countries does not in itself explain the low political and social relevance given to the protection of the environment (Aguilar Fernández 1997: 170).

The long end of the previous authoritarian regime contributed to an overflow of social conflicts and demands. The political institutions, overwhelmed by this quantitative and qualitative increase in various social demands, also showed their receptivity to environmentalist ones at the end of the 1970s. Since the political institutions were not capable of formulating, and even less so, of implementing rigorous and coherent environmental policies, the movement itself had to voice a wide expanse of claims and demands. A favorable climate was also created for environmental mobilizations by the lack of cohesion and of a unified approach amongst the elites. Thus, the political opportunity structure (POS) was quite favorable to the movement.[3]

Perhaps the only unfavorable aspect for the development of the environmental movement is the system of alliances. The political parties of the left (especially the majority Socialist Party) adopted from the outset a policy of at least distancing itself from, if not being hostile toward social movements. As a result, the movement started up with a few competent militants, and without the necessary networks of solidarity needed for its initial impetus. With the democratic transition underway, the development of networks of solidarity that had been generated during the Francoist dictatorship, took two directions. Many disappeared when their demands were taken up institutionally or when a certain 'generational tiredness' was produced by the 'excess' of anti-Francoist mobilizations. The rest have been integrated into the - now legal - political parties, the new protagonists of collective action.

Thus, the movement arose and developed with a notable organizational weakness. Only a network of activists, originating from political parties of the extreme left, made it possible for the movement to maintain a minimum of presence and continuity. Without them, the ecologist movement would not exist. Its positions were excessively doctrinaire, not very pragmatic, and at the same time did not incorporate the usual post-materialist and ecological views. During the transition, Spain was mainly interested in continuing the economic growth and social progress achieved since the 1960s, in defending

2 For an analysis of the Spanish ecological movement as a whole see Fernandez (1999).

3 For an analysis of the political opportunity structure of the Spanish ecological movement see Jiménez (1999).

itself from an emergent and swiftly increasing level of unemployment in the late 1970s, as well as in ensuring the recently obtained democratic freedoms. At the start of the 1980s, Spanish culture differed from that of North Europe in the 1970s. Surveys referring to Spain show indexes of post-materialist culture that, even at the end of the 1980s, are appreciably lower than the European average. It is during the later half of the 1980s that environmentalist concerns increase.[4]

In spite of the difficulties in its genesis, the ecologist movement of the Spanish state has in recent years (1998-1999) undergone a period of reorganization and restructuring, a process in which an important part of the BEM was implicated; this has given rise to the 'umbrella' organization Ecologistas en Acción (Ecologists in Action). The aim of this organization is to overcome localism and dispersion and to seek efficiency, dialogue with the administration and other sectors of society involved in the environmental debate and access to the mass media, aims that are being achieved to a notable extent.

3. EMERGENCE AND CONSOLIDATION OF THE BASQUE ECOLOGIST MOVEMENT

The evolution of the ecologist movement of the Basque Country can be analysed by distinguishing two phases, the first relating to the Lemoiz antinuclear mobilizations, between 1977 and 1982, and the second, from 1982 to the present, covering mainly opposition to construction, infrastructure and waste projects, including the cases of the Leizarán motorway (1989-1992) and, from 1995 onward, the Itoiz reservoir.[5]

3.1 The Lemoiz case: A nationalist struggle

Towards the end of the 1970s, the project to build a nuclear power station at a village on the coast of Bizkaia, Lemoiz, gave rise to an important antinuclear protest movement in the Basque country. The struggle led by the Anti-nuclear Committees coincided with a period of strong social mobiliza-

4 In 1986 only 1.3% of those surveyed knew that the Ministry of Public Works was responsible for the environment. By the end of the 1980s concern for protection of the environment was beginning to increase. 16% between 1987 and 1989. Data collected in Aguilar Fernández (1997: 266, 267).

5 As previously stated, protest-event data cover the Leizaran and Itoiz conflicts, while more qualitative data are used to describe the Lemoiz protest.

tion and is considered to be the first step in structuring a broader Basque ecologist movement.

Throughout the period of the conflict a strong dynamic confrontation took place between the economic and political interests that supported the nuclear power station and the anti-nuclear activists. The kidnapping and killing of the chief engineer from the nuclear power station in 1981 by ETA led to the freeze of the project. Since then, the Basque ecologist movement has continued to demand annually that the infrastructure be dismantled.

The number of protest actions against the nuclear power station was very high. During the early 1980s there was an action against the power station virtually every week. Although the conflict was local in its origins, its level of mobilization was national in scope. In the Lemoiz case, it was Basque society as a whole that, in one way or another, experienced this protest. The slogan, in Basque language, chosen by the movement Euskadi ala Lemoiz (the Basque Country or Lemoiz) expresses that feeling. The Lemoiz anti-nuclear claims, which inaugurated a strategic option maintained to this day, were also directed against the construction of big public works.

The forms of action were very radical. The movement gave priority to public demonstrations, which on occasion achieved a truly massive number of participants (over 50,000 people), and did not always end in a peaceful way. In addition, ETA, the violent organization linked to radical Basque nationalism, carried out numerous acts of sabotage against the power station, including actions which resulted in loss of life.

There was basically only one ecologist organization that led the conflict at Lemoiz: the Anti-nuclear Committees. With the incorporation of other environmental groups in the mid-1980s, they became the ecologist organization 'Eguzki.' During those years other ecologist groups were practically non-existent. The Committees had mainly political allies - principally linked to radical Basque nationalism.

3.2 From the Leizarán to the Itoiz conflicts: Localist encounters

At the end of the 1980s, the Government of Navarra and the Diputación of Gipuzkoa (Provincial Institutions of the Basque Country) undertook to construct a motorway between the locality of Irurzun in Navarra and the locality of Andoain in Gipuzkoa, aiming to improve the system of communications between the two provinces. The 'Lurraldea Coordinator,' an organization with a highly innovative form of activism, mounted opposition against the motorway given its expected negative environmental impacts for the

Leizarán valley. Lurraldea (Territory in the Basque language) was the name of the alternative motorway route proposed by this environmental coordinator.

Following a phase of broad social mobilization, the armed, Basque nationalist organization ETA, decided to participate in the conflict, putting pressure on companies and on the provincial governments involved in the motorway. By 1992, an agreement was reached in which the administration (Diputacion de Gipuzkoa) and the Lurraldea Coordinator agreed on a route for the motorway different from that previously planned.

The Itoiz reservoir, located in Navarre[6], is a hydraulic project of the Government of Navarra that aims to take waters from the mountainous zone of this province towards areas requiring irrigation or to the larger urban nuclei situated further to the South. The reservoir flooded a rural population area and of an important landscape enclave, thus leading to the creation of the Itoiz Coordinator. The intense activity of this group has run up against the Navarra administration which is determined to build this infrastructure and is implicated in several related cases of corruption.

In the mid-1990s, following the government's persistent determination to continue with the project in spite of judicial sentences decreeing its illegality, Solidarios con Itoiz (those in solidarity with Itoiz) managed to paralyze the works with an act of sabotage that gave rise to an important debate within the ecologist movement and Basque society. At present, the works continue to operate, although legally the Government of Navarra is not permitted to fill the reservoir above a specific level. By the time this article is published the conflict will have adopted a new course. The Spanish Constitutional Tribunal will shortly dictate a decisive and definitive ruling on the legality of the reservoir works.

This conflict expresses a certain tendency in the Basque ecologist movement. In 1985, having no faith in the ability of the local councils to fight back, the 'Itoiz Coordinator', a local ad hoc group (Casado da Rocha and Perez 1996), brings together residents councilors and ecologists from the affected area (9 villages would be flooded, while the land of six more would

[6] It is essential to clear up a 'geopolitical' question. The conflict around the Itoiz Reservoir is taking place in Navarra, and we are referring to the Basque ecologist movement. To resolve this apparent contradiction it must be pointed out that although Navarra, in juridical terms, constitutes a different Autonomous Community from Euskadi, politically (by Basque nationalism) it is considered to be part of the Basque nation. Of particular concern is the fact that the region of Navarra, where the reservoir is to be found, is substantially influenced by Basque culture, while simultaneously, the ecological movement against the project conforms, to a certain degree, with the present tendencies of the Basque ecology movement.

be inundated). It was founded in order to defend their land and their communities. From the start, they collaborated with people opposed to other hydraulic projects in the Pyrenees. They also received political and/or economic support from radical Basque nationalist groups, some left-wing parties, and environmental groups and organizations from the Basque Country and Spain, as well as from international and European environmental organizations. But the main burden of the struggle against the Navarrese Government was carried out by the leaders of the mobilization, both through their activism and the courts (Beaumont et al. 1997).

The autonomous government of Navarra and the central government of Madrid have never recognized the Itoiz coordinator as an interlocutor and have refused any kind of contact, dialogue or concessions. On April 6th 1996, the group Soldarios con Itoiz, already well-known for over 15 earlier non-violent direct actions, managed to cut the cables of the equipment that supplied building material to the reservoir dam. As a result of their radical action, work on the reservoir was stopped for almost a year, leading to an intense social and political debate between all parties involved in the conflict. The violent, nationalist, and political-conflict context has been used by the institutions to reject and disqualify any conflict generating demands arguing that it is immersed in and forms part of the destabilization strategy of the 'violent ones' of ETA and their civilian partners in radical nationalist groups.

In this context, the actions of the Soldarios, although evidently having nothing to do with the violence of ETA, have given rise to dissimilar opinions. As expected, both institutions and a part of the mass media found the opportunity to incriminate the entire coordinator. By contrast, aside from some internal criticisms, the coordinator and most of its allies have supported the Soldarios.

3.3 Data

Based on data obtained from empirical research in the TEA project we will now single out those protest events that have a special connection with the two conflicts we have mentioned.

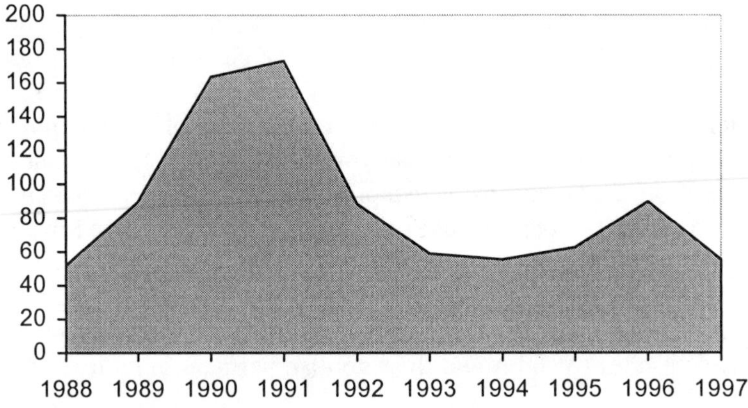

Figure 1. Number of environmental protest-events

In reference to protest-event data (from 1988 to 1997), there were a total of 887 protest events or demand-related actions in defense of the environment. The climax of such events took place in 1990 (163 protests) and 1991 (173) during the Leizarán conflict. Following a period of gradual decrease of protest-events, an upsurge (90 actions) is noted in 1996 due to opposition to the Itoiz reservoir. Figure 1 above, depicts protest events of these two conflicts.

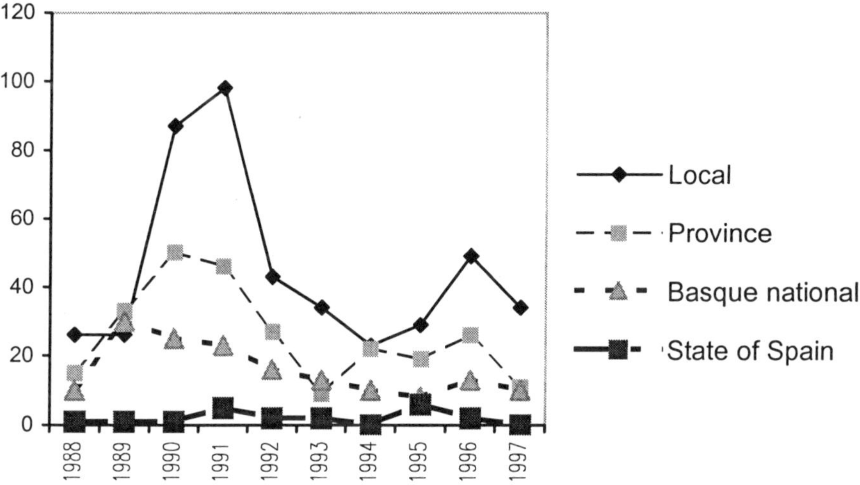

Figure 2. Level of mobilization

Regarding the level of protest-event mobilization during the above period, Figure 2 depicts the importance of protests at the local level for the Basque ecologist protest since the late eighties. Approximately half of all the protests (449) have been local in character. Local in our classification means a protest in which the call to the people to participate remains at village, town or valley level. This presupposes a lower level than provincial protest, which is also important due to the Basque and Spanish administrative division.

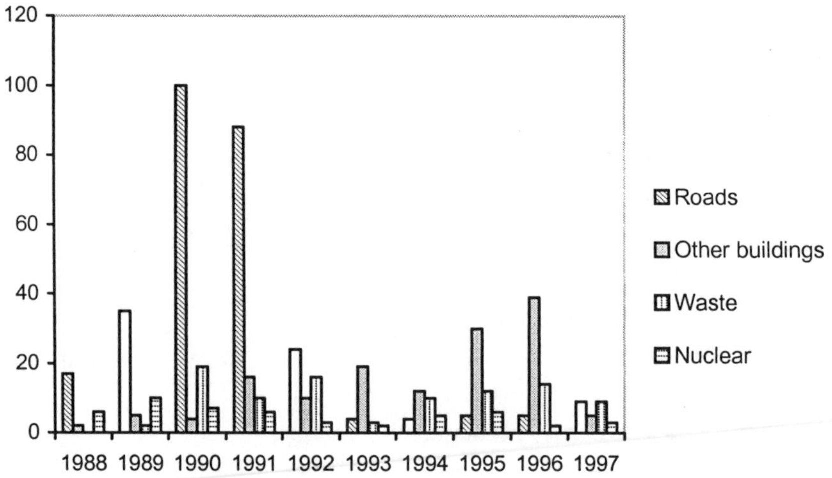

Figure 3. Claims

With respect to the mobilizers' claims, Figure 3 illustrates the importance of claims against road infrastructural projects (291 protest-events), which flourished until 1993 and are best reflected in opposition to the Leizarán motorway. Second in importance are claims against other construction related projects, including recreational harbors, commercial ports or reservoirs. This type of protest climaxed between 1995 and 1996, the period of greatest tension in the Itoiz problem.

Looking at the forms of protest shown in Figure 4, it is worth underlining the importance of demonstrative actions, especially public protest rallies (186) and demonstrations (176). Both forms underwent similar oscillations or fluctuations during 1990 and 1991, followed by a period of decline and a new peak in 1996. Second in importance are press conferences (109) and formal procedures such as gathering signatures, petitions, resolutions or letters (82). Both have undergone an evolution similar to that of demonstrative protests. It should be noted that, most press conferences held by the anti-motorway Lurraldea coordinator or the Itoiz coordinator, took place when they were involved in intense battles of information with the political institutions.

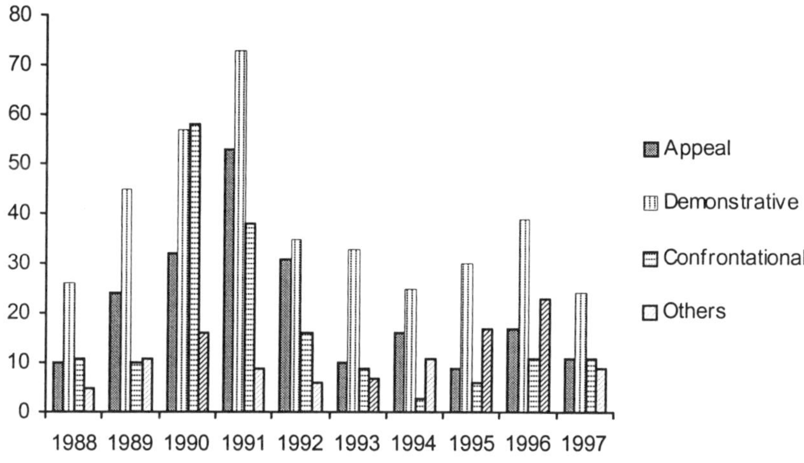

Figure 4. Forms

In this analysis of the evolution of action forms, the disappearance of violent protests, throughout the latest phase (1994-1997) is clearly evident. Nevertheless, there is a tendency during the same years towards an increase of the 'other' forms of action (114 over the whole decade), implying more 'innovative' extensions of conventional protest types such as demonstrative, peaceful and, on some occasions, involving civil disobedience. In total, 105 innovative protests have been counted. Between 1995 and 1996 this type of innovative intense action, generated by the group Soldarios con Itoiz, is most prominent.

Referring to the types of groups that organize protest, Figure 5 shows that informal organizations predominate, most of which are represented by the 'big coordinators' (Lurraldea/Leizarán and Itoiz). In the early 1990s an important role was played by the Lurraldea coordinator. With regard to stable or formal organizations, their evolution is fairly regular, without notable ups and downs. The evolution of the activity of the groups and coordinators as a whole has been linked to the periods of greater or lesser effervescence in those conflicts in which they have been involved. This parallelism is clearly evident in the case of the coordinators, some of which disappear once the project has been completed. A similar pattern is evident for the traditional ecologist groups when involved in a specific dispute.

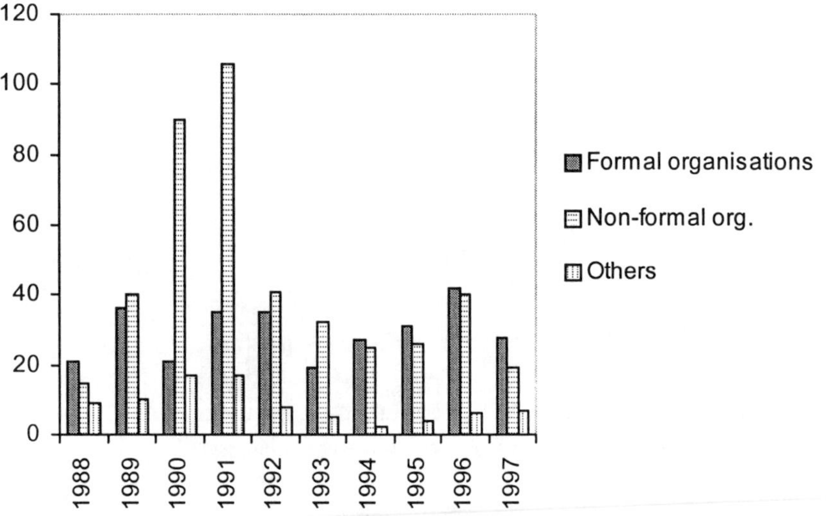

Figure 5. Organizations

4. DISCUSSION

In what follows, we discuss the main causes for the emergence and the consolidation of the BEM, namely its political context, its public discourse and its organizational forms.

4.1 The changing political context

The ecologist movement in the Basque Country was born at the end of the 1970s in order to prevent the Lemoiz nuclear power station from being built. Aside from uncertainties about the mechanisms for resolving social conflicts, both Basque and Spanish political institutions were determined that in the specific case of the nuclear power station, both because of the scale of the project and because of the numerous interests at stake, they were going to ignore social demands. This, in combination with the reduction and moderation of repressive policies (the dictatorship had ended), facilitated the spread and radicalization of the movement's mobilizations.

A position of initial inflexibility was also maintained in the next large scale conflict involving the movement opposed to the construction of a motorway (the Leizarán conflict) during the years 1990 to 1992. Although in this second case the political institutions did, at last, reach an agreement - alteration of the route of the motorway- with the ecologist coordinator that had emerged in an ad hoc fashion, in the case of Lemoiz no agreement was reached; the construction of the nuclear power station was halted both because of the spectacular popular mobilizations and because of the armed actions of ETA. And more recently, the political institutions that defended the building of the motorway have had serious initial difficulties taking the demands of ecologists into consideration; and this is because, amongst other reasons, the ecologists were supported by the BNLM[7], the Basque radical nationalist movement.

While in the Lemoiz case the political transition weakened the new administration's capacity for carrying out the projects inherited from the previous regime, in the Leizarán conjuncture the institutions had recovered their capacity for putting public policies into effect. On the other hand, the lack of co-ordination between the plurality of institutions existing within the Basque political space, together with their relative inexperience, reduced that capacity for putting legislation into practice. Thus, BEM's success in the case of Lemoiz was due to a combination of 1) a policy of closure towards the demands of the antinuclear movement, 2) weakness in applying and implementing the institutional plans, 3) a policy of alliances that reinforced and broadened positions contrary to the nuclear project and 4) elites who were in the process of adapting from dictatorship to democracy and thus poorly legitimized.

Political elites maintained a common line of confrontation, more forceful in the Lemoiz conflict, and less so in the case of Leizarán. In the case of the Leizarán motorway it was possible to establish a dialogue, negotiate and reach an agreement. Yet, even in the later case, the institutions of Navarra and Gipuzkoa never met with the ecologists to talk about their alternative plan, the so called 'Lurraldea' .

Since 1992 BEM has been distancing itself from its traditional allies, the radical left nationalists (the MLNV), as political institutions adopt decidedly more flexible attitudes, to such an extent that on occasion it has been the institutions themselves that have called in leaders or experts from the movements for consultation on – and eventually negotiation of – their public environmental policies. However, in the Itoiz case, the policy of the institutions stands in contradiction to this trend and this has led to a radicalization of that conflict. Aside from the fact that the public institutions continue to

[7] Basque National Liberation Movement (Movimiento Vasco de Liberación Nacional); the ensemble of organizations and groups which are linked to radical nationalism and which support ETA.

maintain an uncoordinated policy based more on conjunctural actions than on strategic planning, in practice the Basque political institutions make environmental decisions. They do so in many cases while paying no heed to the opposition of the movements. Besides, these institutions are perceived by public opinion as actors concerned with environmental problems and as protagonists in their resolution. As a result, the BEM has to compete, although not always successfully, with the policies of the political institutions.

4.2 Alliances and organizational issues

Considering the question of alliances, it must be said that the MLNV has on many occasions conditioned the dynamics of the social movements. In Lemoiz, the emergence of the anti-nuclear movement was closely connected with radical nationalism, and in the Leizarán case the movement later sought the support of the MLNV. These processes of absorption of the ecologist movement and, on occasion, anything but peaceful co-operation between nationalists and ecologists (above all when ETA made its appearance on the scene)[8] should not prevent us from reiterating another assertion. The alliance with the radical nationalists appreciably increased the ecologist movement's capacity of mobilization in the Basque Country.

Aside from the BEM's notable initial organizational strength, its greater mobilizing capacity (and its substantive and procedural successes in the two conflicts we have mentioned) basically responds to the same pattern: to a highly adequate combination of discursive strategies and political opportunity structure. As Diani (1996) points out, political opportunities can be related to the framing processes in different ways.

A significant change has appeared since 1992 in BEM's strategy of alliances. The movement does not appear to be interested in giving priority to alliances with the MLNV, either because on occasions they are too powerful as allies (which finally absorb the original demands of the BEM, as in the Leizarán case), or because they are a handicap as allies when it comes to attempting to talk with the institutions; or, alternatively, because of the growing social discredit of ETA.

Also since 1992, the national organizations are showing a certain inability in promoting campaigns of mobilization at a national level and their activities are being focused above all on giving support to local mobiliza-

[8] The presence of ETA was decisive in the case of Lemoiz and the nationalist/ecologist conglomeration (rather than alliance at that time) was barely affected by its violent actions. In Leizaran, ETA's actions were marginal, hardly influencing the process, although they did create tensions within the alliance of nationalists and ecologists.

tions, led in the majority of cases by local or ad hoc groups, or to activities of environmental consciousness raising.

Compared to the Basque, the Spanish movement is different in terms of its organizational structure. The initial nucleus of BEM was able to incorporate a notable number of activists and sympathizers. This was because solidarity networks were available to social activists who were not incorporated into the political parties, as was the case for militants of political parties of the extreme left who did not wish to, or could not, integrate themselves into an electoral/institutional dynamic, and above all because of the political space of radical nationalism. In this latter case it must be pointed out that the whole of this large movement did not incorporate itself into the competition between parties. By maintaining collective action of the social movement type and by considering that ecologism was also a form of defending the national physical space, the nationalist left supported and in many cases joined the emergent ecologist movement.

4.3 The discourse of the BEM

The BEM encountered and used to its advantage both a broad culture of protest proceeding from radical nationalism and a culture of resistance -an anti-institutional culture of a communitarian type- that had emerged in the struggles against Francoist authoritarianism and -given the difficulties of consolidation of the political transition in the Basque Country- that are still present in the society.

Since its origins in the 1970s, the BEM has maintained a markedly anti-developmentalist discourse; this was a product of the anti-nuclear consciousness that permeated the broad and successful campaign against the Lemoiz nuclear power station, the first of six nuclear reactors that the Francoist administration and the electricity monopolies had planned to install in Euskadi. Together with the anti-nuclear protest, there was an ensemble of fights against the infrastructural plans brought on by economic growth and industrial development. Motorways and roads, recreational harbors and the superport of Bilbao, incinerators and the High Speed Train, plants for energy production, reservoirs and airports have been the central referents of the BEM. It is a movement whose character has been more one of urban ecologism than conservationism, and it has thus been principally linked to political ecology, having a deeply anti-developmentalist character that is opposed to the present Basque development model, the unsustainability of which it works to expose.

Since 1992, it has moved from large anti-institutional and national confrontations to more local conflicts, as well as to more dialogues with the institutions. These readjustments, not always planned, have given rise to difficulties, perplexities, and internal tensions within the movements. Sustainable development is implied in BEM's demands and proposed measures. The movement does not have the capacity to mobilize sufficient resources in order to propose global plans to raise the sustainability of the country. For the time being it has a local orientation and can only respond to specific projects.

The localist framing strategy of 'defense of the land' is articulated against industrialist aggression, against the macroprojects that destroy both the habitat and the small and harmonious Basque community; in the Leizarán conflict, the BEM also used the democratic frame, to the point of abandoning to a certain extent its own contents, its specific environmental demands (Ibarra and Rivas 1996). Thus it was argued that the lack of dialogue and negotiation was the great problem in Basque society and that this lack also affected ecologist demands. From this perspective, the struggle for negotiation resonated favorably within the specific democratic frame of the MLNV, which had as its central strategy the demand for a national political negotiation.

BEM's discourse has changed since 1992. With some exceptions, it no longer appears framed within the Basque national conflict. It has distanced itself from the latter and uses frames that are more autonomous, more exclusively environmentalist. In this way they perhaps succeed in better connecting themselves with environmentalist frames, which are today established and dominant - although with less mobilizing potential than the national frames - in Basque society as a whole.

What is most striking about recent years is a growing tendency towards localism; a tendency that is best expressed in the Itoiz case; without any doubt, the Itoiz conflict continues to have strong Basque nationalist connotations, and as a result continues to be supported by the ecological (and also political) organizations that are more or less close to Basque nationalism. Basque environmental groups appear closer to Preston's first type of political cultural identity, i.e., their identity is immersed in the local community.

5. CONCLUSION

Given its cultural-political context, the Basque ecologist movement appears to be more cohesive and better consolidated than the Spanish one.

Greater confrontation and less fluidity in the relations between ecologists and the administration or business sectors as well as a different political culture in Euskadi have led to better conditions for social mobilization. In addition, more accessible social networks were available for expanding protest activities.

In spite of the fact that Greenpeace is the ecologist organization with the greatest number of affiliates in Euskadi (about 6,000), neither Greenpeace nor other international organizations, such as Friends of the Earth or the WWF, have offices on Basque soil, which means that their discourse and environmental action reaches the Basque public almost exclusively through the mass media. And the differences are notable in this field too, given that in Euskadi we find a spectrum of public and private means of communication that is wider and more varied than in the rest of the Spanish State, with greater opportunities for the ecologist organizations to spread their messages through these media.

In a country where the national conflict expresses itself so clearly as the first and foremost social and political contradiction and where it maintains such a high profile in the mass media and in everyday life, the BEM could not be an exception, remaining on the margins of such a 'cleavage'. Thus, Basque left-wing nationalism has been the central political referent of the BEM, which has found in its complex socio-political framework both a channel to make its demands heard and the foundation for eliciting support for ecologist mobilization.

Nonetheless, the not always peaceful relations amongst the organizations of the BEM itself and with its principal ally (the nationalist left) have generated a localist dynamic, which is the expression of a search by the BEM for its own space of action, autonomy and readjustment within its forms of mobilization.

Regarding the relationship between political opportunity structure and framing process (Diani 1996), in our case the influence has been reciprocal. On the one hand, the discourse of the movements aligned to their advantage those potentially favorable elements of the political context. Their discourse – anti-institutional and proto-nationalist – encouraged a strategy of collaboration by the MLNV, the greatest anti-system ally of the BEM; with this alliance, the movement's possibilities of success were strengthened.

On the other hand, the opportunity structures simultaneously defined and gave priority to – and thus expanded the mobilizing capacity of – a certain type of discourse. The rigidities of the input in the institutions facing non-conventional mobilizations became material for the discourse of the movements. The latter framed that hostile political opportunity structure for its own ends of mobilization; and in this way the ecologist movement confronted the closure of the institutions with proposals for 'dialogue and negotiation'.

We believe, in agreement with the theoretical proposal of Gamson (1992), that the success of these movements in the Basque Country is the result of an interpenetration of discursive strategies and political opportunity structures, in which the interactive processes we have just cited are expressed in one way or another.

Autonomous causes exist for the resurgence of local movements. But in many cases they appreciably increase their mobilizing capacity for external reasons, because of external support. In effect, sometimes the national organizations are unable to carry out large-scale campaigns; but they are able to focus their activities on local actions. From this it follows that the weakness of the national organizations increases the mobilization of the local groups in two ways: on the one hand, local groups are unable to rely on the national organizations and are thus obliged to act themselves; and, on the other hand, they receive the support of the national organizations which are only able to work on limited campaigns.

The tendency towards localism and protests of the NIMBY type (Not In My Back Yard) is common in the Basque and Spanish environmental movements. However, in the Basque case the proliferation of groups and coordinators that confront environmental aggressions or infrastructural projects is more numerous, we could almost say it is exaggerated for a country as small as Euskadi. Some of these local conflicts, as in the case of the Itoiz reservoir, have become the ecologist campaigns or protests that have mobilized the majority of people, that have occupied the most space and time and given rise to the most polemics in the news, and that have brought together the most ecologist organizations in their activities. Itoiz ceased being a problem of local scope and become the most important, widely known and controversial environmental demand not only in the Basque Country, but also in Spain, reaching the European Commission and the European Ombudsman. The protest against the criminalization of, and the prison sentences given to the 8 Solidarios with Itoiz reached the streets of Brussels and, more recently, London.

After more than a decade (1988-1999) of localism and varied forms of organization by Basque ecologism, we are at present witnessing a process of restructuring that responds to the global scope of the problems faced by the BEM. It is conscious of the need to act locally and globally, and of the impossibility of doing this from exclusively local dimensions. Recently, two developments have taken place that mark a change of direction. Firstly, a new national 'umbrella' organization Ekologistak Martxan (1999) has been formed, which brings together the majority of the organizations of the BEM.

Secondly, Eguzki has adopted a proposal to build a new network that will co-ordinate all of the struggles 'in defense of the land' that are raised by ecologist, trade union, fishing, farming and other sectors in Euskadi. From the point of view of the BEM, these two developments are an attempt to give a global perspective and efficiency to the varied local responses to neo-liberalism. Alternatively, they could be consequences of, and reactions to the end of a previous cycle. Another probable third hypothesis would be that they are a combination of the two.

In conclusion, a local orientation may be seen as a rejection of globalization, i.e., resistance to external decision-making which affects their community. This process is complex as it combines, almost never in an explicit way[9], two lines of rejection. There is one line of direct rejection, and another based on a reconstruction which poses as an alternative to the negative consequences of globalization. The former aims to put a stop to the growing distance and globalization in question; the latter is based on the search for solutions to those processes linked to cultural homogenization/ globalization that result in a loss of identity. Another dimension of the local approach is of a strategic nature and refers to the affected group's perception of ecological organizations as incapable of successfully solving local level problems. Local groups believe that these bodies do not have a genuine interest in their demands. They also see a risk that these demands will be negotiated – and reduced – without the consent of those directly affected, and thus sacrificed to the long-term strategy of the ecologist organization.

This mistrust also leads to a specific way of approaching environmental issues. Local groups are not likely to hold face-to-face discussions of their demands with other collective entities affected by the conflict, or to compromise on their demands. This practice differs from that of stable ecologist organizations which usually negotiate with political institutions. What the local groups simply want is for the environmental aggression to disappear from their area. This leads them to chose direct action strategies of high popular participation, organized in an anti-hierarchical way. Long term actions, led by professionals from the large ecologist organizations and based on criteria of supposedly efficient rational planning, are relegated to a second plane. This permanent adaptation to circumstances creates difficulties for the emergence of another parallel process in the strategy of some social movements, the strategy of institutionalization (Eder 1998).

[9] In effect, it is highly risky to affirm that ecological movements (and especially the local ones) have an elaborated 'anti-globalizing' consciousness. For the relationship between social movements and globalization, see Ibarra (1996), McAdam, Tarrow and Tilly (1996), Castells (1997), della Porta and Kriesi (1999), Rootes (1996), Rucht (1999).

REFERENCES

Aguilar Fernández, S. (1997) El Reto del Medio Ambiente: Conflictos e Intereses en la política medioambiental Europea, Alianza Editorial, Madrid.

Barcena, I., and Ibarra, P. (1997) New Forms of Collective Action in the Ecologist Movement in the Basque Country, Paper prepared for the First Regional Conference on Social Movements, Tel Aviv, September 1997.

Barcena, I., Ibarra, P., Torre, P., and Guarrotxena, E. (1999) The Ecologist Protest in the Basque Country: From Nationalism to Localism, Paper for the EPCR Joint Sessions, Mannheim, 26-31 March, 1999.

Barcena, I., Ibarra, P., and Zubiaga, M. (1995) Nacionalismo y ecología. Conflicto e institucionalización del movimiento ecologista vasco, Los Libros de la Catarata, Madrid.

Barcena, I., Ibarra, P., and Zubiaga, M. (1996) Nationalism and Social Movements: Estonia, Slovenia and the Basque Country. Discursive Resonance and Harmonisation, Institute of Political Science of The Hungarian Academy of Sciences, Budapest.

Barcena, I., Ibarra, P., and Zubiaga, M. (1997) The evolution of the relationship between ecologism and nationalism, in M. Redclift and G. Woodgate, The International Book of Environmental Sociology, Edward Elgar, Cheltenham, pp.300-315.

Barcena, I., Ibarra, P., and Zubiaga, M. (1998) Movimientos sociales y democracía en Euskadi. Insumision y ecologismo, in P. Ibarra and B. Terejina, Movimientos sociales, transformaciones políticas y cambio cultural, Trotta, Madrid, pp.43-64.

Beaumont, M.A., Beaumont, J.L., Arrojo, P., and Bernal, J.M. (1997) El embalse de Itoiz, la razóno o el poder, Bakeaz-COAGRET, Bilbao.

Casado da Rocha, A., and Perez, J.A. (1996) ITOIZ. Del deber de la Disobediencia Civil al Ecosabotaje, Pamiela, Pamplona.

Castells, M. (1997) El poder de la identidad (vol.2). La era de informacion, Alianza, Madrid.

della Porta, D., and Kriesi, H.-P. (1999) Social movements in globalizing world: An introduction, in D. della Porta, H.-P. Kriesi and D. Rucht, Social Movements in a Globalizing World, MacMillan, Basingstoke, pp.3-22.

della Porta, D., and Rucht, D. (1995) Left-libertarian movements in context: A comparison of Italy and West Germany, 1965-1990, in J.C. Jenkins and B. Klandermans, The Politics of Social Protest. Comparative Perspectives on States and Social Movements, University of Minnesota Press, Minneapolis, MN, pp.299-272.

Diani, M. (1996) Linking mobilization frames and political opportunities: Insights from regional populism in Italy, American Sociological Review **61**, 1053-1069.

Diani, M. (1997) Social movements and social capital: A network perspective on movement outcomes, Mobilization **2**, 129-148.

Doherty, B. (1997) Tactical Innovation and the Protest Repertoire in the Radical Ecology Movement in Britain, Paper prepared for the European Sociological Association Conference, Essex, August 1997.

Eder, K. (1998) La insitucionalización de la acción colectiva. Hacia una nueva problemática teorica en el análisis de los movimientos sociales, in P. Ibarra and B. Tejerina, Movimientos sociales, transformaciones políticas y cambio cultural, Trotta, Madrid, pp.307-331.

Eguzki (1987) Lemoiz 1972-1989. Selfedition, Bilbao.

Fernandez, J.M. (1999) El ecologismo español, Alianza, Madrid.

Gamson, W.A. (1992) Talking Politics, Cambridge University Press, Cambridge, MA.

Ibarra, P. (1996) Globality and Difference in Three Social Movements, Paper prepared for the Conference Research Committee 47, ISA, Santa Cruz, USA, May 1996.
Ibarra, P., and Rivas, A. (1996) Environmental public discourse in the Basque Country. The conflict of Leizarán motorway, Comparative Social Research **Suppl. 2**, 139-151.
Jiménez, M. (1999) Consolidation through institutionalisation? Dilemmas of the Spanish environmental movement in the 1990s, Environmental Politics **8**, 149-171.
Kousis, M. (1998) An exposition of environmental movement research worldwide: the challengers, the challenged and "sustainable development", in A. Gijswift , F. Buttel, P. Dickens, R. Dunlap, A. Mol and G. Spaagaren, Sociological Theory and Environment. Part II. Procedures of the Second Woundschoten Conference (RC 24-ISA), SISWO, University of Amsterdam, Amsterdam, pp.85-97.
Kousis, M. (1999) Sustaining local environmental mobilisations: Groups, actions and claims in Southern Europe, in C. Rootes, Environmental Politics, Special issue: Environmental Movements: Local, National and Global, pp.172-198.
Kriesi, H.-P. (1995) The political opportunity structure of new social movements: Its impact on their mobilization, in J.C. Jenkins and B. Klandermans, The Politics of Social Protest. Comparative Perspectives on States and Social Movements, University of Minnesota Press, Minneapolis, MN, pp.167-198.
McAdam, D., McCarthy, J.D., and Zald, M.N. (1996) Introduction: Opportunities, mobilizing structures and framing processes - towards a synthetic, comparative perspective on social movements, in D. McAdam, J.D.McCarthy and M.N. Zald, Comparative Perspectives on Social Movements, Cambridge University Press, New York, NY, pp.1-20.
McAdam, D., Tarrow, S., and Tilly, C. (1996) To map contentious policies, Mobilization **1**, 17-32.
McCarthy, J.D., and Zald, M.N. (eds.) (1987) Social Movements in an Organizational Society, Transaction Book, New Brunswick, NJ.
Neidhardt, F., and Rucht, D. (1992) Towards a Movement Society? On the Possibilities of Institutionalizing Social Movements, Paper prepared for the First Conference on Social Movements and Societies in Transition, Berlin, October 1992.
Pastor, J. (1998) La evolución de los nuevos movimientos sociales en el Estado Español, in P. Ibarra and B. Tejerina, Movimientos sociales, transformaciones políticas y cambio cultural, Trotta, Madrid, pp.69-89.
Preston, P.W. (1997) Political/Cultural Identity: Nations and Citizens in a Global Era, Sage, London.
Rivas, A. (1999) Un modelo para el análisis de la dimension ideatica de los movimientos sociales: El discurso del movimiento ecologista vasco, Ph.D. Thesis, Universidad País Vasco.
Rootes, C.A. (1996) Environmental Movements from the local to the global? Paper prepared for the Second European Conference on Social Movements, Vitoria, Spain, 2-5 October, 1996.
Rootes, C.A. (1997) The Transformation of Environmental Activism, Paper prepared for the 6th IRNES Conference, London, September 1997.
Rucht, D. (1999) The transnationalization of social movements: Trends, Causes, Problems, in D. della Porta, H.-P. Kriesi and D. Rucht, Social Movements in a Globalizing World, Macmillan, London, pp.206-223.
Smyth, J. (1998) Nacionalismo, globalizacion y movimientos sociales, in P. Ibarra and B. Tejerina, Movimientos sociales, transformaciones políticas y cambio cultural, Trotta, Madrid, pp.321-336.

Taylor, B.R. (1995) Popular ecological resistance and radical environmentalism, in B.R. Taylor, Ecological Resistance Movements: The Global Emergence of Radical and Popular Environmentalism, State of New York University Press, Albany, NY, pp.334-355.

Tejerina, B., Fernandez Sobrado, J.M., and Aierdi, X. (1995) Sociedad civil, protesta y movimientos sociales en el País Vasco, Servicio Publicaciones del Gobierno Vasco, Vitoria.

Chapter 8

Grassroots Environmental Action in Portugal (1974-1994)

Elisabete Figueiredo, Teresa Fidélis and Artur da Rosa Pires

1. INTRODUCTION

There is increasing evidence that citizens of non-core nations are as (if not more) concerned about the environment than citizens of core countries, especially regarding health, economic impacts, and private property (Dunlap 1996; Figueiredo 1999; Kousis 1999b).[1] While environmental attitudes and movements have been relatively well studied with reference to the more developed countries (Dunlap and Mertig 1995, Dunlap 1996), these works have just began for Portugal, a non-core nation (Rodrigues 1995; Fidélis et al. 1996; Gil Nave 2000). During the first years of democracy in the seventies, Portugal experienced a relatively large number of protests, while in the eighties their numbers decreased. The effective consolidation of democracy, social stability, both economic and political, as well as a change in social values was strongly felt after the mid-80's, when notable increases in participation in civil society and public life occurred.

The integration of Portugal in the European Community in 1986 has forced the central state to accelerate the development of environmental legislation and administration already underway in order to meet new environmental requirements and quality standards. Growing environmental problems, together with increased local information networks, awareness of new legislation, and pressure for government action may have been factors which, since the late eighties, have support the new growth of environmental action.

[1] Some have characterized these as linked to materialistic values.

K. Eder and M. Kousis (eds.), Environmental Politics in Southern Europe, 197–221.

When examining the environmental movement, recent studies distinguish between formal organizations and informal community-based or grassroots groups (Kousis 1999b). While the first tend to be professional organizations with a bureaucratic structure, community-based groups usually concentrate their actions on local issues and mobilize using informal networks.

Studies of grassroots environmental activism emphasize the large variety of social groups and socio-professional categories involved in the protests (Freudenberg and Steinsapir 1992; Taylor 1995; Kousis 1999b). Women are amongst the most represented social categories, both in terms of participation or leadership. Activists strongly believe in the citizen's right of participation in the decision-making process (Fidélis et al. 1996). The above studies have shown that the majority of the grassroots movements are concerned with the protection of public health and the negative impacts on property values, incomes and the local economy as well as on the local ecosystem.

The majority of local environmental protests tend to be non-violent (Taylor 1995; Freudenberg and Steinsapir 1992; Fidélis et al. 1996; Kousis 1999b). In general, as these studies show, activists have been using alternative forms of public participation, such as road blockades, picketing and demonstrations, outside traditional institutional channels. It has also been documented that activists usually approach various agencies or bodies such as state agencies at various levels, formal environmental organizations, or large environmental organizations in order to get assistance from them. The bodies approached are, in general, local entities or local representatives of national or international bodies or associations.

The main purpose of this chapter is to provide a global vision of grassroots environmental action in Portugal between 1974 and 1994. It will describe the character of grassroots environmental protest cases in terms of the types of participating groups, the actions they engage in, the claims they make regarding the source of environmental intervention and the resulting offences, as well as in terms of the bodies they approach for assistance and the responses they receive, the groups they challenge and the replies they face. At the same time, this chapter investigates the context of the regional distribution of environmental problems and economic growth, as well as environmental attitudes and formal environmental organizations.[2]

[2] The data originate from Project EV5V-CT94-0393 - Grassroots Environmental Action and Sustainable Development in Southern European Union, European Commission, DGXII.

2. THE REGIONAL CONTEXT

For environmental and planning purposes, Portugal[3] is divided in five main regions: Norte, Centro, Lisboa e Vale do Tejo, Alentejo and Algarve. The population density is highly differentiated. The regions including the two largest cities – the Northern region around Oporto and the region of Lisbon including Setubal and Almada – are most densely populated. In addition, it must also be stressed that nearly two thirds of the population are concentrated in the littoral areas, whereas the interior is increasingly being abandoned in spite investment efforts to stabilize the local population. Significant growth in terms of production can be found in the industrial, construction and service sectors of the economy, whereas fisheries have forfeited a large share of the Portuguese economy. A decrease in employment figures is evident in almost every sector, with the exception of construction and services. The regional differentiation of development within the Portuguese mainland is as follows.

The Norte region is strongly marked by the differentiation between the littoral and the interior areas, both in terms of demographic as well as economic and development levels. In the littoral areas, with a population density of 329/Km2, industry – especially labor intensive industry – is the main economic activity. Most of the education and development infrastructures are concentrated in this area. By contrast, in the interior, where the population density is 30/Km2, agriculture still dominates economic activity. In environmental terms, the littoral areas, like Barrinha de Esmoriz, Esposende and the area of the National Park of Gerês, can be cited as a set of important and sensitive areas under pressure by development trends.

In the Centro region, the working population has been shifting from agricultural into industrial activity, mainly concentrated in the littoral areas such as Aveiro and Leiria and in some interior main cities, like Viseu, Guarda and Castelo Branco. The development of the urban centers associated with these areas has also attracted a large number of related services, contributing to the social and economic strength of the central region. In addition, it is a region that has strongly benefited from recent investments in transport infrastructure, mainly main roads or highways, which are important factors in attracting new economic activity. The littoral areas are under greater pressure from development.

In economic terms, the Lisboa e Vale do Tejo region is considered the most developed of the country. It constitutes 13.5% of the total national area

[3] It has an area of approximately 88.683km2 and a total population of 9.853.896 inhabitants (INE 1991).

and has 35% of the total population. In spite, the progressive decline in importance of the Lisbon metropolitan area in terms of industrial employment, this region is still dominated by industrial activity, mainly in Setúbal and in Lisbon, along the Northern margin of the river Tejo. The location of the national capital in this region means that the tertiary sector is a center of major economic activity. The natural or environmentally sensitive areas which are threatened by development pressures and which generate controversy as well as attracting public and institutional concern, are mainly concentrated in the area of the Natural Park of Sintra and Arrabida, in the littoral areas of Estoril and Cascais and in the valleys of the rivers Tejo and Sado.

The Alentejo region is characterized by a low levels of population density and development. The most important economic activity is agriculture and extractive industry. The agricultural activity is mainly focused on cereal production, but during the last decades increasing erosion and consequent abandonment of the farmland has been observed. The main industrial area is located in the littoral city of Sines. Up until the last decade, the littoral area of this region had, probably, been under the least amount of pressure in terms of development project location, like tourism or agro-industry.

The region of Algarve is characterized, in economic terms, by the tourism sector and related equipment and infrastructures. The only industrial centers are located in Faro, Loulé and Olhão (MARN 1995). The typical Mediterranean agricultural activity has been progressively disappearing and is being replaced by intensive agriculture. Pressures of development are mainly felt in littoral areas, where environmental balance has clearly been disturbed.

Portugal has an important natural heritage, highly differentiated within its territory and, in general, a good physical environment. Severe environmental problems relate mainly to water and waste management as well as nature conservancy issues. A large part of the funding dedicated to the environmental sector is still allocated to built or update water and waste management infrastructures. Agriculture and industrial activities are responsible, in large part, for water and soil pollution. In addition urban activities and tourism are responsible for significant and burdensome amounts of water pollution and solid wastes.

3. ENVIRONMENTAL ATTITUDES AND ENVIRONMENTAL ORGANIZATIONS

One way of looking at the cultural context within which grassroots environmental activism appears is through an examination of environmental attitudes and environmental organizations. The brief presentation that follows aims to provide this context.

3.1 Environmental attitudes of Portuguese public opinion

In Portuguese public opinion, concern about the environment currently ranks 5th after drugs, unemployment, health (AIDS), poverty and social exclusion and is followed by problems such as the cost of living, safety/violence, housing, schools and education (Observa 1998).[4] For the majority of the Portuguese, the environment is an immediate and urgent problem, its protection being a priority in relation to the promotion of economic development (EU surveys 1986, 1988, 1992, 1995, 1999).

Closer examination (EEC 1988) reveals that the Portuguese identify environmental problems with materialistic values and tend to see other activities, such as fighting inflation, maintaining economic growth and stability, and the maintenance of order, as higher priorities. The first 'post-materialistic' value referred to (i.e., the possibility of public participation) ranks fifth amongst a materialistic and post-materialistic set of proposals. The priority given to materialistic and environmental values can be explained by the historical, political, social, cultural and economic context

[4] Portuguese – and international – public opinion on environmental issues has been influenced by factors such as advertising, mass media, political discourse, environmental protection organizations, among others (Figueiredo and Martins 1996). A 'media revolution' occurred in Portugal at the beginning of the 90's, mainly due to the advent of private television, brining with it growing competitiveness between the various television stations, as they search for higher ratings and develop means and programs designed to capture the public's attention. Thus, environmental awareness has generally developed within this environment of media-hype and is less the product of education and training (Figueiredo and Martins 1996, 1999). Portuguese point to the media as being one of the most trustworthy sources of information on the environment and consequently, it is an important means of socialization in that area. As stated by Mansinho and Schmidt (1994) "the media have played a prominent progressive role in the mobilization of public opinion, denouncing the problems and providing information with an educational component ... With the new newspapers started in 1990, the new radio stations and private television, the environment acquired a definitive place in the media's lineup".

particular to this country. As stated by Lowe and Goyder (1983), the growth of post-materialistic values has developed out of the peace, freedom of expression, and economic prosperity that our country only recently acquired. Due to these developments, it is not surprising to find Portuguese public opinion valuing physical and economic safety more than before.

The relatively strong presence of environmental values may be a consequence of the need to make the level of economic national growth compatible with European levels in the wake of Portugal's entrance into the European Community. By the same token, membership in the EU also subjects the country to practical obligations of an environmental nature. This conjunction creates the conditions for the appearance of a series of environmental demands, propagated by the mass media, public and private entities, environmental groups, and others. These demands also create the possibility for greater citizen participation and expression with respect to environmental issues. (Figueiredo and Martins 1996).

While, in general, environmental attitudes are strong, environmental behavior is weak (EEC 1986, 1988, 1992, 1995, 1999); thus, *green* attitudes in Portugal are still rather fragile. This fragility is particularly visible in the kinds of environmental problems that the Portuguese consider to be most worrisome, as well as in the articulation of the relationship between attitudes and behavior (Figueiredo and Martins 1999).

Although 58% of the Portuguese (Observa 1998) consider themselves to sympathize with organized environmental movements, only 14% declare themselves to be active members of an association. Therefore, even though the environmental protection agencies are held in esteem by Portuguese citizens, or in other words – to borrow an expression from Observa (1998) – although they enjoy *'a great capital of sympathy'*, this capital is not invested in concrete actions and in active associations.

Another important issue, also revealed by the data in the surveys mentioned, is that environmental sensibility in Portugal increases with urban complexity. This means that people from larger urban areas have a greater perception of environmental problems, because they are also more salient to these areas. The level of social engagement also increases in the higher social strata as well as at lower age levels (Mansinho and Schmidt 1994; Figueiredo and Martins 1999). Furthermore, the inhabitants of rural areas in Portugal tend to exhibit different sensibilities than do their urban counterparts. Although further empirical confirmation of this needed, according to the surveys consulted (EEC 1986, 1988, 1992, 1995, 1999), the rural inhabitants generally point to a lack of basic sanitation and waste removal as the leading environmental problems in their areas of residency, while those in urban areas complain of problems related to atmospheric pollution and a lack of green spaces.

3.2 Formal environmental organizations in Portugal

In Portugal, organized environmental activism only began to influence public opinion towards the end of the '80s. Before this time, in spite of their increasing number, activist organizations were limited in scope and had very little ability to influence either public opinion or administrative and political agencies. Until 1974, the environmental movement was all but absent. The Group for Environmental Protection (Liga para a Protecção da Natureza (LPN) was the first environmental movement in Portugal (its creation dates from 1948). It was founded by scientists associated with conservative international movements interested in the protection of rural life (Mansinho and Schmidt 1994). The LPN produced studies related to nature conservancy of high scientific quality. However, it kept a low public profile and exerted little political influence. The absence of environmental groups prior to 1974 is frequently explained by the existence of an authoritarian regime (Melo and Pimenta 1993) which effectively eliminated any attempt at public organization, and which imposed limits on the circulation of information, multiple censorship mechanisms, and low levels of education and salaries (Figueiredo and Martins 1996). Before 1974, environmental problems still evoked only limited resonance and were very localized due to the country's low levels of industrialization and urbanization (Melo and Pimenta 1993; Mansinho and Schmidt 1994).

After 1974 and until the middle of the '80s, the number of formal organizations increased (Melo and Pimenta 1993) among the numerous revolutionary popular movements. These environmental movements did not limit their scope to environmental causes, but instead also made claims on social and political issues (Rodrigues 1995). The Portuguese Ecologist Movement (Movimento Ecologista Português (MEP), for example, argued against industrial and consumer oriented development models.

The evolution of environmental movements in Portugal can be divided into three phases (Rodrigues 1995). The first phase can be characterized by the rise and subsequent disappearance of numerous environmental groups, which lacked serious organizational capacity due both to there being hardly any tradition of associative behavior in the Portuguese society and to the general public's unwillingness to accept the ecologist message. That public was still more concerned with bread, work, housing, health and education. The second phase spans the period from 1976 to 1983 and is associated with organized protests against nuclear power. The environmentalist discourse was marked by the political radicalism dominant in Portuguese society. In 1979 the Portuguese Association of Ecologists / Friends of the Earth (Associação Portuguesa de Ecologistas / Amigos da Terra) was created, becoming the predecessor of one of the most important existing environmental groups. In 1982 the first political pro-environment organization, the green party (Partido os Verdes), appeared. This party attracted severe opposition from other environmental groups due to its integration into the Portuguese Communist party. The third phase encompasses the evolution of the organized environmental movement in Portugal and extends from 1984 to 1988. During this period, growing economic and social stability opened up greater space for public awareness of environmental issues (Figueiredo and Martins 1996). The entrance into the European Community and the progressive opening of Portuguese society to the international agenda has also facilitated new organizational strategies. The most important event in this phase, however, is considered to have be the publication in 1987 of the Environmental Framework Law (Lei de Bases do Ambiente) and especially the Environmental Associations Law (Lei das Associações de Defesa do Ambiente) which, according to Melo and Pimenta (1993), made environmental issues institutionally and socially relevant. 1987 was also the European Year of the Environment which encouraged, among other things, the organization of several local and regional groups as well as environmental education. It was during this phase that most of the important Portuguese groups consolidated themselves, articulating their positions within the wider spectrum of environmental groups and acquiring almost hegemonic status (Rodrigues 1995).

After 1989 a new era began, characterized by the creation of the National Association of Portuguese Environmental Groups and increasing public and media interest in environmental issues. Until 1993 there were approximately 111 environmental groups[5], of both national and local scope. Many of these have direct connections with similar organizations at the international level (such as Greenpeace and Friends of the Earth).

Table 1 shows the number of formal environmental groups by region, as well as the number of local environmental protest cases.[6] Interestingly, the regions of Lisbon and North show the largest numbers in both formal organizations and grassroots actions (68.5% and 67.4% respectively). A detailed analysis of the regional distribution of environmental groups shows that they concentrate more in the littoral areas, mainly near Lisbon and Oporto

[5] Groups registered at the National Institute of Environmental Promotion, Ministry of Environment and Natural Resources.

[6] To be presented in the next section.

(Rodrigues 1995). This concentration relates to the higher concentration of urban areas and industrial activity, leading to increased social, economic and environmental problems and, thus, to the need for more active public intervention. The protest case analysis also illustrates major concentrations in littoral areas, such as Lisboa, Braga, Aveiro, Porto and Setúbal. As regards membership size, most formal groups have between 100 to 300 members.

Table 1. Regional distribution of regional and local environmental groups

Regions	Number of formal environmental groups*	%	Number of local environmental protest cases	%
Norte	27	2.5	221	40.4
Centro	18	16.7	136	24.8
Lisboa e Vale do Tejo	47	43.5	148	27
Alentejo	10	9.3	21	3.8
Algarve	6	5.5	22	4
Total	108	100	548	100

*Ipamb (1996); **Kousis, Aguilar Fernández and Fidélis (1996)

The main actions and functions of these groups include protest, publication and education related campaigns, preparation of studies, projects and data collection, protection of local resources and cultural traditions, and the promotion of conferences and debates. Their main publication networks are in the media, although they also use direct involvement with local populations through information sessions, auditions and mediation processes.

4. GRASSROOTS ENVIRONMENTAL PROTESTS IN PORTUGAL (1974-94)

This section analyses the type of participants, forms, and claims of local environmental protest cases[7], as well as the role of the state and other agencies affected by the protest. Protest case analysis was applied in a survey of selected newspapers, Jornal de Noticias and Publico, from 1974 to 1994.[8] A

[7] The project from which the data for this section has been taken – *Grassroots Environmental Action and Sustainable Development in Southern EC* - involved the University of Crete, the University of Salamanca and the University of Aveiro and aggregated data from Greece, Spain and Portugal. A comparison of the results for the three countries is presented by Kousis (1999a, 1999b).

[8] For a detailed presentation of protest case analysis, i.e., the specific type of content analysis, see Kousis (1998, 1999a).

large majority (46.7%) of newspaper articles came from Publico, about one third (34.9%) from Jornal de Noticias and 18.4% of the cases have been picked up from both newspapers. In addition, several articles were collected from a national periodical (2%).[9] Concerning the number of articles by cases, more than half are based on information from one article (57.8%). The remaining 15.5% include two articles, followed by 7.5% cases with three articles. The biggest case identified in the Portuguese data set has 53 articles, which dealt with the siting of an industrial landfill. Most cases involve public action forms which last a few days. It is important to note that more than one tenth (13.5%) of the cases last between one and two years, while only 1.6% last for more than twenty years. As expected, concerning the temporal distribution of the cases, more than half (53.5%) occurred between 1991-1994 – i.e., within the last four years of the twenty year period, once greater importance came to be attached to public participation under the growing social, economic and political stability in the mid-eighties. We also notice that between 1974 and 1980 14.2% and between 1980 and 1985 only 5.1% of the total cases occurred. The relatively significant percentage observed in the first years of democracy may be explained by the fact that it was the first period in which, after several decades of authoritarianism and subordination, the Portuguese people had the opportunity to participate in the political life of the country, leading, therefore, to an explosion in terms of social participation. However, this explosion of public participation was more a social and political de-compression than a real change in values and social behavior in the direction of effective citizenship (Braga da Cruz 1994).

A large majority of cases (77.3%) have involved only one community, whereas 22.7% of the cases have involved two or more communities. The interrelation between this data and the type of project/activity leading to protests would shed additional light on the geographically widespread features of the mobilization processes. In fact, a majority of the projects or activities exert only local influence. Rural environmental protest at the local level is more frequent (53%) than urban protest at that level (31.5%). These results concur with the findings of Freudenberg and Steinsapir (1992), Schnaiberg (1992) and Kousis (1999a) with regard to community based features of the environmental grassroots movements. The rural features of the majority of protest cases distinguish Portugal from other Southern European Countries (Kousis 1999a). This may naturally be attributed to the signi-

[9] The reason for the short number of cases identified here relates to the fact that this magazine has only been published since the middle of 1994, the last year considered in the time span of this research project.

ficant number of development projects located outside urban areas and rural areas' greater exposure to related effects. They may also be due to lower levels of education and civic participation or to cultural factors, namely, to a fear to change, to suspicion of new technologies, to sentimental values attributed to the land and, of course, to the salience of land in economic activities in these areas – more so than environmental sensitivity and local resource valorization.[10]

In the majority of cases (83.3%) specific information was not available on the number of participants. For cases where this was available, 8% involved between 100 and 1000 participants and 3.8% involved between 1001 and 2000 participants. Only three cases have involved more than 5000 participants. These figures are in agreement with the previous information showing smaller values for cases involving more than one community. The estimated number of participants gives an additional clue as to the dimension of the mobilization processes. Approximately half of the cases (52.1%) have involved less than 100 participants, 22.2% involved between 100 and 500 participants and only 4,5% of the cases seem to have involved more than 1000 participants.

As illustrated in Figure 1, residents (84%) and local governments (33.5%)[11] are the local participating groups most frequently referred to in the cases. Local environmental groups (24.7%) and labor and trade unions (18.5%) also illustrate an important degree of involvement and the predominance of community based protests. The heterogeneity revealed by the type of participants is surprising due to the traditional features of Portuguese rural communities. The low level of involvement of women's groups suggests the limited influence of feminist groups in Portuguese society. In particular cases, however, qualitative data analysis from the field (Fidélis et al. 1996) showed high involvement of women in the mobilization processes.

[10] As shown by Kousis (1999b) the rural character of grassroots environmental activism is more prevalent in Portugal than in Spain and Greece – something which needs further investigation.

[11] Reference here is mainly to the presidents of the 'freguesias' (communities).

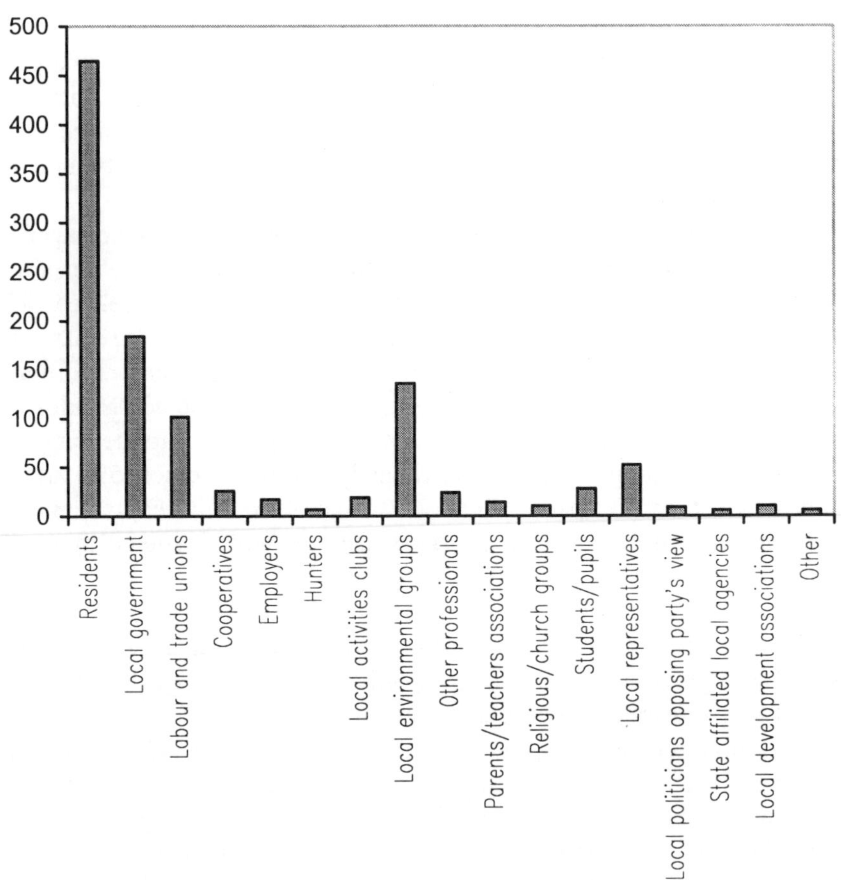

Figure 1. Local participating socio-economic, political and cultural groups [12]

The most frequent action forms were claims or demands (94.7%), as shown in Figure 2. In addition, the two other action forms most used in the protests were complaints to authority (74.7%) and signatures (45.6%). These numbers reflect that the majority of cases in Portugal may by characterized as passive and non-radical, confirming the statements of Freudenberg and Steinsapir (1992). By contrast, there were some cases where action forms reflect higher levels of protest. Among them demonstrations and public protest assemblies (16.4%) or the court route (7.5%). Protests that are more

[12] The source of all figures is Kousis, Aguilar Fernández and Fidélis (1996).

radical have involved source blockage (5.5%), or road blockades (7.1%). Violent episodes are rare.[13]

Regarding the regional distribution of local environmental protest, in general, the Northern region is the one recording the most cases (40%), followed by the Lisboa e Vale do Tejo region (27%), perhaps due to the existence of the two metropolitan areas and their high population density. The high level of industrialization and urbanization can also explain some of these figures. The regions with a lower number of cases are Alentejo (3,8%) and Algarve (4%). The development trends observed in Alentejo and briefly summarized in point 3 explain this result. What is surprising is the result of the region of Algarve, where the impacts of tourism on the environment assumed enormous proportions. Portuguese society, however, tends to identify environmental problems (EEC 1986, 1988, 1992, 1995 and 1999) more with industrial activity and the removal and treatment of wastes, be they domestic or industrial. This data helps to explain also the low number of observed protests in the Algarve region. On the other hand, the fact of having used a newspaper - Jornal de Noticias -, which focuses on Northern and central regions of the country, may have influenced the results for Alentejo and Algarve regions.

The most commonly referenced non-local participating groups are the environmental groups (9.6% of the cases) followed by citizens groups from adjacent areas (5.3%), political party representatives (4%) and local governments from adjacent areas (3.6%). These values may, however, be considered very low. They most likely reflect the overall low level of environmental awareness and social solidarity, as well as the existence of more immediate and daily individual or private interests.

[13] Compared to Spain and Greece, Portugal has the lowest number of violent episodes (Kousis, Aguilar Fernández and Fidélis 1996).

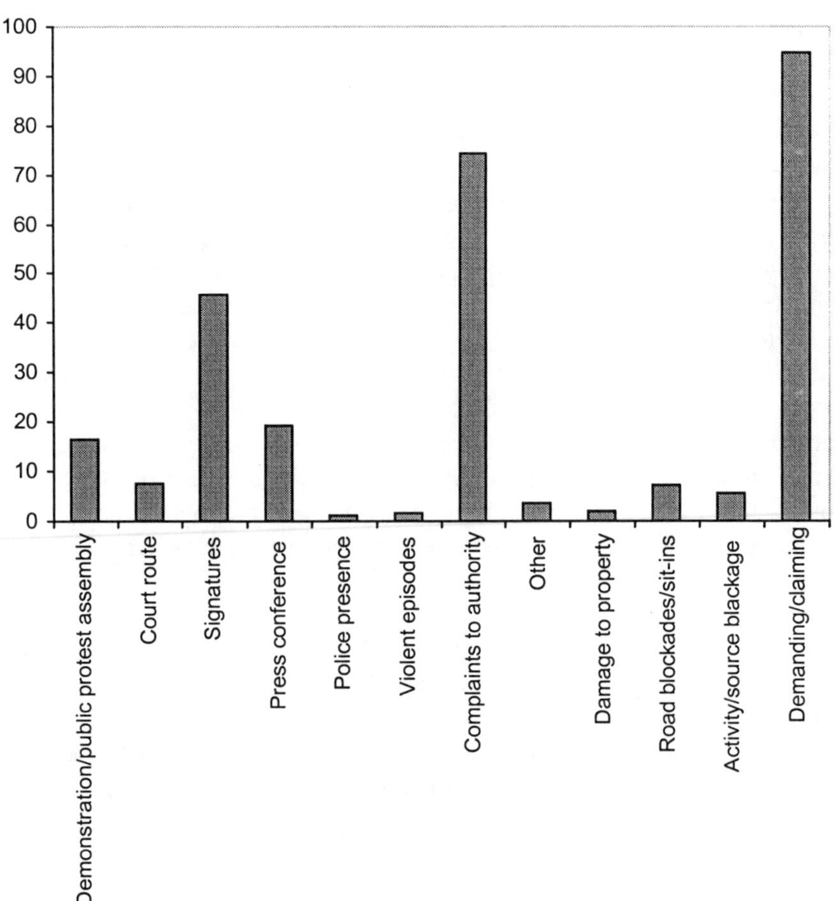

Figure 2. Action forms taken during the protests

Turning now to the sources/activities leading to mobilization, Figure 3 illustrates that the most controversial ecosystem-intervening sources related to waste disposal and treatment (34%) followed by construction and infrastructures (18.4%), manufacturing activities (9.8%) and finally, wildlife areas and energy installations (9%). The first three figures reflect the economic profile summarized in point 3, which means that they deal with aspects visibly associated with the economic growth of the country in the last twenty years. This finding may be related either to the significant growth of industrial activity during the last decades, or to the identification of only those environmental problems directly interfering with people's daily life.

Concerning the type of source, the most prevalent were fully functional sources already in operation (57%) while a minority were siting sources (32%). The high level of exposure sources highlights the seriousness of existing environmental problems in Portugal. The smaller number related to siting reflects traces of the secrecy of decision-making processes as well as limited environmental awareness and education enabling the public to foresee potential impacts from development projects.

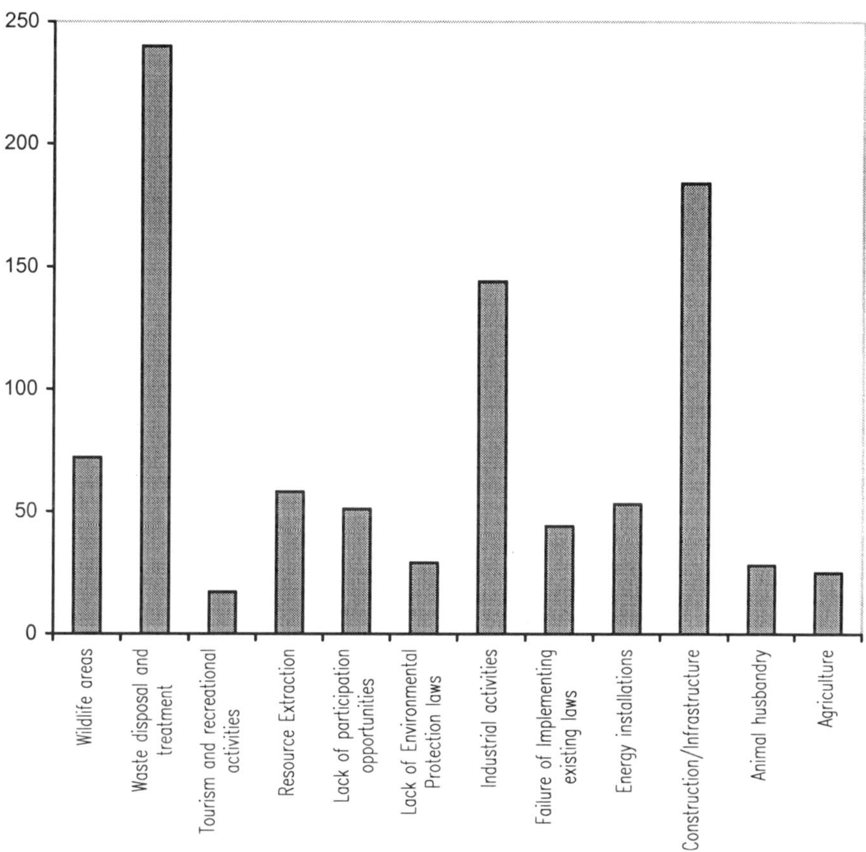

Figure 3. Primary source or activity leading to mobilization

An examination of the regional features of local environmental protest reveals the following aspects. The Northern region of Portugal holds more than half of the cases related to open landfill/free waste disposal (57%), such as the Canedo (Feira) and Padim da Graça (Braga) cases. This result is strengthened by the current waste management situation in Portugal, where a significant number of municipalities do not have proper waste treatment systems. In addition, almost half (48%) of the cases relating to untreated disposal of non-toxic industrial waste were identified in this region. This is a problematic area, with an industry location pattern spread along some important rivers and using them as open sewage channels: the Ave, with the textile, metals, chemical and agriculture-food industries; the rivers Douro, Cavado and Leça with various industries along their banks.

The Centro region also shows a significant number of cases (29%) related to the untreated disposal of non-toxic industrial waste. We can mention the Vouga river and especially the industrial zone of Estarreja with its paper and chemical industry, as well as the Certima river with metal industries of Agueda, and the food-agriculture industries of Bairrada. The cases related to roads, highways, tunnels, bridges are spread all over the various regions – the North region (36%), Lisboa e Vale do Tejo region (31%) and Centro region (21%). These mobilizations demonstrate the impact of the implementation of the National Road Plan (Principal Routes, Complementary Routes and other roads).

The primary source leading to one of the main mobilizations in the Alentejo region was the storage of toxic/hazardous waste (23%) as foreseen by the Industrial National Waste Treatment System. This was followed by the agriculture infrastructure (20%) in connection with the Alqueva dam project. In the Algarve region the primary sources leading to mobilization were roads, highways, tunnels, and bridges (10%) associated with the Via do Infante. This mobilization included multiple communities along the coastal zone. The mobilization occurred mainly for economic reasons (destruction of important agricultural land). There were also some cases related to stone extraction (11%) resulting from the construction activity for projects in the tourism industry. The majority of environmental degradation problems result from tourism activity (inadequate urban planning, human pressure on the littoral areas, and coastal erosion), however, there were no mobilization cases directly related to these aspects.

In 34% of the cases the mobilizers considered the source/activities as fully responsible for eco-disturbances and demanded that some mitigating measures be undertaken. There are also an important number of mobilizers (31.3%) who consider that source/activities are fully responsible for eco-disturbances and that they should not be put into operation. These may be seen as the most radical attitudes assumed by the mobilizers. They may reveal mistrust of authorities as well as doubts about the efficiency of environmental legislation, state intervention, and related control systems. Alternatively, they may reflect low levels of environmental education and a lack of information on existing mechanisms to minimize or even solve some environmental problems. As we saw in the fourth part of this chapter, both characteristics (mistrust of public authorities as well as the absence of information) are important features of Portuguese society with regard to environmental matters.

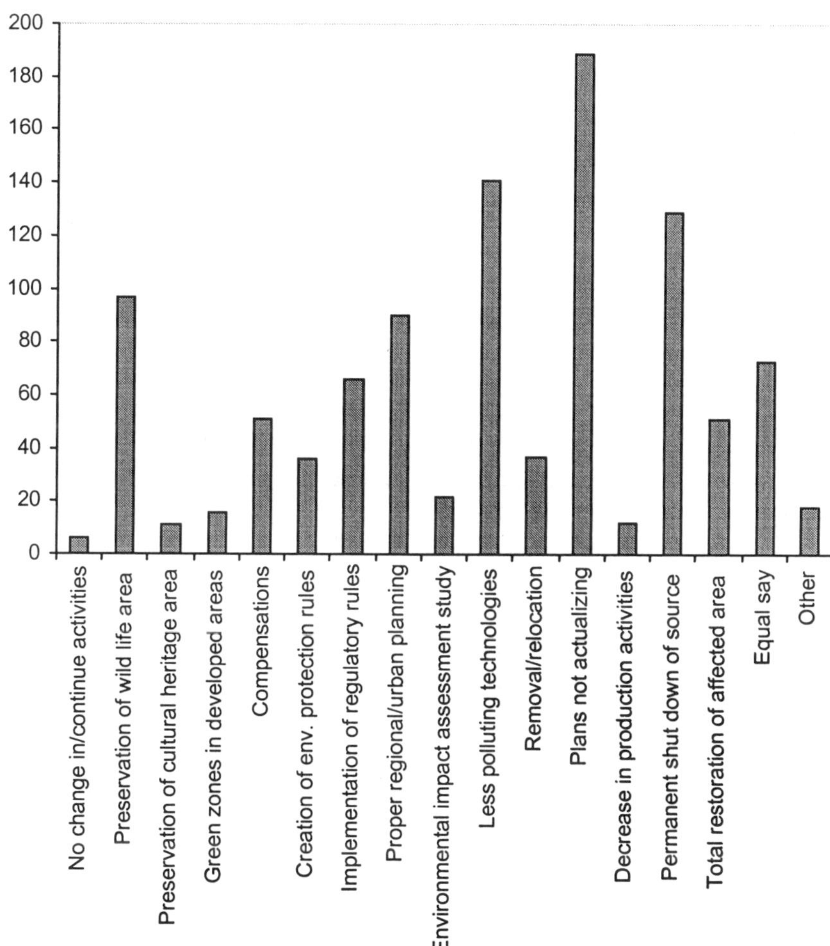

Figure 4. Resolutions proposed by the mobilizers

As we can see in Figure 4, non-actualization of planning is the most common resolution proposed by the mobilizers (34,4%) and it may be related to siting project cases. This choice may be seen as the most direct and easily applicable and is also related to the above discussion. The second type of resolution related to better non-polluting technologies (non-main processes related, 25%) and proposals for permanent shut down (24%). Apparently, there are more mobilizers proposing resolutions for the siting situation than for the exposure one.

The data show that the main environmental offense identified by the mobilizers was the destruction of an ecosystem (60%), followed by water pollution or contamination (48%) and atmospheric pollution (44%). These results can easily be associated with the figures obtained about the primary source and with the current environmental problems summarized above. In general terms the mobilizes identified more ecosystem impacts (51.3%) followed by economic impacts (17.6%), health impacts (11.9%) and negative political impacts (8%). Among the ecosystem impacts identified, the most frequently referred to were fresh water (54%) and airsheds (43%). Suspected negative public health impacts (33%) were also frequently pointed out.

Figure 5 depicts the bodies to whom the mobilizers went for help. Local government was the agency that was more frequently approached (14%) followed by the state (12.4%) and the sub-local government (freguesias) (12%). The interrelation of this data and the time periods considered in the project would allow one to identify the extent to which appeals to the state for support are declining, thus reflecting a decline in the importance attached to centralized features of decision-making, and an increased valorization of local government.

As Kousis (1999a) states, an increase in protest cases is clearly evident before or during the electoral periods. Such an increase is justified by the perception of greater receptivity and support for social demands which, in general, the political parties on the one hand and the governments on the other, demonstrate in these moments.

Information on the above bodies' responses to the protesters is not complete. Based on the available information it appears that for the most part these responses are positive, while fewer are neutral and negative responses are rare. Positive responses come mainly from the state (at all levels), local government (including freguesias), and large environmental groups.

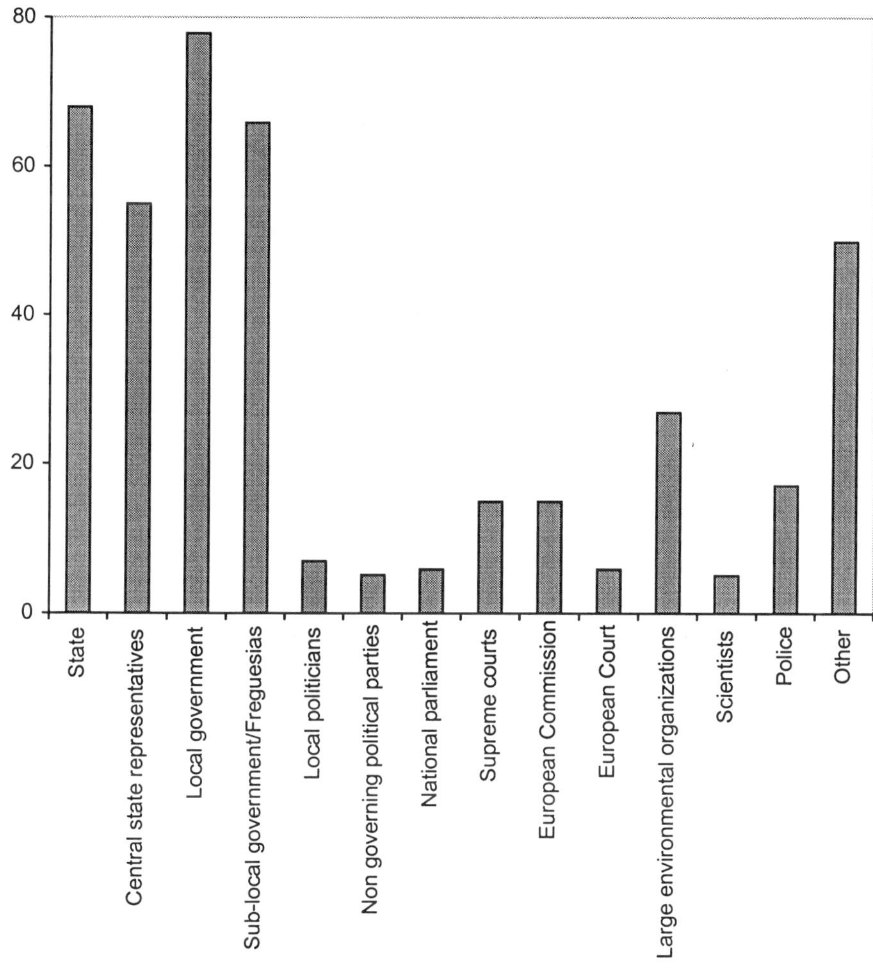

Figure 5. Bodies approached by the mobilizers for assistance

The major groups challenged by the mobilizers are the state (all levels), producers (private and semi-private) and local government, as seen in Figure 6. These results reflect the importance of the main actors involved in the current development process. On the one hand, state and local agencies are responsible for infrastructures such as roads, water supply, wastewater collection and treatment, urban solid waste collection, disposal and final treatment. On the other hand, development proposals in fields such as industrial activity come mainly from different types of producers.

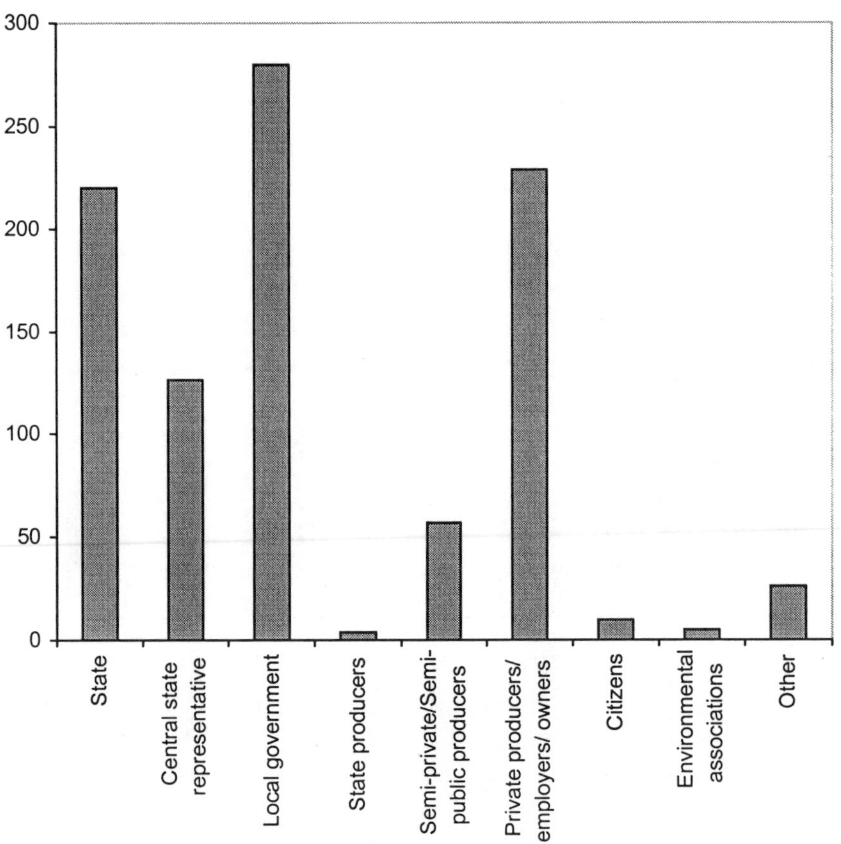

Figure 6. Groups challenged by the mobilizers

Again, information is limited on the challenged groups' responses to the mobilizers[14] The data at hand show that overall, when information is available, there are more positive responses, followed by neutral and negative ones.[15] Positive answers come mainly from the state (all levels), local gov-

[14] Information is available only for forty percent of the cases.

[15] Negative Response includes the following categories: Ceased Action by Force, Ceased Actions by Law, Pleaded for Stop of Actions, Job Loss Threat, Refuse Talks, Continued Operations. Neutral Response includes the following categories: Asked for Evidence of Problem, Eco Protection Exists, Promised Jobs, Eco Problem Insignificant, Indifference/ No Response, Too Costly to Correct, Compensations. Positive Response includes the following categories: Temporary Ceasing of Operations, Formal Recognition, Opened Negotiations, Measures/ Alternatives Promised, Control Tech/ EIA/ Some Measures Taken,

ernment and producers. Contrary to what one might expect, private producers give more positive than negative responses.[16] However, they often remain neutral in their responses to mobilizers claims. While local government performs a mediating role, it is the state that has decision-making power over various development projects.

5. CONCLUSION

The evidence provided in this chapter illustrates the increasing trend toward grassroots environmental protest cases in Portugal. This trend is related to the transformations which occurred in the social, economic and political context of the nation after 1974. Portugal has, in the last two decades, known unprecedented economic and social development, the consequences of which can be seen in the construction of diverse infrastructures and equipment, as well as in the growth of environmental problems. The transformation in the political context from a dictatorial regime to a democracy and the political, economic and social changes and instability witnessed in the decade after the Revolution of April 1974, have, on the one hand, aggravated environmental problems. On the other hand, however, they have also allowed for an increase in the intervention of the civil society in various areas, including the environment.

Since the early nineties the number of grassroots protests due to environmental problems also witnessed a sharp increase. The brief presentation in section three shows that among the Portuguese there is a relatively strong sense of environmental values. However, this characteristic stands in contradiction to the fragility of public opinion when it comes to assessing the relationship between attitudes and behaviors. The discrepancy verified between attitudes and practices suggests that Portuguese society is sensitive to environmental issues, but does not have an effective environmental conscience (EEC 1995; Figueiredo and Martins 1996, 1999). It is also worth noting in recent years the growing involvement and participation of Portuguese citizens in environmental issues, particularly when local interests or people are involved. We are therefore looking at a sporadic, specific, and locally oriented initiatives which are directed at problems that threaten the everyday live and concerns of the people (Figueiredo and Martins 1996, 1999). To

---

Decrease Production/ Most Measures Taken, Permanent Stop of Eco Disturbance, Better Alternative Chosen.

16 This too differs from results for Spain and Greece – something needing further study (Kousis, Aguilar Fernández and Fidélis 1996).

reinforce these findings it is also important to stress the limited involvement of Portuguese citizens in formal environmental organizations.

It is also in the early nineties that legislation on environmental issues became more abundant and, as we have demonstrated in section three, the role of the environmental protection agencies was institutionally recognized, due in part to a change in the formal Portuguese environmental movement, that is, its passage from a discourse that appealed for radical social changes to one more institutionalized and less radical. As stated by Melo and Pimenta (1993), in this period most of these agencies "became more pragmatic, abandoning the philosophy of a radical rupture with the system ... The agencies began to make changes to the system from within." It is due in large part to the work of the environmental protection agencies and on a larger scale to the definitive appropriation by the media of environmental issues, that Portuguese society has acquired a heightened sensibility for these issues.

Environmental protest cases are usually tied to the community where the implementation of the project/activity takes place. This, as well as the type of arguments used in the protests – often focusing heavily on health risks and negative economic impacts –, apparently encourage the NIMBY attitude of Portuguese public opinion regarding environmental issues (Figueiredo and Martins 1996, 1999).

Most of the environmental protests in Portugal analysed here reveal striking similarities with similar cases in other countries, for example: the emphasis on the need for more information on the projects and activities at issue; the demand for greater involvement and participation of the local population in the decision-making process, as well as challenges put to public authorities and producers – the promoters of projects/activities – to take the necessary actions (Freudenberg and Steinsapir 1992; Kousis 1999b, 1999b). The response of the state is in most cases rather neutral. Even when its response is positive, its intervention in the protesting process usually does not go beyond promising a solution, which in the majority of the cases never materializes into concrete results.

As regards the regional distribution of grassroots environmental activism, the Northern region is the one recording most cases followed by Lisboa, Vale do Tejo, and the central region. The high levels of urbanization and industrialization in the littoral areas of those regions, as well as the fact that the majority of the country's population is concentrated there, help explain these figures. In addition, the majority of development projects and infrastructures tend to be implemented precisely in these regions. In contrast with these figures, the Alentejo and Algarve regions have almost insignificant percentages of protest cases, given that they are sparsely populated and less developed. Alentejo is a region marked by the absence of development projects, with the exception of the Alqueva dam project and the Sines industrial area. Yet one would expect more protest in the Algarve region, with the highest tourism development in Portugal and its related negative impacts on the environment. Here, the economic dependence of the locals upon tourism may help explain, to a certain extent, the low number of protest cases. The regional distribution of protest cases in association with the development features of each region highlights the close relationship that exists between the level of socio-economic development and the levels of environmental concern.

The principal sources of grassroots environmental protests in Portugal are related with the removal and treatment of garbage waste, the construction of infrastructures, and industrial activities – all aspects that themselves are closely linked both to the country's path towards economic development as well as to the decidedly rural nature of the grassroots environmental movements in this country.

The forms of protest most used include public appeal for closure, stopping construction or planning, complaints to public authorities, petitions, press conferences and public demonstrations. Although the intensity of the protests increases with their duration and with the absence of results, as elsewhere, violence is not characteristic of these movements in our country (Freudenberg and Steinsapir 1992; Kousis 1999a).

Further research is needed to understand the strong rural character of the protest cases, their community-based and nonviolent features, as well as the state's responses to the mobilizers' claims and demands. These should also be studied in relation to the social, economical and political context. An interesting comparison could be made between grassroots environmental movements with non-environmental grassroots movements in terms of participating and challenged groups.

Given the unavailability of certain types of data and especially the type of approach used – content analysis of newspapers – it is not possible to evaluate completely the responses of approached and challenged groups to these movements. Additional sources of data are necessary to carry out a more careful investigation as regards the cause of the protests, the proposed solutions, the achieved solutions, and the role of the state in this process.

These additional sources or methods could also provide more information and thus shed more light on the character and dynamics of local environmental activism. The global picture supplemented by content-analysis could thus be enriched with qualitative studies bringing out details on specific cases missed by the press, such as women's involvement or the ambivalent stance of local government.

REFERENCES

Dunlap, R.E. (1996) A sociological perspective on environmental problems, in C. Borrego et al., *Actas da V Conferência Nacional sobre a Qualidade do Ambiente*, DAO/ CCRC, Aveiro, pp.99-120.

Dunlap, R.E., and Mertig, A.G. (1995) Global concern for the environment: Is affluence a prerequisite? *Journal of Social Issues* **51**, 121-137.

EEC (1986) *Europeans and the Environment in 1986*, Brussels.

EEC (1988) *Europeans and the Environment in 1988*, Brussels.

EEC (1992) *Europeans and the Environment in 1992. Eurobarometer 37.0*, Brussels.

EEC (1995) *Europeans and the Environment in 1995. Eurobarometer 37.0*, Brussels.

EEC (1999) *What Do Europeans Think about the Environment*, Brussels.

Fidélis, T., Figueiredo, E., Bastos, S., and Rosa Pires, A. (1996) *Grassroots Environmental Action in Portugal - A Case Study on a Landfill Sitting*, Presentation to the Euro-Conference Environment and Innovation, Vienna.

Figueiredo, E. (1999) O Norte e o Sul das Questões Ambientais - Breve Reflexão Acera do Valor Social do Ambiente, in Solidários / OIKOS, *Projecto Interdependência Norte Sul - Cultura e Desenvolvimento; Ambiente e Desenvolvimento*, Solidários / OIKOS, Oliveira do Bairro.

Figueiredo, E., and Martins, F. (1996) Pensar verde - Contributos para o Estudo Formacao de uma Consciencia Ambiental em Portugal, in I.C.S/ UL, *Dinâmicas Multiculturais, Novas Faces, Outros Olhares*, ICS, Lisboa, pp.417-436.

Figueiredo, E., and Martins, F. (1999) *O Cidadão na Construçãode um Ambiente Melhor*, Comunicação apresentada ao Seminário sobre Higiene Ambiental, Coimbra, HUC.

Freudenberg, N., and Steinsapir, C. (1992) Not in our backyards: The grassroots environmental movement, in R.E. Dunlap and A. Mertig, *American Environmentalism*, Taylor & Francis, Washington, pp.27-35.

Gil Nave, J. (2000) *Environmental Politics in Portugal?* Ph.D. Thesis, European University Institute, Florence.

INE (1991) *Recenseamento Geral da População e Habitação*, Instituto Nacional de Estatistica, Lisbon.

Kousis, M. (1998) Protest Case Analysis: A Methodological Approach to the Study of Local Environmental Mobilizations, Working Paper No. 570, Center for Research on Social Organization, Department of Sociology Ann Arbor, MI, University of Michigan.

Kousis, M. (1999a) Environmental protest cases: The city, the countryside and the grassroots in Southern Europe, *Mobilization* **4**, 223-238.

Kousis, M. (1999b) Sustaining local environmental mobilisations: Groups, actions and claims in Southern Europe, in C. Rootes, *Environmental Politics, Special issue: Environmental Movements: Local, National and Global*, pp.172-198.

Kousis, M., Aguilar Fernández, S., and Fidélis, T. (1996) *Grassroots Environmental Action and Sustainable Development in Southern European Union*, Final Report, European Commission, DG XII, Contract No. EV5V-CT94-0393.

Lowe, P.D., and Goyder, J.M. (1983) *Environmental Groups in Politics*, Allen & Unwin, London.

Mansinho, I., and Schmidt, L. (1994) A Emergencia do Ambiente nas Ciencias Sociais, *Análise Social* **39**, 441-481.

MARN (Ministerio do Ambiente e Recursos Naturais) (1995) *Plano Nacional de Politica de Ambiente*, Ministerio do Ambiente e Recursos Naturais, Lisboa.

Melo, J.J., and Pimenta, C. (1993) *Ecologia e Ambiente*, Difusão Cultural, Lisboa.

Observa (1998) *Folha Informativa No 4 & 5*, ICS, Lisboa.

Rodrigues, E. (1995) *Os novos Movimentos Sociais e o Associativismo Ambientalista em Portugal*, CES, Coimbra.

Schnaiberg, A. (1992) The hazards of production: How do communities react? *Science* **255**, 1586-1587.

Taylor, B.R. (1995) *Ecological Resistance Movements: The Global Emergence of Radical and Popular Environmentalism*, State of New York University Press, Albany, NY.

III

SUSTAINABILITY DISCOURSES AND THE EMERGENCE OF INSTITUTIONS FOR COLLECTIVE ACTION IN SOUTHERN EUROPE

Chapter 9

Sustainable Development and the Participation of Environmental NGOs in Spanish Environmental Policy

The Case of Industrial Waste Policy[*]

Manuel Jiménez

1. INTRODUCTION

In the early 1990s, environmental issues acquired an unprecedented level of prominence in Spanish public opinion and on the political agenda. Social and political conflicts over the implementation of policies such as water planning, road infrastructure, or waste management plans led to a change in the traditional, closed style of policy-making in favor of open public discussion in the search for a social and political consensus on the plans to be implemented. Environmental NGOs played a crucial role when it came to mobilizing public opinion in order to introduce environmental concerns into decision-making processes. At the local level, and despite some incidents of violent confrontation, they managed to redirect NIMBY protests over the location of specific projects towards broader debates on policy contents and to generate unprecedented levels of citizen involvement in local politics. At the state level, the confrontational strategy of decision-makers towards environmental NGOs shifted to one of negotiation and problem-solving. The tendency of state authorities to describe environmentalists as radicals and troublemakers was replaced by a new willingness to give them a voice in the decision-making process through new advisory institutions such as the Advi-

* I am grateful to Chris Rootes for his support and suggestions on previous drafts of this article, as well as for the comments from the editors of this volume. I would like also to thank *Ecologistas en Acción* for providing empirical information and Justin Byrne for reading this text critically.

K. Eder and M. Kousis (eds.), Environmental Politics in Southern Europe, 225–253.

sory Committee for the Environment *(Consejo Asesor de Medio Ambiente)* or the National Water Council (*Consejo Nacional del Agua*). This development coincided with the incorporation of the vocabulary of sustainable development as a central concept shaping policy planning in Spain.

The objective of this chapter is to offer a preliminary approach to the study the relationship between the participation of environmental NGOs and the development of Spanish environmental policy. The significance of the Spanish case resides, on the one hand, in the underdeveloped character of the country's environmental policies within the context of the European Union (EU) and the 'European pressure' to move forward through the different policy stages in a limited period of time. On the other, Spain constitutes an interesting case because of the particular democratic character of the Spanish political system, as reflected in deficient legal provisions for rights of the defense and promotion of public interests through standard administrative and judicial procedures.

In the first section, I outline the nature of the participation of environmental groups in advanced industrial states as part of the evolution of environmental policy towards the paradigm of sustainability. The second section focuses on explanations of policy change and the impact of environmental NGOs' participation as a factor in policy development processes. In the remaining sections, the case of the development of Spanish environmental policy is analysed in these terms, with special reference to changes in industrial waste policy over the last decade.

2. ENVIRONMENTAL POLICY PARADIGMS AND THE PARTICIPATION OF ENVIRONMENTAL NGOS

Environmental policy development has evolved through three stages, characterized by the prevalence of different types of state response or approach to the environmental issue. Each of these responses is located in a specific policy paradigm or belief system, which defines the environmental issue as a political problem, identifying the goals to be pursued as well as influencing the specific institutional and organizational configuration of the decision-making process and the use of a particular set of policy instruments.[1] Such belief systems incorporate normative assumptions about the

[1] "Policymakers customarily work within a framework of ideas and standards that specifies not only the goals of the policy and the kind of instruments that can be used to attain

desirability of the involvement of environmental NGOs in the decision-making process as well as about the instrumental value of their participation in the policy process. In this sense, each stage in the process of environmental policy development can be considered to offer a set of resources and incentives for the participation of environmental NGOs as part of the rules of the policy game (Pierson 1993).

In some cases, the presence of environmentalists in the network of actors participating in the policy process might be considered undesirable. Decision-makers might regard pressure from environmental groups as a potential threat to their policy goals. Thus, decision-makers might follow a strategy of exclusion, for example by denying environmentalists' access to conventional channels of public participation in democratic systems and/or by using repressive measures. In other cases, policy-makers might view the pressure of environmentalists as a necessary input in order to achieve their policy goals. Decision-makers then follow a strategy of facilitating the participation of environmentalists through those mechanisms of public participation for which provision is made in the administrative and judicial arrangements of the state, as well as creating specific policy instruments designed to foster participation.

Table 1 relates the three stages of environmental policy development to different opportunities for participating and influencing the policy process.

---

them, but also the very nature of the problems they are meant to be addressing." (Hall 1993: 279)

Table1. Environmental policy paradigms and environmental NGOs' participation

UNDERSTANDING OF THE PARTICIPATION OF ENVIRONMENTAL NGOs	OPPORTUNITY STRUCTURE		STATE-ENVIRONMENTALISTS INTERACTIONS				POLICY PARADIGM (type of state response to environment)
	Legitimacy/ Status	Level of access	Environmentalist strategy	State strategy	Type of interaction	Level of influence	
REJECTED	Not granted	Limited	Socialization	Repressive	Confrontation	Limited/ sporadic	NON-DECISION (reactive)
LEGITIMATE	Partially granted	Peripheral	Institutional changing strategies/ Socialization	Institutionalization at the margins	Confrontation / problem-solving	Medium (peripheral policy change)	SECTORAL
INSTRUMENTALIST	Widely granted	Central	Policy proposals/ implementation	Increasing inclusion in decision-making	Increasing consensus on basic goals	Great (paradigm change)	INTEGRATIVE (towards sustainability)

2.1 The first stage: 'Non-decision'

The first policy stage is characterized by the absence of a specific state response to the environmental issue. This was the prevailing situation in industrialized countries until the progressive incorporation of the environment into the political agenda during the 1970s. The natural environment is, at this stage, seen in terms of its utility to the economic system (Baker et al. 1997). State intervention takes the form of health or sanitary measures in those cases where the negative impact of economic production takes the form of accidents or leads to emergency situations in industrial areas. The causal link between the nature of production systems and these 'incidents' is not clearly identified, nor is the relationship between environmental deterioration and other policy decisions clearly traced. These incidents are either seen fatalistically, as inevitable 'natural' events or as consequences of the haphazard functioning of the production system, to be overcome by greater progress in the technical reduction of randomness (Baker et al. 1997). In the meantime, state intervention is confined to isolated initiatives, for example the introduction of special plans for emission control measures in some industrial areas, following strategies of dilution by means, for example, of policies favoring the construction of tall chimneys, which merely shift the problem to other areas.

These measures are carried out by existing sectoral departments of the administration (industry, health, agriculture) where environmental concerns are subordinated to other, generally economic interests, and usually on the basis of command and control instruments related to health standards in industrial activities or municipal regulations.

In this 'non-decision' stage, the demands of environmental NGOs are ignored. The absence of legislation (in constitutions or legal codes) defining the environment as a public or collective good usually precludes the recognition of environmental NGOs as having legitimate access to conventional mechanisms for citizen participation (access to administrative procedures of norms adoption and judicial review). Furthermore, the deficiencies of the administrative and judicial norms concerning the environment curtail the legal basis of the demands of NGOs and reduce the efficacy of litigation against polluters.

However, the opportunities for participation will vary according to the quality and extent of the general legal provisions related to citizens' participation in any given political system. In general terms, the access of environmental NGOs to policy arenas will be limited and restricted to the latter stages of the policy process. Environmentalists' influence depends on their capacity to mobilize public opinion over topical issues and critical events which are usually framed in terms of local environmental grievances. Conflict (and repression) is the prevailing pattern of interaction with decision-makers. At this stage, environmental movements might be expected to have a strong grassroots component, an organizational base characterized by a lack of professional resources, weak territorial coordination, and limited relationships with scientific communities.

By the 1970s, the visibility of environmental degradation, the experience of the failure of the dilution approach, scientific progress, and the spread of environmental awareness among the population, among other factors, modified the response of the state to environmental problems.

2.2 The second stage: The sectoral response

The second stage is characterized by the recognition of the causal links between environmental degradation and economic production. Environmental problems are defined as a by-product of economic activities, so that "state intervention involved a zero-sum game between the protection of the environment and the reduction of the cost of economic production" (Weale 1992). As Baker et al. (1997) point out, the dominant discourse establishes economic development as a precondition for environmental protection. This discourse is especially prevalent in those countries immersed in the process of industrialization. Therefore, the new state response to the environment consists of a willingness to cope with the problematic externalities of economic growth, but it does not involve the redefinition of the objectives of economic growth (Jänicke 1996).

As Paehlke (1990) has noted, the incorporation of the environment into the state's agenda does not imply any administrative innovation in the compartmentalized, closed, centralized, and technocratic administrative state, but merely a new specific area of decision-making. This is usually headed by a high- status national administrative entity, with limited powers to coordinate the actions of other departments of the state that retain environmental powers, and with limited political autonomy to influence outcomes in other policy areas. The solution to environmental problems is sought in extending policy instruments on an incremental basis (extension of command and control techniques of administrative regulation (Weale 1992)) in combination with management plans for specific problems such as water or air pollution. In this institutional setting, environmental policy focuses on mobilizing financial resources and formulating standards relating to particular media such as water or the air (Jänicke 1996). These measures are usually demand-orientated, and intended to provide infrastructure and a legal framework to meet the expected increasing demand (e.g., for water or for waste disposal facilities), with a limited impact on production processes. In the case of industrial pollution, the generalization of end-of-pipe solutions and the management of waste are typical initiatives of this stage.

The incorporation of the environmental issue into the state agenda is accompanied by the formal recognition of environmental organizations as legitimate representatives of environmental interests. At best, recognition may figure in a specific general law on the environment; in other cases, recognition might be progressively incorporated through sectoral environmental legislation.

At the same time, the institutionalization of environmental arenas of decision-making around new environmental departments usually increases the opportunities of environmentalists to lobby and participate from the earliest decision-making stages in policy processes onwards. Their capacity to do so depends on the powers and political autonomy of the environmental authorities vis-à-vis other state departments. The nature of the interaction between environmental organizations and environmental departments will vary according to the policy arena in which the policy process takes place, as well as the particular issue at stake.

The sectoral approach to environmental issues is reflected in the scant weight of environmental departments in such well-established policy domains as agriculture, public works, or industry. On some occasions, environmental departments, in alliance with environmental groups, may try to influence the policy process in these policy domains, while on others, they will keep a rather low profile or even support other departments' decisions. A department's attitude will depend on the level of government support for different policy issues at any given time, as well as the structure of power across state departments. But even with respect to those issues on which environment departments hold a great deal of power and, therefore, occupy a central position in the policy process, the environmental organizations' access to, and influence on the decision-making process does not necessarily increase. Again, the pattern of interaction will depend on the political autonomy of environmental departments to implement environmental measures which might be opposed by other state departments.

This makes the institutionalization of environmental NGOs' access to decision-making processes extremely problematic, as reflected in the creation of advisory committees promoted by environmental authorities. These bodies are often characterized by the predominance of institutional actors, their scant political leverage, and their irregular pattern of functioning. Environmental organizations, especially those dealing with crucial policy issues (such as energy, transport, industry, etc.), see these institutionalized fora as a means of legitimating environmental policy, rather than as a channel through which they can effectively promote their proposals.

In short, in this second stage, sectoral environmental policy signifies the concession of access to new arenas, in which environmental organizations are formally recognized as legitimate representatives of environmental interests. However, these channels of interaction with authorities are usually conceived, or used in practice, as a means of confining the participation of environmentalists to the periphery of the policy process (advisory committees, public inquiry procedures, etc.). In contrast, the capacity of initiative and the power to establish the contents of the debate, remains in the hands of established state actors.

The pattern of interaction between environmentalists and authorities therefore wavers between conflict and collaboration, depending on the issue at stake and the state's need for legitimacy. The influence of environmental NGOs in the policy process depends on their capacity to politicize (or socialize) these arenas (usually resorting to non-conventional forms of participation), and to construct alliances (adversarial advocacy coalitions) around their policy alternatives.

At this stage, environmental organizations usually show a high degree of coordination at the state level, and at the regional level in decentralized systems, and marshal resources to work through the state level and to coordinate environmental movements around specific campaigns, as well as to propose alternative policy solutions in collaboration with scientific communities.

2.3 The third stage: The integrative policy approach

The third stage is characterized by the consolidation of the idea that a high level of environmental protection is a precondition for long-term economic development: "By the late 1980s, it had become clear that the shortcomings involved in the environmental policy strategies of the 1980s left many problems of pollution unresolved or growing worse" (Weale 1992: 23). These ideas crystallized into specific environmental policy proposals in advanced industrial countries (the EU, the Netherlands, Germany), where "the debate has started to switch to the overall resources input to industrial production" (Jänicke 1996: 307).

This stage is usually defined as a transitional phase towards sustainability (the Fifth European Action Plan on the Environment) or ecological modernization (Weale 1992), implying recognition of the need progressively to integrate the environment into the economic model of development in order to achieve sustainability in the medium to long term. The precautionary principle guides state intervention through the integration of environmental concerns into other policies (a progressive greening of sectoral policies). Policy goals focus on the demand side of the production process, aiming at reducing the consumption of inputs (water, energy) and the production of waste.

The promotion of the conditions for the greening of the administration involves the re-organization of institutions for integrated environmental planning: environmental authorities acquire full powers and autonomy, at the same time as sectoral departments also include environmental considerations in their policies (taking sustainable development as their objective). As Weale (1992) points out, this implies the pursuit of a change in "the relationship between the state, its citizens and private corporations, as well as changes in the relationship between states". Changes in the configuration, institutionalization and organization of the decision-making process involve a democratic turn towards the decentralization and the integration of economic and social interest groups as partners in the policy process (as indicated by the notion of shared responsibility in the Fifth European Action Plan). Policy instruments include the improvement of existing measures (regulations, judicial control, financial incentives, market mechanisms) and the development of new instruments (voluntary agreements, thc creation of deliberative fora comprising social interests, etc.)

The recognition of environmental NGOs extends in this stage into new areas of the decision-making process (institutionalization of advisory committees in the fields of energy, town-planning, industry, agriculture, etc.), where they gain greater control and influence over the content of discussions. Participation in itself is an objective of environmental policy, not only for legitimization purposes, but also as an instrument for efficiently achieving results. Access to environmental information, to the courts and to the decision-making process is encouraged by the state. As stated in Rio Summit's Agenda 21 (art. 27), 'governments, in consultation with NGOs, should initiate a process to review formal procedures and mechanisms for the involvement of these organizations at all levels, from policy-making and decision-making to implementation.'

Environmental NGOs are characterized in this phase by their high levels of professionalization and issue specialization within a division of labor across networks that transcend national frontiers. Alternative proposals are related to central issues of economic policies and policy planning.

This section has highlighted the fact that the openness of decision-making processes to environmentalists increases as state responses to the environment move towards the paradigm of sustainability. This scheme for classifying the relationship between environmental NGOs and different environmental policy approaches will be used to analyse the role of environmental NGOs in the process of environmental policy development in Spain.

3. THE PARTICIPATION OF ENVIRONMENTAL NGOS AND POLICY CHANGE AND CONTINUITY IN SPAIN

The evolution of Spanish environmental policy constitutes an interesting case for examining the relationship between exogenous (the Europeanization dynamics) and endogenous (the pressure of environmental NGOs) factors of policy change toward sustainability. The peculiar situation of Spain is not only seen in the pressure to develop (in a short period of time) an advanced environmental policy meeting EU guidelines, but also in the lack of certain 'democratic prerequisites' for the deployment of sustainable development strategies. In this sense, environmentalist pressure does not only center on substantial environmental problems, but also involves struggles to overcome the shortcomings of the Spanish democratic system (above all the scant legal provision for citizens' rights with respect to the collective defense or promotion of public interests through standard administrative and judicial procedures).

In fact, a number of studies have explained the development of Spanish environmental policy in terms of the tension between European pressure and the resistance of domestic actors. Indeed, it would be difficult to account for Spanish environmental policy development and progress on the environmental issue without taking the EU into account as a crucial exogenous factor for change (Font 1996; Pridham and Konstadakopoulos 1997). However, the influence of European environmental policy has been weakened by the national context which worked as an obstacle to implementation (Jänicke 1996). These studies have not usually identified the participation of environmental groups as an important factor in policy change. When considered, environmental NGOs have been depicted as prisoners of the institutional context (which has limited their access and influence), the political milieu (ignoring their demands), or that old favorite, cultural determinism which claims a "Mediterranean syndrome" associated with low environmental awareness (La Spina and Sciortino 1993). In contrast, I offer another interpretation which highlights the crucial role of these actors in the development (or 'Europeanization') of Spanish environmental policy.

In 1992, when the EU approved the Fifth European Action Plan on the Environment, the process of institutionalization of the environment as an area of state intervention in Spain was still incomplete. In the mid-1980s, and after the Department of the Environment (a small unit within the Ministry of Public Works) had produced numerous drafts, the Socialist government definitively abandoned the idea of enacting a general law on the environment. This (non)decision was taken to satisfy the Ministry of Economy, which had stressed the negative consequences that such law would have for economic development.[1]

Despite the absence of a general law, the incorporation of EU directives led to notable progress in environmental legislation. In many cases the new set of norms on water (1985), the coast (1988), industrial waste (1986), environmental impact assessment (1986), and natural areas (1989) constituted the first initiatives in these fields. Nevertheless, the practical impact of this legislation was significantly weakened in the process of adoption and application (not to mention enforcement) (Pridham 1994; Font 1996; Aguilar Fernández 1997).

Even when incorporated into the Spanish legislation, EU environmental standards were hardly applied. This situation was highlighted by the practice of bilateral bargaining between the Ministry of Industry and polluters over the level of emissions, or the omission of the EIA report in the authorization procedures of many public works projects. Practical application deficits were also due to the environmental authorities lack of (technical) resources to fulfil their obligations, their limited political leverage to impose compliance with environmental regulations on other departments' activities, as well as their scant capacity to engage different actors affected by their policy decisions (i.e., to create their own policy public, providing incentives for industry to negotiate and accept environmental measures).

The main cause of these practical application (and enforcement) deficits has been the prevailing notion among state decision-makers that environmental measures obstruct economic growth. According to this view, Spain's relatively lower level of economic development prevents application of European environmental standards since many basic communication infrastructure still had to be built and politicians (and citizens) were very sensi-

[1] The complexities added by the process of decentralization and the incorporation of EU environmental directives were also used to justify this decision.

tive to manufactures' threats to close plants, thereby fuelling employment. The idea that economic development is a precondition for environmental protection shaped the Spanish position in the negotiations on EU environmental measures. This is exemplified by Felipe González's speech to the Spanish Congress on his government's position on the CO^2 restrictions discussed at the meeting of the European Council in Dublin in 1990:

> We have made a map with good data but suitably colored to discuss with the Community ... we have greened our territory when actually it is quite dry ... we have greened the territory because Spain produces 1.2 tonnes of CO^2 per inhabitant a year, while France produces 1.7, Holland 2.9, Germany 3.2 ... We should certainly have an environmental policy, but it should be recognized that the growth we need in energy production is going to require, at the same time, a greater contribution by our country to the so-called green-house effect, even if environmental measures are implemented. Obviously it is very easy for the Dutch government to say: 'We must all stop at the current level of emissions. That is the limit'. That measure could be imposed on us by a qualified majority. Yet they are two or three times above our level, with a completely different level of development and capacity for energy production ... Meanwhile in that way, our possibilities for industrial growth and economic growth will be completely limited. (p.31)

The low priority attached to the environment in the government's agenda and the reactive nature of state responses is also reflected in the absence of a national plan (or apparently national strategy) for the environment establishing environmental policy goals in the medium term.

Many of Spain's sectoral laws not only involved regulatory (coercive) measures but also envisaged the public promotion of pollution abatement or the rational use of environmental resources through management plans. Sectoral plans, however, were subject to continuous delays in formulation and, when approved, they subordinated environmental considerations to the goal of growth. Those plans in which environmental concerns acquired any prominence in the discussion, as in the case of the Hydrologic Plan, or in the various Waste Management Plans (as well as many plans for the management of natural resources in protected spaces), have proved extremely difficult to elaborate or implement. In 1992, Jiménez Beltrám, the head of the state environmental department, defined the environmental policy context as one in which "rather than the integration of environmental aspects into other policies, there was a process whereby environmental projects, programmes and proposals were being conditioned by those other policies" (Información de Medio Ambiente 1992). In short, in the 1980s and early 1990s, Spanish environmental policy constituted a field in which elements of non-response and sectoral types of responses were intermingled (Pridham and Konstadakopoulos 1997; Aguilar Fernández 1997).

Around the mid-1990s, symptomatic changes took place in the response to environmental problems. At the level of discourse, this process was clearly influenced by the Fifth EU plan and Agenda 21. These symptoms can be observed in the way that the environment is now framed as an object of public intervention, and in a number of modifications to the institutional and organizational configuration of the policy process and the use of new (in the Spanish context) policy instruments.

The Socialist government established the concept of sustainability as the reference point for environmental policy in 1993. After recognizing the absence of any clear guidelines in past environmental policies, the Ministry of Public Works and the Environment announced its decision to draw up a National Environmental Plan.[2] At the institutional level, this policy change was reflected in the completion of the institutionalization process of environmental bodies and/or authorities (through the creation of a State Secretariat in 1993 and the Ministry of the Environment in 1996) as well as in the progress on new inter-ministerial and inter-territorial coordination efforts and the encouragement given to the involvement of socio-economic and environmental interest groups in decision-making. These changes have also been reflected in changes in environmental policy instruments (the implementation of coercive measures, financial incentives, participatory instruments and voluntary agreements and pacts).

At this point, two interrelated questions can be posed: What factors have promoted this change? And: To what extent does it imply a shift toward a new policy paradigm approaching the parameters of sustainability? To put it in Majone's (1989) terms: To what extent do these changes in the discourse of decision-makers, in the institutional and organizational configuration of the policy process, and in the application of policy instruments affect the core belief system of the dominant policy paradigm? Or should they be seen as marginal/peripheral modifications?

[2] Declarations of the Ministry of Public Works, Transport and the Environment, to the Environment Commission of the Congress. The plan envisaged the creation of a national environmental fund as one of the European Cohesion Funds. Although this national plan never saw the light of day, the general guidelines announced by the ministry represented a step forward in environmental policy; sustainability also entered the vocabulary used in reports, guidelines, and papers on agriculture, tourism, energy, industrial, and other sectoral policies); for example, in the 1995 white paper *An industrial policy for Spain*, sustainable development figures as the core concept of the entire industrial strategy and was discussed by social and political interest groups (Quercus, 1995: 48).

I have depicted Spanish environmental policy in the 1980s as an amalgam, combining elements of 'non-decision' and 'sectoral' responses to the environmental problem. In this policy context, the participation of environmental organizations in the policy process through conventional means was very limited. The absence of a general environmental law (or equivalent) explicitly recognizing the legitimacy of environmental organizations to defend environmental interests left their participation in administrative procedures to the discretion of decision-makers, who could deny them the right to initiate judicial reviews of administrative decisions, or ignore their petitions for information. Equally, the lack of penal regulation of the environment (as well as the judicial system's lack resources, expertise, and ecological awareness) made the strategy of litigation unattractive in terms of cost-efficiency.

By the mid-1990s, the process of the incorporation of EU environmental law (and new general dispositions on administrative procedures) paved the way for the formal legitimization of environmental NGOs' participation in administrative procedures. The approval of the EIA regulations in 1988, and of the law on free access to environmental information in 1995, illustrate the changes that have taken place in the administrative context faced by environmental NGOs. Similarly, the reform of the penal code in 1994, and state prosecutors' and judges' improved handling of environmental demands have increased the efficiency of recourse to the courts.[3] At the executive level, the process of institutionalization of environmental departments has also given NGOs greater formal access to the decision-making processes. This was reflected in the configuration of the Advisory Committee for the Environment in 1994, in which environmental NGOs were in a majority, and which was originally designed to favor their participation in environmental policy-making.

These changes have brought an increase in the number of arenas in (or channels through) which environmental NGOs have access to the environmental policy process. The overall explanation for this process would stress the dynamics of the Europeanization of environmental policy as the main driving force behind these changes. However, a more detailed analysis of the policy process behind each of the above-mentioned changes would reveal the importance of the institutional-change strategy of environmental groups trying to modify the rules of the policy game in their favor, as well as

[3] Despite the substantial improvements in the penal law, the main constraint on environmental groups' access to judicial proceedings remain unchanged: the great cost (the need to pay for expert witness and to provide financial guarantees in the case of administrative law).

contributing to changes in the content of environmental policies. For instance, the recognition of environmental groups as legitimate representatives of environmental interests in administrative procedures and courts is to a large extent attributable to jurisprudence established by the Spanish high courts, itself a result of legal action initiated by the NGOs themselves. Similarly, the incorporation of a new set of crimes against the environment into the 1994 penal code, or the transposing of a directive on free access to information in 1995, are best explained in the light of campaigns waged by environmental NGOs.

The efforts of the environmental NGOs were not only confined to the creation of new policy arenas in which they could have a voice, but also implied striving for control of these arenas. Increasing formal opportunities for access to the policy process in administrative, executive and judicial arenas were relativized by the state authorities' strong control over activation (or, more accurately, deactivation), so as to fix the content and the terms of the discussion in such a way that they could marginalize the role of environmental NGOs in the policy process. Again, the case of the EIA procedures illustrates this point well.

It has already been shown that decision-makers can void the participatory philosophy of policy instruments, marginalizing the EIA procedure in the process of developing projects; equally, the difficulties the EU faces in ensuring correct enforcement of these policy instruments have been highlighted above. In this sense, in Spain, despite the fact that EIA procedures have been mandatory since 1988, until the 1990s the authorities' prevailing strategy was not to apply (activate) this norm (by resorting to legal prerogatives such as special urgent procedures or by infringing the law by ignoring the requirements relating to public scrutiny of the process), or to control the content of the process (e.g., by disregarding the discussion of alternatives options or carrying out the EIA in the final stages of the authorization process).[4]

These enforcement deficits provoked numerous complaints to the EU (Font 1996), but despite the initiation of an infringement procedure by the Commission in 1992, the Spanish authorities were politically able to withstand EU pressures and postpone practical application of the EIA norm. In fact, the Commission was mainly concerned with the incomplete transposition of Annex II of the directive. As a result, the infringement procedure was

4 Furthermore, the environmental authorities' lack of political leverage meant that they were unable to demand compliance with this procedure. Such was the situation that, the first time the environmental authorities demanded compliance with the EIA, they made the news (Quercus 1989, 46: 43).

suspended when Spain agreed to amend its national legislation – something that it has still not yet done. Nonetheless, since 1992, a change has taken place with respect to formal compliance with EIA procedure.[5]

A crucial factor behind this change has been the ability of Spanish environmentalists to link the realization of EIA studies to access to European structural funds. Criticisms addressed to the Commission concerning Regional Development Plans (a prerequisite for applying for European funds) revolved around the alleged infraction of EU environmental norms and the lack of EIA studies of projects liable to receive EU funding. In coordination with other European environmental NGOs, Spanish organizations initiated a campaign that led to modification of the rules for the concession of European funds which incorporate, among other prerequisites, the carrying out of EIA studies.[6]

Therefore, improvements with respect to the formal compliance with EIA regulations in Spain have been due, among other possible causal factors, to a combination of pressure from the Commission and from environmental organizations, and more specifically, to the ability of environmentalists to link the lack of application of this policy instrument to the structural funds policy process. In this way, environmental NGOs assured the activation of this arena as the state could no longer ignore the requirement to carry out EIA studies.

Another key opportunity to gain access to the policy process arose in the late 1980s and early 1990s as the state response moved towards the sectoral paradigm. This came through implementing various environmental management plans that involved the construction of infrastructures (such as dumps in the case of the National Hydrologic Plan, or waste facilities in the case of the Waste Management Plans), projects in which state intervention gained visibility and the potential to politicize the authorization process (including the EIA procedures) were greater due to the multilevel government dimension of the process and public resistance to many of the projects envisaged in these plans.

The role of EU environmental policy and increasing emphasis put on implementation have doubtless played a decisive role. But as shown here, their role cannot be understood without considering the capacity of environmental NGOs to activate new arenas and to gain control over them. As a consequence of their efforts to modify the constraints of the policy process

[5] The technical resources available to state environmental authorities were increased in order to undertake this task (Escobar 1994).

[6] The lack of EIA studies led to the EU's denial of 238 million ECUs in 1994 (Econoticias, vol. 24, 1994).

in their favor, they have also been able to influence the content of environmental policy.

4. THE PARTICIPATION OF ENVIRONMENTAL NGOS AND CHANGES IN SPANISH HAZARDOUS WASTE POLICY

Over last fifteen years, industrial waste policy in Spain has evolved through the three different environmental policy stages outlined in section 1. The early 1980s was marked by the prevalence of the 'non-decision' paradigm, as industrial waste was hardly identified as a political problem. Dilution into the environment, usually by transferring pollution to water or soil (through direct discharge or unsafe disposal methods), was common practice. Dumping was scarcely regulated or controlled, and only large firms employed end-of-pipe solutions and obtained authorization to incinerate or directly dump their waste into the sea. In 1985, the environmental authorities recognized that 87% of hazardous waste was 'eliminated' without any control (DGMA 1985). The scant enforcement of coercive measures (a loose regime of sanctions and discharge taxes) had little impact on producers' behavior or technological innovation. Moreover, the laxity of Spanish regulations encouraged imports of waste from Northern European countries, where stricter regulations were enforced.

Although these problems continued unabated (especially in terms of the weak control of waste producers and limited enforcement of sanctions), the context changed significantly in the late 1980s, when industrial waste was defined as a policy problem and incorporated into the agenda. The state response initially took the form of implementation of pollution abatement plans in industrial areas, where the visibility of environmental degradation first highlighted the failure of dilution practices.

At this point, the intense work of local environmental groups in industrialized areas proved critical. They launched press campaigns and, in some cases, legal actions against illegal dumping by manufacturers, and at the same time denounced the connivance of environmental authorities. Although environmentalists were able to establish alliances with affected groups (fishermen, residents, etc.), these campaigns were restricted to the local level. Nonetheless, this pressure often favored negotiations on corrective measures[7] between the administration and polluters.

The creation of the Spanish branch of Greenpeace in 1984 significantly increased the media impact of these local incidents of environmental degradation, giving state-wide resonance to the local actions (within the context of internationally-designed campaigns). The role of Greenpeace would become increasingly important, as the organization provided the expertise on which the environmental movements based their alternative solutions to the problem of industrial waste.

These local corrective measures involved the generalization of end-of-pipe solutions which led the state environmental authorities to adopt managerial measures to rationalize, coordinate and safely dispose of the increasing quantities of industrial waste. The elaboration of the national plan on industrial waste in 1989 marked the adoption at state level of a new sectoral approach to the problem.[8] Through the approval of this plan, the Spanish state fulfilled, albeit after a two year delay, the EU environmental guidelines in this field.[9]

7 The *Plan Corrector de Vertidos de la Bahía de Algeciras* in 1989 illustrated this tendency: the continuous action of the local ecologist groups AGADEN and VERDEMAR forced the authorities to draw up a series of corrective measures, to initiate the first legal proceedings against polluters, and to create a monitoring commission with environmentalist participation. In turn, the petrochemical industry in the area began to show some environmental awareness, commissioning studies on the pollution of the bay.

8 In January 1990 the Catalan Government approved a regional plan. However, this was never implemented because of the strength of the opposition in those localities which had been intended to house waste facilities. Resistance was especially strong in the rural area of Conca del Barberá in Tarragona. Although there are regional variations in the development of industrial waste policy in Spain, this chapter examines the overall evolution of state-wide policy.

9 The obligation to formulate national plans was envisaged in the national law on industrial waste (*Ley 20/1986, de 14 de mayo, Básica de Residuos Tóxicos y Peligrosos*) that transposed the content of Directive 78/319 on this topic into Spanish law. However, effective transposition was delayed until mid-1988, when the regulations for its application were approved, specifying among other things the sanction regime (*RD 833/1988, de 20 de julio*). Together with the pressure from the environmental movement and the need to implement EU environmental legislation, the agreement of the Convention of London prohibiting incineration at sea after 1995 (and, reducing it significantly after 1991) made elaboration of a plan especially urgent. The Convention of London agreement was, in part at least, the result of the international campaign launched by Greenpeace in 1985, which local environmental groups in Santander, as ARCA, joined in 1987 when the Spanish government authorized the incineration at sea of waste from the Spanish petro-chemical industry. Similarly, the campaign of environmental groups in Huelva (CEH) against discharges of Titanium Dioxide in the Gulf of Cádiz also raised the visibility of the problem, highlighting the need to find solutions within Spanish territory. The need for a plan be-

Since then, two different national plans have been formulated, both by Socialist governments. The first national plan initially covered the period 1989-1993. However, it was hardly implemented and only in 1995 was it replaced by the current plan, which will be in force until 2000.[10]

An analysis of some of the essential components of the two plans revels the re-orientation of the state's response to the problem of industrial waste. This has consisted of a shift (at least at the level of policy formulation) from a sectoral approach based on the provision of infrastructure for ever greater production of waste towards a more integrated approach based of the principle of prevention (reduction of the production of waste) through the promotion of clean industrial processes (i.e., a movement towards the third stage in the model outlined in section 1).

Table 2 offers a comparison of the two plans, as well as of the (aborted) draft of the first national plan review in 1993, in terms of the priority given to prevention rather than provision of treatment infrastructure. The table also refers to the role played by environmental actors in the elaboration of the two plans as well as their role in process of policy re-orientation.

---

come even more urgent in the 1990s as the principle of national self-sufficiency in waste treatment and the criteria of proximity of waste infrastructure (to waste production areas) were progressively incorporated into the European policy as a consequence of the EU's adherence to international agreements (Convention of Basilea in 1989-1994).

10 *Plan Nacional de Residuos Industriales* and *Plan Nacional de Residuos Peligrosos* approved by the Spanish Council of Ministers in March 1989 and January 1995, respectively.

Table2. Evolution of policy approach to the problem of hazardous wastes in different National Plans

	First National Plan (1989-93)	1992-3 National Plan Review (1994-98) (aborted draft)	Second National Plan (1995-2000)
Estimated waste production (millions of tons per year)	1.7	2.0	3.6
Targeted waste reduction	15% (generic guidelines)	15%-25%	40% (objective)
Estimated budget for waste infrastructure (% of the total)	36%	38.5%	16.5%
Support for incineration	Critical (envisaged four facilities to burn 100,000 tons per year)	Critical (four integral plans including, one state initiative, to burn 150,000 tons per year)	Discouraged in terms of central subsidies (security landfills are promoted instead)
Participation of environmental groups in the decision-making process of the different documents and level of access to information	The process was completely closed. Ad hoc information: once the plan had been elaborated, the Director General of the Environment met with environmentalists to inform them about its contents and ask for their support for the construction of waste facilities *no access to information	Opaque process of formulation; the draft was circulated in September 1993 to diverse groups along with a request for their comments *no access to information	Drafts took environmentalists' position into account; diverse monographic meetings; participation in the final draft through the CAMA. Public participation is mentioned in the plan: to be guaranteed through the CAMA *commitment to improve public access to information

Sources: compiled by the author, percentages taken from the various official documents cited in the bibliography

The radical opposition of the environmental movement to the first plan was grounded on their lack of participation in the decision-making process[11], and the plan's failure to address the need to transform industrial production processes to solve the problem of waste.[12] At the level of general principles (without making it a clear priority), the plan highlights the need to reduce the volume of industrial waste and established a 15% waste reduction goal for the period. The largest item of envisaged investments was for the promotion of clean technologies. In practice, however, the priority was to solve the lack of treatment infrastructure. The prevention programs considered in the plan (i.e., those affecting the production process rather than the management of waste) were sidelined when it came to implementation, as public investment prioritized the promotion of management infrastructure and service projects (Santamaría Arinas1996: 371). Hence, the plan emphasized waste treatment through state promotion of facilities to ensure the development of a minimal capacity to handle the waste generated in Spain, which was seen as a pre-requisite to curb the previous lack of waste control. The plan envisaged the construction, on the initiative of the state, of some ten sites (including three or four incinerators, physical-and-chemical treatment plants and landfills) along with other projects sponsored by the private sector or regional administrations.

However, this first experience of state management failed to meet its initial objectives.[13] This was due to the lack of coordination with regional governments, funding problems[14], the lack of collaboration from waste producers, and massive public opposition to the local siting of waste disposal facilities. In the case of industrial incinerators alone, at least 25 different projects – at different phases of planning – were abandoned between 1990

[11] The decision-making process on the first draft was closed to environmental NGOs and dominated by the recently created environmental authority and the waste management infrastructure and services lobby (a new industrial sector in Spain with a pronounced multinational character). It was probably the environmental authorities' lack of capacity (or opportunities) to include waste producers and the representatives of the autonomous communities in this process that reduced their capacity to enforce the plan.

[12] Environmental organizations also questioned the reliability of the waste inventory used by the administration, demanding that it be made public, as well as the creation of an independent National Chemical Security Council, with powers of sanction and normative powers to control the production and management of industrial waste (CODA 1995).

[13] Except in the case of physical-chemical treatment, the planned facilities were not actually built.

[14] One of the environmental authorities responsible for the 1989 plan emphasized funding problems as the main cause of the failure of the plan, due to the changing investment priorities of the Ministry of Industry in the context of the economic recession in the early 1990s.

and 1995 as a result of local opposition.[15] While business organizations and the public authorities saw the construction of infrastructure as a precondition for compliance with legislation and as a prerequisite for the adoption of preventive programs, environmentalists demanded a moratorium[16] on the construction of infrastructure, the number and type of which they were only willing to discuss after the state designed and implemented alternative industrial policies (to progressively replace toxic substances and eliminate processes producing toxic wastes). Only then, would environmentalists be prepared to accept some infrastructure, but not incinerators, as a transitory solution. Environmental NGOs also denounced their lack of access to information and decision-making processes, in contrast to the privileged access of the representatives of the industrial sector. This state-wide campaign was coordinated by CODA (*Coordinadora de Organizaciones de Defensa Ambiental*) and Greenpeace, and in the case of the incinerators, by the State Network of Citizens Against Incineration, in which the environmental movement was able to define its common demands and strategies.[17]

Despite this lack of access to information and the decision-making process, there was an increasing tendency towards formal compliance with EIA procedures (as argued above, this in itself was partly due to pressure from environmental groups). Independent of the quality of the assessment, environmental groups found opportunities to politicize decision-making processes and to mobilize public opinion (the ongoing process of political decentralization, along with the gatekeeper role played by local authorities in the administrative process, also increased the opportunities for support within the administration). The rapid upsurge in local disputes about the same issue created an opportunity for environmental NGOs to transcend the

15 Information gathered from various sources, specially from *El País*, Greenpeace 1994 and CODA 1995.

16 To some extent the movement against incineration adopted strategies similar to those deployed by the antinuclear movement in the late 1970s. The promotion of local and regional government declarations against industrial waste facilities, above all, incinerators, recalled the 'nuclear free zone' declarations that had proliferated two decades earlier. Similarities can also be seen in the industrial interest groups this movement confronted: promoters of incineration resemble those which had defended nuclear energy (the energy sector), and their strong relation with the industry and some environmental authorities seems to have reproduced the closed pro-nuclear policy community composed of the energy administration and corporate interests. For the economic interests behind incineration, see Greenpeace (1995); for the continuity of the energy policy community after the democratic transition, see Lancaster (1989).

17 This state-wide network included, besides environmental NGOs, trade unions, neighborhood organizations, and many municipalities. See 'Guidelines for an alternative waste policy' (*Red Estatal Ciudadanos contra la Incineración* 1992).

debate on the location of specific sites in order to challenge the definition of the policy itself, and to push forward alternative policy conceptions based on the precautionary principle and the reduction of waste through the promotion of clean technologies. Backed by high levels of mobilization[18], and the pressure of the costs of delay caused by this opposition, environmentalists were able to activate an alternative policy coalition, not only in the courts and arenas related to the administrative authorization process (in which, through their strategy of institutional change mentioned above, they were gaining power), but also in ad hoc town hall plenary sessions, special commissions of regional parliaments, or conciliatory roundtables, where open discussion of alternative solutions to the industrial waste problem publicly exposed this issues to local and regional authorities as well as to citizens.

Figure 1 traces the evolution of conflicts over industrial waste in Spain as reported in *El País* from 1988 to 1997.[19] The peak in 1992 coincided with the preparations for the review of the plan for the period 1994-1998, which maintained, and even accentuated, the existing emphasis on waste management.[20]

Although the review established waste reduction as the main objective, it also emphasized the urgent need to provide basic treatment capacity, "postponing other main objectives to a favorable economic conjuncture that would allow the required investment for any process of industrial re-conversion". The main goal of the plan was the construction of four integral waste facilities (which would include incinerators) across Spain. The incineration (of 120,000 tons per year) was considered an "urgent environmental necessity" on which the survival of Spanish industry depended. This discourse was widely disseminated in specialized fora organized by the waste industry (the new *Club Español de los Residuos*, the Spanish counterpart of the European Waste Club, played a crucial role in this respect), as part of an information campaign designed to establish a political consensus among the different administrations and political parties. This included an aggressive campaign against environmental groups, as well as consideration of the possibility of resorting to the use of the 'general interests' legal prerogatives to

18 In fact, the turn of the decade saw a peak in the overall level of mobilization on environmental issues (Jiménez 1999). The 1990s also saw the organizational consolidation of the environmental movement (Jiménez 1999).

19 Data on environmental protest has been obtained from media analysis of reports in *El País* as part of the ongoing EU-funded comparative research 'The Transformation of Environmental Activism' (EC Contract number ENV4-CT97-0514).

20 After 1992, conflicts around waste infrastructures have focused on landfills (Somozas, Subiza, Nerva, etc.) and incinerators of biosanitary waste (Agost).

overcome the resistance of local administrations. This publicity campaign gave environmental NGOs an opportunity to formulate their comments and suggestions on the draft, although the elaboration process proved quite closed to environmentalist participation (CC.OO 1993). The review was never completed, as in 1994 the new Secretary of the Sate definitively decided to draw up a new plan for the period 1995-2000.

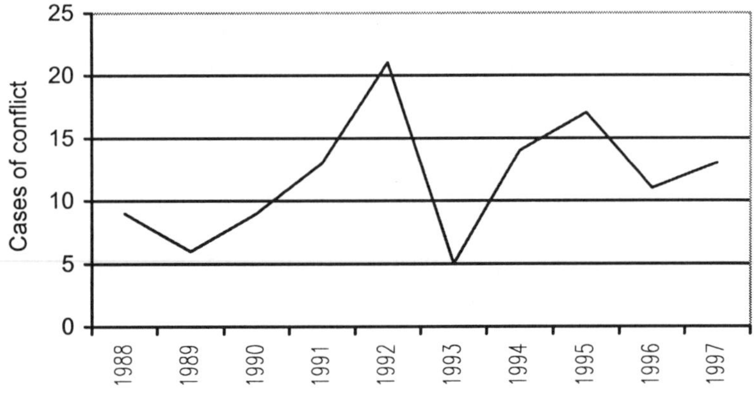

Figure 1. Evolution of conflicts over industrial wastes in Spain 1988 - 1997

The new plan in 1995 involved a re-orientation of the state's response along the lines of the EU environmental policy on sustainability and, more specifically, directive 91/156 on industrial waste. The previous emphasis on waste management was replaced by prioritizing preventive approaches aimed at the reorientation of producers' behavior to adapt their activities to the need for waste minimization.

The elaboration of this second national plan not only came closer to meeting environmentalists' demands[21], but also implied a shift in the state environmental authorities' strategy for dealing with environmentalists. The confrontational approach seen in the past gave way to one of negotiation, in which NGO representatives were included in the decision-making process

[21] In general, this plan was assessed positively by environmental groups, but not considered completely satisfactory. Their criticism centered on the lack of an integral perspective – which would include other sources of pollution –, the lack of regulatory, economic and technical mechanisms to achieve a reduction in waste and the promotion of clean technologies, the absence of progress on public access to environmental information (CODA 1995).

from the very beginning through their participation in the elaboration of the new document in monographic meetings organized by the ministry.[22] The plan was also discussed in the working group on waste of the Advisory Committee for the Environment (CAMA), created in 1994 explicitly to allow environmental NGOs (and other social interests groups) to participate in this and similar decision-making processes. The plan established the goal of a 40% reduction of waste generation by the year 2000 and the eventual abandonment of incineration of industrial waste (only one incinerator, already approved by the Catalan government, was considered), despite the opposition of the industry.[23] The content of this plan was influenced by the EU's strategy and legislation on industrial waste (directives that at that time had not been transposed into Spanish law). This influence also derived from the obligation of the member-states to formulate and enforce national and regional waste plans, an obligation that led the EU to initiate the procedure of law infringement against Spain. The abandonment of eight incinerator projects in 1992-3[24] and the total opposition to the new draft plan from environmentalists (and trade unions) was a crucial element in the policy (learning) process that led to the adoption of a new approach in 1995.

5. CONCLUSIONS

In the previous section I have highlighted some aspects of the relationship between the participation of environmental NGOs and the development of Spanish environmental policy. First, the nature of these groups' participation (the opportunities for access and influence in the policy process) has changed according to the nature of the state's response to environmental problems. In this way, the shift in Spanish environmental policy towards the parameters of sustainability, as the case of industrial waste policy seems to indicate, has been accompanied by the increasing participation of environmental NGOs in decision-making processes.

22 The new Socialist legislature, which brought about the renewal of the environmental authorities and the administrative upgrading of the environmental department (with the designation of Cristina Narbona as State Secretary for the Environment and Housing), was also a decisive factor in the development of this new attitude towards environmentalists.

23 Representatives of the industrial sector criticized the decision to abandon incineration, the potential limitation of toxic substances in industrial processes, and the limited investment in waste infrastructure (CEOE 1994).

24 The most significant developments were the failure of the projects in Medina Sidonia and Almadén, but other projects were also rejected in Toledo, Soria, Madrid, Guadalajara, Murcia, Alicante, etc.

Second, policy development has also been related to the capacity of Spanish environmental NGOs to pursue strategies for institutional change, to create and gain control of alternative policy arenas from which they have advanced their demands with greater chances of success. Given the constraints that the administrative, judicial, and political systems have placed on environmental groups, these strategies can be interpreted in terms of struggles for the democratization of the political system. That is, as part of the struggle for a democratic opening that can also be seen as an institutional precondition for the successful implementation of advanced environmental policies.

The contribution of environmental NGOs to policy development has been highlighted by almost every study analysing environmental policy development in Northern European countries. So much so, that their involvement in decision-making has been incorporated as a condition for successful environmental policy and has thus been considered, in both international and national environmental policy programs, as a policy goal in itself on the road to achieving sustainability. However, because of the backwardness of Spanish environmental policy, the role of the EU as a major force for policy change has tended to be the center of attention in analyses of the Spanish case. In this respect, and in accordance with policy process theory, I have shown, thirdly, how this pressure has been reinforced by the pressure of environmentalists in their interactions with state decision-makers. In this sense, I have argued that in most cases it is necessary to take the activities of the Spanish environmental movement into account in order to understand the Europeanization, and, more generally, the evolution of the environmental policy in Spain.

The evolution of waste policy since 1995 suggests that the prospects for the fulfillment of the goals envisaged in the 1995 National Plan are at best uncertain.[25] This plan incorporated the subsidiarity principle, in accordance with the federal-like structure of the Spanish state. Indeed, the actual programmatic definition and implementation of waste policy depend on regional plans elaborated by the governments of the respective autonomous communities. In 1997, at least half of these plans had still to be approved. Moreover, the new waste law (updating Spanish norms to European regulations on waste, Directive 91/156/CEE) was not passed until 1998. Even then the new law was heavily criticized by environmental NGOs for its orientation towards waste management rather than waste prevention, the consideration of incineration as a valid system of treatment, the limited responsi-

[25] Another twelve countries also became the object of European Commission infringement procedures for their failure to adopt industrial waste standards.

bility of waste producers, the lenient sanctions regime for infringements, and the scant definition of economic and financial measures, while social participation and access to information are only vaguely referred to (Aedenat 1997). This situation seems to reflect the conservative government's and new Environment Ministry's lack of interest in implementing the (Socialist) plan, as well as their unwillingness to put forward any environmental policy that runs against industrial interests.

As far as changes in the prevailing policy understanding of environmentalists' involvement in the policy process are concerned, it seems that the institutional-change strategy pursued by environmental NGOs has had limited success, except in the particular context which gave rise to the 1995 industrial waste plan. Although the reform of the penal code and the (incomplete) transposition of the directive on free access to information (both of which occurred under the previous Socialist government) have expanded the potential opportunities for environmentalist pressure, the control that environmental NGOs gained over other arenas during the discussions on industrial waste policy has subsequently diminished.

The advisory committee mentioned above did not lead to the institutionalization of environmental NGOs' participation in decision-making, as the environmental authorities were suggesting it would in 1994. When, at the time of its creation, the environmental authorities accepted that the committee would have a majority of environmentalists as a condition for their participation, this implied an initial period of intense involvement in policy-making. However, many environmental NGOs soon saw their expectations frustrated, since the committee did not develop into a platform for participation from which they could influence the decision-making process, but rather as a means of legitimating state environmental policy. This led the NGOs to gradually leave the committee before the last legislative elections. The new Minister of the Environment in the conservative government elected in 1996, soon showed her scant interest in this committee, encouraging the environmental movement organizations to withdraw again from the committee after just two preparatory meetings has been held in one year.[26]

[26] The environmental groups' criticism of the CAMA during the first Socialist government focused on the lack of economic and technical support for its activities (provision of resources for its operation), and the scant leverage or echo of its resolutions in the government, (ultimately a consequence of the limited political leverage of the Secretary of the State for the Environment). However, the CAMA functioned as a channel of communication, in which environmental groups obtained information about new legislative actions and discussed problematic issues in special sessions open to different interlocutors (even those not directly represented), resembling small parliaments for interest groups

It seems that the activation and control of policy arenas still depend on the capacity of environmental NGOs to generate social conflict, and that the state's prevailing image of environmental participation is linked to strategies of legitimization rather than to a new, shared view of NGO involvement in decision-making as a necessary input into the process of environmental planning.

REFERENCES

Aedenat, (1997) *Comentarios Generales sobre el Anteproyecto de Ley de Residuos*, http://nodo50.ix.apc.org/aedenat/residuos/rsu/htm.

Aguilar Fernández, S. (1997) *El Reto del Medio Ambiente: Conflictos e Intereses en la política medioambiental Europea*, Alianza Editorial, Madrid.

Baker, S., Kousis, M., Richardson, D., and Young, S. (1997) Introduction: The theory and practice of sustainable development, in S. Baker, M. Kousis, D. Richardson and S. Young, *The Politics of Sustainable Development: Theory, Policy and Practice within the European Union*, Routledge, London, pp.1-42.

CC.OO (Departamento Confederal de Ecología y Medio Ambiente) (1993) *La problemática de los Residuos Tóxicos y Peligrosos*, Confederación Sindical de CC.OO, Madrid.

CEOE (Comisión de Medio Ambiente) (1994) *Observaciones a las Líneas Básicas del Plan Nacional de Residuos Peligrosos (1995-2000)*, Unpublished Document.

CODA (1995) *El Plan Nacional de Residuos Peligrosos*, Unpublished Document.

DGMA (1985) *Medio Ambiente en España*, MOPT, Madrid.

Escobar, G. (1994) Evaluación de Impacto Ambiental en España: Resultados Prácticos, *CyTET* **11**, 585-595.

Font, N. (1996) *La europeización de la política ambiental en España. Un estudio de implementación de la directiva de evaluación de impacto ambiental*, Ph.D. Thesis, Universidad Autónoma de Barcelona.

Greenpeace (1994) Incineración de Basuras: Razonas para un No, Report of Greenpeace.

Greenpeace (1995) La "guerra sucia" de la Incineración, Report of Greenpeace.

Hall, P.A. (1993) Policy paradigms, social learning and the state. The case of economic policy-making in Britain, *Comparative Politics* **25**, 275-296.

Inforamción de Medio Ambiente (1992) *Oportunidad y necesidad de una estrategía nacional de medio ambiente*, MOPTMA-MIMAM, Madrid.

Jänicke, M. (1996) The political system's capacity for environmental policy, in M. Jänicke and H. Weidner, *National Environmental Policies. A Comparative Study of Capacity-Building*, Springer, Berlin, pp.1-24.

Jiménez, M. (1999) Consolidation through institutionalisation? Dilemmas of the Spanish environmental movement in the 1990s, *Environmental Politics* **8**, 149-171.

---

presided over by the Ministry of Public Works, Transport and the Environment. In that context, the CAMA brought environmental issues and environmental groups unprecedented media attention at the national level. In fact, quite probably in order to avoid this level of public prominence, the current ministry has tried to void the committee of any political character, reducing its functions to a minimum.

La Spina, A., and Sciortino, G. (1993) Common agenda, Southern rules: European integration and environmental change in the Mediterranean states, in J.D. Liefferink, P.D. Lowe and A.P.J. Moll, *European Integration and Environmental Policy*, Belhaven Press, London, pp.216-234.

Lancaster, T.D. (1989) *Policy Stability and Democratic Change. Energy in Spain's Transition*, Pennsylvania State University Press, University Park/London.

Majone, G. (1989) *Evidence, Argument and Persuasion in Policy Process*, Yale University Press, New Haven, NJ.

Paehlke, R. (1990) Democracy and environmentalism: Opening a door to the administrative state, in R. Paehlke and T. Torgerson, *Managing Leviathan*, Broadview Press, Lewinston, NY, pp.35-55.

Pierson, P. (1993) When effect become cause. Policy feedback and political change, *World Politics* **45**, 595-628.

Pridham, G. (1994) National environmental policy-making in the European framework: Spain, Greece and Italy in comparison, in S. Baker, K. Milton and S. Yearly, *Protecting the Periphery: Environmental Policy in Peripheral Regions of the European Union*, Frank Cass, London, pp.80-101.

Pridham, G., and Konstadakopulos, D. (1997) Sustainable development in Southern Europe? Interactions between European, national and sub-national levels, in S. Baker, M. Kousis, D. Richardson and S. Young, *The Politics of Sustainable Development: Theory, Policy and Practice within the European Union*, Routledge, London, pp.127-151.

Quercus (1989) *El MOPU exige por primera vez formular una declaración de Impacto Ambiental.*

Quercus (1995) *Estrategia para reducir el impacto de la industria.*

Red Estatal de Ciudadanos contra la Incineración (1992) *Líneas básicas de una política alternativa de Residuos*, Unpublished manuscript.

Santamaría Arinas, R.J. (1996) *Administración Pública y Prevención Ambiental: Régimen Jurídico de la Producción de Residuos Peligrosos*, IVAP, Bilbao.

Weale, A. (1992) *The New Politics of Pollution*, Manchester University Press, Manchester.

Chapter 10

Is Spanish Environmental Policy Becoming More Participatory?

Institution Building Versus Current Experiences of Participation

Susana Aguilar Fernández

1. INTRODUCTION

Over the last few years, the Spanish state administration for the environment (as well as many of the regional ones) has been gradually allowing social groups to partake in the making and management of certain conservation policies. This move towards a more participatory policy style is exemplified by the creation of new advisory institutions, the organization of regular seminars, and the setting up of specific workshops for environmentalists and other interest groups in order to discuss controversial issues. Furthermore, the growth of sustainable development-related local action since the mid-70s, the need for outside expertise in the administration and a new international mood of democratization have also played a role in the gradual establishment of this participatory approach. It is still too soon to determine whether this approach implies a real change in the attitude of public authorities or whether it is simply a way of paying lip-service to the 'ecological cause' while trying to come to terms with European developments. Undoubtedly, some figures in the past socialist government (1982-96) attempted to bring green issues to the fore and entered into heated debates over different topics with environmentally reluctant business associations. It is also beyond question that the new conservative government has upgraded the administration of the environment with the creation of a Minister for the Environment in 1996. Yet alongside these promising developments, the current policy appears to suffer from the same

K. Eder and M. Kousis (eds.), Environmental Politics in Southern Europe, 255–275.

old problems as before: the second-rate importance of environmental topics within the cabinet, a problematic intergovernmental relationship between the state and the regions, and the lack of relevance of social groups in the policy-making process. On the one hand, environmental protection in Spain clearly shows that old habits die hard but, on the other hand, it seems to be facilitating some room for novel political experimentation.

2. A TRADITION OF PUBLIC INTERVENTION IN SPANISH POLITICS

In the 1980s some authors described the Spanish political style as one in which, to the detriment of society as a whole, top bureaucrats and civil servants had concentrated a disproportionately large amount of power in their hands (Linz 1981; Martín Rebollo 1984). This situation, which can be traced back to the turn of the century, was reinforced by the Franco-regime, especially after a generation of Opus Dei technocrats came to power in the 1950s. One of the few groups which could have counteracted this situation, the business community, was either not interested in making its public presence as an organized group more visible (because tycoon-like entrepreneurs already enjoyed privileged personal contacts with individuals from the administration and government), or lacked the capacity and resources to set up strong national-based interest groups.

After the economic boom of the '60s there gradually emerged several new social movements which sought to modernize Spanish society. Modernization was understood in a two-fold sense: More equitable distribution of the wealth which was being generated so that entrenched social inequalities could be diminished, and greater state investment in certain public services to counter the effects of a regime which, having neglected its role in the promotion of social welfare, lagged well behind most European countries in this respect. The last demand was especially evident in relation to one of the more active of these social movements: the neighbor's movement. Facing a situation of unplanned material growth, this movement, which in the mid-70s managed to establish approximately 150 associations in Barcelona and 120 in Madrid, took to the streets to obtain above all a decent housing policy, a good network of public transportation and schools, and an improvement in public health and the urban environment (Alonso 1991).

During the political transition from dictatorship to democracy (1976-1982) a number of studies were carried out to analyse whether the new regime was being accompanied by the emergence, or purposive establishment, of a new pattern of relationship between the state and society. Although some events – such as the signing of the Moncloa Pacts in 1978 and the conclusion of a number of neo-corporatist-like agreements between the unions (basically UGT and CC.OO) and the business organization (CEOE) in the '80s – might have hinted at a new trend of co-operation between public authorities and interest groups, most scholars concluded that they simply entailed a form of remedial action which did not break with the old tradition of state intervention (Foweraker 1985). Consequently, the impact of the interest groups was uneven and lacked an institutional basis in this period and, aside from the existence of rare deliberations between private and public actors, a regular consultation pattern did not exist (Gunther 1996).[1]

The peculiar features of the Spanish political transition have been frequently referred to when explaining the low degree of social involvement in the process as well as the weak role of social movements in promoting the final goal of democratization. During the last stage of Francoism political parties and social movements worked together, in a symbiotic way which sometimes blurred the differences between them, in the fight against the regime. However, once the first general elections took place in 1977, leftist parties (basically the Communist Party) began to reveal some reluctance in using their social leverage to the full in order to accelerate political changes for fear of bringing about a counter-reaction on the part of the army and the immobile forces from the old regime. This fear led them to adopt a cautious strategy in 'street politics'.[2] Social movements strongly linked to these parties were closely controlled and discouraged from openly and frequently demonstrating to press home their demands. Thus, parties went from being promoters of social agitation to 'respectable' organizations which, being inclined to political consensus with other interest groups (at the top), tried to justify the new 'demobilization mood' to their rank and file (at the bottom).

1 The budgetary policy-making process, for instance, was still in the hands of the same old public agencies and bureaucrats while the input of the business sector was negligible because its lobbying centered merely on individual ministers (Gunther 1991).

2 That fear is somehow reflected in the fact that the Spanish constitution – as well as the regional statutes and many municipal ordinances – allows for the use of referenda and popular initiative techniques, but does not facilitate their enforcement (Subirats 1997).

3. EUROPEAN MEMBERSHIP AND PRESSURES FOR SOCIAL PARTICIPATION IN THE PUBLIC SPHERE

After Spain entered the European Union in 1986, a number of studies revealed that the participation and relevance of social groups in most policy areas remained elusive.[3] Yet in environmental policy, a field in which the regulatory activity of the EU was comparatively intense, pressure to change the non-participatory system of intermediation was more sustained. By and large, this could be accounted for by the fact that deficient implementation was more visible in light of obligations to comply with a large number of highly complicated technical directives. More recently, the salience attached to the concept of sustainable development has brought to the fore the fact that the compatibility between economic growth (or the continuation of traditional productive activities) and environmental protection can only succeed if locals are incorporated into the decision-making process. Furthermore, recent experiences (in Spain, the United States and elsewhere) have shown that whenever locals have been excluded from participatory procedures, sustainable development policies have resulted either in failure or flawed solutions. Last but not the least, basic EU principles of subsidiarity, shared responsibility and partnership – all of them linked to sustainable development – emphasize letting locals partake in environment-related decisions.

All this may explain the emergence of some political experimentation and institution building which principally aimed at getting public and private actors to collaborate more closely in order to improve the quality of the policy-making process. These new developments have, for instance, taken the shape of environmental pacts between certain large firms and regional governments in highly polluted areas, and voluntary agreements between industrial sectors and the state administration (Aguilar Fernández 1997).

Although it might still be too soon to conclude that this new 'participatory mood' has replaced the old 'statist' approach, a number of factors seem to indicate that these novel practices will not easily wither away – though of course, whether they will end up functioning effectively or simply paying lip service to popular mottoes such as 'the call for a democratic and open administration' is an altogether different matter.

The factors which favor a gradual process of embedding participatory practices in Spanish environmental policy are:

[3] This, which was true of, amongst others, environmental protection and gender politics (Valiente 1997), led some authors to use the term 'statism' or 'dirigiste' to define the Spanish political style (Aguilar Fernández 1993).

- the growing recognition by public authorities that non-collaborative policies are increasingly doomed to failure[4]. Clearly, the growth from the mid-'70s up to 1990 of sustainable development-related grassroots movements, which mounted more than 3.000 protest events during the period 1976-1994[5] (Kousis, Aguilar Fernández and Fidelis 1996), has to do with this recognition. This local action, which in many cases entailed fierce opposition to certain projects sponsored by public authorities, is not only economically costly (in terms of delays, missed deadlines, and the loss of Community financial support), but has frequently ended up in political crises as well: highly visible failures to implement decisions and the consequent need to abandon putting certain plans into practice have eroded the legitimacy and authority of political parties in the public's eye.
- the need for 'external and independent expertise' by over-burdened (and sometimes insufficiently informed) public officials. This factor, together with the new political opportunities which environmental groups have been given by entry into the EU, is undermining the traditional reluctance shown by the administration towards social participation in the public sphere.
- a new international political mood (which is also reflected in the emergence of new terms in the scholarly world, such as 'transformational politics', 'risk societies', 'empowerment'...) that stresses the need for the state to withdraw from certain policy areas – mainly related to the provision of welfare and judicial arbitration – so that local communities can play a larger role in restructuring the social fabric, thus promoting novel models of participatory democracy.

Once these factors have been analysed, a brief discussion of the actual functioning of two new institutions in Spanish environmental policy will follow: the Advisory Council for the Environment (CAMA) and the National Council of the Climate. Likewise, some attention will be paid to the participatory opportunities opened up by impact studies (EIAs) and the recently transposed directive on access to environmental information.

[4] "In the last several years, government agencies [in the United States] have recognized that they can not hope to secure public approval for many activities, such as the siting of waste facilities, without providing sufficient opportunities for citizens to participate in critical judgments about acceptable levels of risk ... ; decision making without public involvement invites ethical lapses and policy failure." (Kraft 1996: 205)

[5] These events were located in approximately 10,874 articles from a national newspaper, *El País*, although 94 case articles from the magazine *Quercus* were added. This and further information related to these environmental protest can be found in Kousis, Aguilar Fernández and Fidélis (1996).

This discussion will provide arguments for the debate about whether or not what is being said in public matches what is actually being done. More precisely, it will allow one to assess the real importance of the developments analysed in terms of their effecting (or simply easing) real political change.

4. THE GROWTH OF GRASSROOTS MOVEMENTS IN DEMOCRATIC SPAIN

Local environmental protests increased in Spain from 1974 to 1990. This 'disruptive situation', not controlled by national organizations, can be connected with an initial democratic and spontaneous upsurge which peaked by the end of the 1980s and which coincided with a period of extensive public investment in infrastructures. Yet this upsurge slowed in the final period from 1991 to 1994, probably as a result of the experience (and achievements) of democratic local politics and a 'saturation effect'.

Table 1. Local environmental protest in Spain (1974-1990)

	Frequency	Percent
1976-80	191	26,2
1981-85	197	21.1
1986-90	281	38,6
1991-94	59	8.1
TOTAL	728	100

Some of the most important features of this dynamic of protest are the following: About one fifth of the actions have taken place in Andalusia and another fifth in Madrid, followed by Valencia (12,3%), Castille-Leon (9,5%), Aragon (6,8%) and Galicia (5,3%).[6] Approximately 72,7% of the cases are based on one community alone whereas 27,3% involve more than one, and only 4,5% affect more than ten. As to the duration of the protests, one third finish in less than a week, another third last between one week and one year, while the rest continue for longer periods of time. The number of participants

[6] Andalusia is a geographical area of high environmental mobilization because, being the second largest region with the most extensive unspoiled surface in Spain, it has developed an impressive policy for the conservation of spaces. Its important primary economic base also explains why a lot of protests revolve around agricultural issues and protected areas. Another important source of action is waste facilities siting since Andalusia's large area and low population density make the region suitable for the construction of these infrastructures. The absence of Catalonia might be explained by the fact that the *El País* supplement for that region was not covered between 1982 and 1985.

was in most cases between 26 and 500. Although 23,8% of the mobilizers have a rural character – their actions being linked to protected spaces or traditional activities such as farming and cattle breeding –, the majority of them relate to urban environments. This is due to the fact that many contentious events revolve around issues concerning industrial pollution. Thus, the principal sources of mobilization are the construction of roads, highways, and the like (5,4%), agricultural infrastructures (4%), weapons and military facilities (3,8%) and nuclear plants (3,6%).[7]

Considering the grassroots nature of the protest, it is not surprising that the main participants are local area residents (58%), followed by local environmental groups (45,6%), unions (24,9%), local political representatives (22,7%) and local governments (17,1%). Usually, the actions take the form of demands and claims (91,6%) as well as complaints to the authorities (69,9%), demonstrations (32,1%), petitions (20,3%), press conferences (19,4%), and the use of judicial procedures (15%). "Only when these 'traditional and accepted' ways of protest fail to have any effect, do people resort to more unconventional ways of expressing their discontent", such as strikes and shop closures (13,6%), blocking an activity or the source of protest (13,6%), violence (13,5%), and road blockades and sit-ins (11,7%) (Kousis, Aguilar Fernández and Fidélis 1996: 4-10). In this respect, violent episodes have been higher in Spain (13.5%) than in Greece (7.3%) and Portugal (1.8%). Mobilizers tend first to address their local governments (12,5%), resorting then to the state government, ministries or governing party (9,6%), regional governments (9,1%), regional courts (7,1%), Community institutions (embracing the Commission, the European Court and the Parliament) (6,4%), central state representatives at the local level (5,5%), and large environmental groups (5,1%). Locals are not generally supported in their demands by non-locals, and only political party representatives (13%) and environmental groups (12,9%) show some interest in promoting their cause. The four basic groups challenged by the mobilizers as being responsible for the eco-disturbance are the state (42,3%), the local government (41,6%), private producers and employers (34,1%), and the regional government (31,3%).

Although the increasing relevance of sustainable-related local protests in Spain is, as previously mentioned, undoubtedly linked to the new democratic political climate dating from the late '70s (or to a "favorable opportunity structure", as Tarrow would put it), it also seems to be even more strongly connected with a feeling of frustration, experienced by locals, at being excluded from policy decisions affecting their immediate environments and ways

[7] Other important sources of mobilization are: the lack of environmental protection laws (4%) and the failure to implement existing laws (3,6%).

of life. Consequently, these new and predominantly single-issue movements might better be explained by an old theory of social movements, that of deprivation-frustration-aggression, than by the newest ones which center on the formation of identities and the construction of shared meanings (Melucci 1988).

The exclusion of local communities from policy-making responds to the statist political style described above and also to the widespread (but increasingly false) belief that decisions based on technical and 'objective' arguments are readily accepted by target groups. As the literature on risk societies as well as certain public policy analyses show, Europeans have begun to view rather fearfully the way technological progress is evolving and to regard experts with mounting suspicion (Giddens 1998; Beck 1998; Aguilar Fernández and Subirats 1998). That scientific legitimacy has gradually become an object of dispute may in part account for the new call for democratic participatory practices in different public fields. Furthermore, the increasing recognition on the part of the authorities "that democratic decision making and public support are crucial to a successful environmental politics, and to the formulation of environmental policy that is both technically sound and politically acceptable" (Kraft 1996: 10), goes a long way to explain this call. However, the 'democratization motto' has hitherto had limited influence everywhere[8], while in Spain its implementation has been practically negligible.

To some extent, the reluctance to promote democratic practices – what one could call the 'democratic prejudice' – might be easier to understand (though not to justify) when dealing with environmental policy as opposed to other public policies. Protecting the environment constitutes a highly conflictive task because it congregates a large number of stakeholders with divergent interests[9]: "international organizations, governments, multinationals, industrial associations, environmental groups, farmers, unions, consumers and citizens are some of the actors who, by means of disparate strategies, fight for playing the lead in a policy area characterized by the absence of agreement about how to evaluate, and to tackle, environmental deterioration" (Aguilar Fernández 1997: 18-19). Alongside this situation, the promotion of ecological causes can also be envisioned as a relatively new endeavor in which political routines and patterns of behavior amongst the different interests are not as established as in other fields of public intervention, thus contributing to the frequent adoption of zero-sum game-like strategies by the contending groups. To complicate things

[8] Even in Germany, a country which has been dealing with citizens' council techniques for more than 30 years, only 30 experiences have ever been implemented (Font 1997).

[9] "Democratic politics ... is made especially difficult when individuals and groups hold sharply divergent perspectives on the issues and are reluctant to compromise" (Kraft 1996: 10).

even further, the balance between expertise and popular control which democratic governments must strike is at once both hardest and most important in areas such as environmental protection, where expertise is essential and where the interests of consumers, labor, industry, and local residents are opposed (Rose-Ackerman 1995).[10]

Not surprisingly, then, experiences of local participation in environmental policy-making are merely anecdotal in Spain. In general, public authorities attempt to implement decisions without allowing for any type of consultation. It is true, nonetheless, that under 'normal' circumstances citizens are not likely to show a great interest in getting involved in time-consuming and tiring decision-making processes, and that only when a conflict is unleashed are they bound to demand their right to participation. Logically, only when the implementation of certain projects get out of hand and decisions are violently contested will authorities try to orchestrate some sort of information campaign or public involvement strategy. However, this being a reactive and 'forced bottom-up approach', it introduces biases and distortions in the participatory process; for instance, when information campaigns do eventually take place, they are conducted without strong conviction because, according to a widespread belief held by officials and even by some environmentalists, locals can be easily brainwashed and manipulated by vested interests. Hasty and badly organized participatory procedures, which only try to justify and facilitate the implementation of decisions previously taken, simply help to compound the problem of increasing numbers of grassroots protests. This situation is perfectly exemplified by the case of the regional park Cuenca Alta del Manzanares (Aguilar Fernández 1996).

4.1 The dispute surrounding the declaration of the regional park Cuenca Alta Del Manzares in Madrid

In the initial phases leading up to the declaration of the regional park, Cuenca Alta del Manzanares, information (apart from the public notice stage which was legally enforceable) was not provided to the locals by regional authorities. The absence of information was criticized by a conservationist group, COMADEN (Madrid Coordinating Committee for the Defence of Nature), which also denounced the fact that only political actors (the regional government and some municipalities) had been given a voice in the process. This would help explain, according to the same source, why at first the re-

[10] According to Rose-Ackerman (1995), "equally fundamental is the issue of making democratic values operational in modern states where hierarchy and expertise cannot be avoided" (Rose-Ackerman 1995: 1).

gional park faced enormous grassroots opposition and why locals were sceptical when information began to be distributed by public authorities later on. Officials from the regional government and the Environmental Agency justified their initial stance by arguing that transparent information is generally of no use because locals only trust their most immediate peer groups who, on the whole, tend to manipulate 'objective' facts. Likewise, experience would allegedly show that, regardless of the number of meetings and round tables organized, the declaration of protected areas usually evolves along conflictive lines. Following this line of argument, experts and technicians have become unreliable sources for the residents[11], while protected areas have only become accepted once they have begun functioning and have proven to be beneficial. Accepting that the information available in the Manzanares case was insufficient, one environmental representative from Aedenat (Ecological Association for the Protection of Nature) claimed, nonetheless, that participatory and informative procedures were not that important since area residents tended to be easily manipulated by local representatives and interest groups and to dispute systematically any declaration for the protection of spaces.[12]

4.2 The conflictive nature of industrial waste policy

Some of the most relevant cases of violently disputed decisions have been related to the construction of large public facilities, such as the examples of the dam in Itoiz and the railcar at the Leizarán show.[13] On the whole, however, the policy area which has experienced the most local protests has been industrial waste management. Three cases which have occurred in Andalusia help to illustrate the conflictive nature of this policy: the attempt to build a landfill for

[11] Despite this skeptical attitude towards participatory procedures, politicians and regional officials seem to have drawn some lessons from the Manzanares case because, concerning a more recent process of declaration (that of the Southeast Park, 29.836 hectares), *ex ante* information was distributed and the period of public notice and allegations was extended. Yet most information seems to be given *a posteriori*: for instance, only when the farmers of a certain municipality initiated a campaign against the park, did the Environmental Agency decide to go to the place to explain the benefits of the project to the locals.

[12] This skeptical opinion can be attributed to the sometimes tense relationship between environmentalists and locals. In 1982, for instance, before the declaration of the regional park, about 3.000 villagers from Manzanares el Real demonstrated against conservationists because of the latter's opposition to the construction of a dam in the natural park of La Pedriza.

[13] The geographical location of these two facilities (the first one in Navarre and the second one in the Basque Country) adds some drama to these cases because, in this Northern region, environmental protests are radicalized by the participation of extreme left-wing nationalist groups, principally *Herri Batasuna* (United People).

hazardous waste in Gibraleón elicited a civil war-like ambience which, after a protracted conflict between supporters and opponents to the facility, ended up with the loss by the PSOE of one of its strongholds in the 1991 municipal elections (Aguilar Fernández and Subirats 1998); the refusal of the village of Nerva to allow the construction of an industrial landfill in its surroundings not only brought about tremendous social and political tension, but also delayed the works enormously; the opposition of the city council of Medina-Sidonia, alongside locals and environmentalists, to an incinerating plant, led to the abandonment of the project and resulted in the EU money assigned to it being wasted.

Leaving aside specific examples, the hard evidence is that, since the beginning of the '90s, different environmental groups and local committees have managed to halt about 25 projects of incinerating plants and to impede the operation of all those related to industrial waste disposal (Aguilar Fernández 1997). The most significant outcome of this situation has been an important change in the public approach to industrial waste management. In 1995, after the administration publicly recognized the failure of the Law 20/86 for Toxic and Hazardous Residues and the 1989 National Plan of Industrial Waste[14], a process to reform the policy was set in motion and finally led officials to recognize that the minimization of waste at its origin should be stressed more strongly. The Plan for Toxic Residues and the Plan for Polluted Soils, both passed in 1995, embody this reform. The first plan reveals the refusal by the previous socialist government to continue promoting incineration as a disposal technique for industrial waste.[15]

5. EXPERTISE DEFICIT IN THE ADMINISTRATION AND POLITICAL OPPORTUNITIES IN THE EUROPEAN SCENARIO

Since environmental protection contains an important element of technicality, expertise is a very valuable political resource for those groups who have a stake in the policy. Although the recent European move towards 'nationalization' and subsidiarity in environmental policy may bring about some slowing down in the influx of directives (Aguilar Fernández 1994), the impressive

[14] Prior to the state administration's failure, the Catalan government likewise did not succeed in implementing its 1983 Law of Industrial Residues through the 1990 Plan of Hazardous Waste. This failed attempt brought about a change in the policy as well.

[15] The first plan also aims to minimize waste generation by 40% in the year 2000 while it envisions the construction of a number of landfills as an alternative to incinerating plants.

number of Community rules enacted hitherto has literally flooded the Spanish bureaucracy. Alongside the problem of the seemingly ever-growing quantity of directives, which has forced officials to work in haste while neglecting formal, legalistic, and time-consuming procedures, there is a quality problem: as the content of the directives is highly technical and complex, the administration frequently lacks the expertise necessary to establish the best position on the issues for Spain as well as to implement the decisions in a later stage. That the Spanish administration fails to draw on enough experts is frequently referred to by officials who, when flying to negotiations in Brussels, see themselves lacking the minimal technical backing which other member-states enjoy.[16] This situation favors the position of those environmental groups which can resort to their expertise in order to make their presence more palatable in the policy-making process. Besides, if the administration persists in not relying on them, experts can always circumvent it and provide Euro-bureaucrats with important information that public authorities try to conceal. In addition, environmentalists have the option of directly addressing the European Commission with their complaints. These strategies were useful during the negotiations leading to the passage of the habitats directive and will also likely be resorted to during its implementation.

5.1 Environmentalists' input into the habitats directive

The habitats directive was negotiated exhaustively for five years (1988-92). The negotiations ground on for so long mainly because the Spanish government rejected the directive on the grounds that Community financial instruments for the creation of the network Natura 2000 were missing (Aguilar Fernández 1998). In the face of government refusal to put forward specific proposals, a conservationist group (SEO/Bird Life) drafted a document which, after being leaked to all the member-states, made the European Council accept the co-financing of the directive. As Spain continued to block the text, claiming that the extent of financial support should be established and that the receipt of funds could not be considered "an exceptional case" (Quercus 1991), the SEO prepared a second proposal which fixed the level of European funding for less developed regions in which habitats of preeminent interest existed.[17] Only when it became clear for the conservationist group that the gov-

[16] This situation has been recently exemplified by two case-studies: exhaust emissions (Aguilar Fernández and Jiménez 1998) and the habitats directive (Aguilar Fernández 1998).

[17] This level was established at approximately 500 million ECUs for at least three years. According to the SEO proposal, the EU would contribute between 50 and 75% of the total

ernment would not submit the text to Brussels and that the habitats issue was being used as a legal precedent to achieve direct Community financing for future directives, did it decide to send the proposal to the rest of the member-states, the Commission and the Dutch Presidency of the EU. The final obstacles on the way to the passing of the directive were eventually removed after concerted European pressure was exerted on the Spanish government.[18]

The implementation of the habitats directive will help to demonstrate the extent to which authorities are really willing to let social interests participate in conservation policy. This is so because the directive commits member-states to elaborate within three years a list of Special Areas of Conservation which will have to be based mostly on negotiations between the different administrations and between the administration and social groups. However, according to Juan Carlos del Olmo (general secretary of WWF/Adena), "a national-based and institutionalized process of consultation is currently missing" and the same applies to information procedures (Adena web page, 12-11-97). Yet it is too early to reach a conclusion about what shape the implementation process will finally take.[19] This process can still be carried out in two opposed ways: a new open way, through consultation, participation and the search for consensual decisions, and a traditional, secretive way, lacking transparency and social involvement. The choice of one or the other will provide a crucial test for the real intentions of the government, but even if it opts for the last one, it will not be without its problems – as it previously used to be the case: Adena has already warned that it will make its own assessment of the Spanish list of Special Areas of Conservation directly available to the Commission by means of its European network and bureau in Brussels. Besides, the Commission, probably under the guidance of some environmentalist group, will intervene if it considers that a member-state has 'forgotten' to include an important area in the list. Likewise, NGOs will also intervene if the biannual reports submitted by the governments abuse annex V of the directive which refers to the species which can be collected for surveillance or productive reasons.

---

expenditure. In exceptional cases, the Community financing could also amount to up to 90% (Quercus 1991).

18 The visit of representatives of the European Commission and the Dutch Presidency of the EU to the State Secretary for Water Policies and the Environment, Vicente Albero, made the government finally accept the passage of the directive.

19 Although a national-based negotiation may be missing, social participation in the implementation of the directive is taking place in several regional projects under way. For example, a pilot project by WWF/Adena which relies on the support of the European Commission and the regional government of Andalusia, is facilitating the application of the directive in the Natural Park of Alcornocales.

6. A NEW INTERNATIONAL PARTICIPATORY MOOD: SOME SPANISH EXPERIENCES

A new international approach fostering communitarianism and a democratization of political life has recently emerged.[20] It entails not only a larger involvement of private individuals in the public sphere through different instruments (deliberative opinion polls, alternative dispute resolution procedures or simply, mediation techniques[21]), citizen's juries (*Plannungszelle* in Germany), citizen's advisory committees, local participation structures, national issues forums and study circles, citizen's panels, and televoting) (Font 1998), but also less public regulation. These instruments serve the purpose of making citizens more aware of the difficulties surrounding policy-making by letting them decide on certain disputed issues.[22]

The ethical imperatives of this approach include, amongst others, "a politics of participation that provides every member of society with full opportunities to influence the political, social, and economic institutions affecting their lives, that fosters collective and personal responsibility to fulfill that task; a politics of conflict resolution and healing that acknowledges and respects differences and goes beyond 'us against them', or 'right versus wrong', in promoting co-operation and community in all matters; a politics of ecology and unitizing consciousness, which understands that we are only part of a seamless web of life, that we are responsible to all life on earth, and that our journey can have sacred meaning" (Schwerin 1995: 3-4).

The Spanish experience in this regard has been very limited, consisting only of the nuclei of participatory intervention: groups of persons (about 25)

[20] Communitarianism is one of the four main pillars of British Prime Minister Tony Blair's New Labour and has also been promoted by President Clinton. Barber, the author of the book Strong Democracy and a vigorous advocate of more participation in modern societies, is advisor to the US President. However, this idea is to some extent interpartisan: it has been equally promoted by the Democrats and the Republicans in the United States. "Both liberal and conservative policy-makers have recently embraced the language and symbolism, if not the substance, of empowerment. This shift is reflected in numerous public statements and the titles of new programs and organizations such as Clinton's Community Empowerment Zones, and the conservative republican organization, Empower America." (Schwerin 1995: 175)

[21] "Mediation is a form of alternative dispute resolution or popular justice, which is viewed as a supplement to, or alternative to, the formal legal system ..., In recent years the practice of mediation has grown rapidly in the United States and in other countries. It is used in a wide variety of disputes, such as ... environmental and public policy conflicts." (Schwerin 1995: 7)

[22] This is the goal of empowerment: "to increase your capacity to define, analyse, and act on your own problems" (Schwerin 1995: 56).

chosen at random which, after a process of plural information and collective deliberation, agree on a position related to a collective problem which demands a public answer (Font 1996). They have been put into practice first in the Basque Country (Guipúzcoa) between 1992 and 1994, precisely because of past traumatic experiences of local violent action such as the one at Leizarán: The first case, which took place in Idiazabal, revolved around the location of a football field; the second one in Astigarraga dealt with the siting of a sports center; and the third one, which was more complicated, concerned the lay-out of the highway Maltzaga-Urbina. Following the generally positive results of these experiences, a citizen's jury in the municipality of Sant Quirze del Vallès, and a citizen's council in the municipality of Rubí, both in Catalonia, were organized – the first case dealt with the General Planning of the municipality whereas the second one sounded out locals about the use of some municipal states (Font 1998).[23]

7. AN ASSESSMENT OF THE INSTITUTION BUILDING PROCESS IN ENVIRONMENTAL POLICY

The most significant example to date of institution building in environmental policy has been the Advisory Council of the Environment (CAMA) set up in 1994. At the beginning of 1995, the State Secretary for Environment and Housing, Cristina Narbona, pointed out that the only occasion when the President of the Government, Felipe González, had shown an interest in the environment was when the CAMA was created (Aguilar Fernández 1997).

The CAMA was set up by the RD 224/94 with the aim of "promoting the participation of prestigious social organizations and individuals in making and monitoring sustainable, development-oriented, environmental policy" (CAMA 1994). The RD acknowledges that the creation of the CAMA responds to the need to comply with measures adopted by the EU (it is explicitly mentioned that the V Environmental Action Programme and the Decision of 7 December 1993 have led to the Advisory Forum for the Environ-

[23] Despite the insistence of organizers, the ecologist group ADENC refused to participate in the first process whereas the second one could not count on the presence of the neighbors' associations either. The first case ended up with a consensus on the need to control the growth of the municipality and to revitalize the center by concentrating different facilities; through the second process it was agreed on that a car park and a green area would be built in a certain area, against the proposal of the city council which favored a commercial center and a petrol station.

ment in the Community being set up) and the international level. Being composed of 36 people belonging to a wide array of groups (administration, environmentalists, farmers, entrepreneurs, consumers, citizens...), the CAMA contains a working group on territory, infrastructures and environmental impact which has debated several environmentally-sensitive projects in 1994: the dam in Itoiz, the extension of the Barajas airport, and the layout of the highway from Madrid to Valencia across the river Cabriel. One year later, aside from Barajas and the former highway, the planning affecting Delta del Llobregat, while the dams in Garita and Riansares, and the General Plan of Infrastructures were subject to scrutiny as well. The relevance of two specific issues, Itoiz and the Madrid-Valencia road scheme, facilitated the organization of two monographic workshops to analyse their environmental repercussions. To date the CAMA has functioned with mixed results: it has been a good forum of debate but, at the same time, it has had little leverage within the state administration.[24] After the creation of the MIMAM, the CAMA has undergone some modifications in order to adjust its structure to the new ministry. Besides being renamed as the Advisory Forum for the Environment, this institution has been given the task of promoting measures for the creation of employment and the reinforcement of environmental education. The Minister responsible for the MIMAM, Isabel Tocino, has also stressed the need to broaden its social basis in order to permit the participation of interests "related to land planning, hunting, fishing, littoral planning, ... and even the mass media" (Información de Medio Ambiente 1996: 5). Along these lines, it has also been pointed out that the Forum will facilitate the participation of economic and social interests with the aim of improving the state of the environment (above all, in the management of natural parks and forest policy), and that – using an approach that emphasizes shared responsibility – it intends to foster a "dialogue with the affected social sectors and NGOs" (Información de Medio Ambiente 1996: 8). However, the frequent clashes between the minister and the environmental associations involved in the institution are at odds with these public statements.

Another example of institution building is the National Council of the Climate which was established last February to comply with the Kyoto agreements. Although its principal aim is to design a Spanish strategy for global warming by means of consensual practices and the participation of citizens, the Council is exclusively composed of public agencies: nine ministries plus the Presidency. This 'homogeneous' composition constitutes a surprising factor if participatory procedures are to be taken seriously.

[24] This argument would account for the abandonment of CAMA by Greenpeace and Aedenat in 1995 (El País, 20-6-1995).

Alongside these institutions, two important European directives, DIR 90/313 and DIR 85/337, which have already been incorporated into Spanish law, are meant to facilitate social involvement in environmental policy issues. The DIR 90/313 on free access to environmental information, transposed by the Law 38/95, guarantees the right to environmental information without previously having justified a specific interest in the matter. It also establishes that requests of this type will have to be answered in a maximum of two months. Yet 80% of the requests formulated by public departments have been denied over the last three years.[25] After the creation of the MIMAM, Isabel Tocino announced, in her first appearance before Parliament, a project whereby an office to provide free environmental information would be established. This office has not yet been put into practice nor have any other plans related to the directive. This situation, combined with the pending conflict with Brussels over the allegedly incorrect transposition of the directive, has led to a stalemate. In the meantime, the number of complaints brought before the European Commission concerning the infringement of this directive is beginning to grow.

Before the DIR 85/337[26] on Environmental Impact Assessment was transposed into Spanish law by the RDL 1302/86 and its reglement RD 1131/88 – which again Brussels considered inappropriate –, references to this instrument could be found in a number of sectoral regulations. Although the EIA is meant to embody two main principles, the precautionary and the participatory (from the bottom-up) principles, its implementation in Spain has simply paid lip service to them. The application of flawed and routinized methodologies and the haste and lack of transparency with which these evaluations have been generally carried out are proof of this disinterest. Thus, thc participation of environmentalists, citizens, and other social actors in the early stages of the elaboration of the EIAs has been very elusive. Confrontational strategies have therefore been prevalent among conservationist groups and residents, in alliance sometimes with political parties, regional and local governments and public agencies. This has created complex and uneasy, ad hoc coalitions based on the exclusion of these actors from the policy-making process.

In Spain EIAs have opened up two new ways for the involvement of social groups in environmental issues: institutional meetings at the central level (this procedure was not envisaged by the directive) and public information processes (included in the directive). The first one begins with the organization of a period of consultation which, run by the DGIA (General Direction of Environ-

[25] Econoticias nº 46 1996

[26] The dir 85/337 has been replaced by the dir 97/11 which introduced some changes related to certain public and private projects. Although this new directive has not been dealt with in Spain yet, a bill on environmental impact assessment is now in preparation.

mental Information and Evaluation), is open to all potentially affected interests or to those that have relevant information about the foreseen projects. The second one allows the submission of allegations to the project and its corresponding EIA. Yet this second possibility takes place at the final phase of the process and promoters do not usually take notice of it.[27]

Until February 1993, 44 out of 107 EIAs elaborated by the state administration entailed significant changes, and only 6 projects (one highway, three quarries, one gold mine and one gravel-site) were denied authorization for environmental reasons (Font 1996). Since 1992 the EIAs formulated by the state administration include an annex which spells out the consulted actors, which of them provided answers and the authors of the allegations. From 1992 to 1994 the number of institutions consulted has grown (13 in 1992, 20 in 1993, and 22 in 1994) although replies have slightly decreased (33% in 1992, and 29% in 1993 and 1994). The number of allegations has evolved in the following way: 2,1 in 1992, 6,9 in 1993 and 4,4 in 1994. Between 1988 and the beginning of 1991, out of 75 complaints on EIAs submitted to the DG XI, only 6 led to infringement procedures.

Table 2. Complaints relating to EIAs before DG XI and the ratio between them and the total number of environmental complaints

	1988	1989	1990	TOTAL	EIA/ environment
EIA	8	34	30	72	28,4%

Source: Font 1996

8. CONCLUSIONS

Participatory experiences in environmental politics do not leave much room for optimism. Firstly, they basically amount to a reactive approach: only after locals demonstrate against the declaration of protected spaces do regional authorities try to appease them by organizing information campaigns and giving them some voice in the process; only after numerous grassroots actions are organized against industrial waste incinerating plants does the state administration take on board the minimization principle; and only after the traumatic experience of Leizarán have nuclei of participatory intervention been arranged in the Basque Country.

[27] Important innovations can be expected as a result of the DIR 90/313 on free access to environmental information which directs the public authorities to make information relevant to EIAs available to any physical or juridical person that requires it.

Secondly, these experiences have not been pursued wholeheartedly: the CAMA could have set a good example in relation to participation but its current influence on the decision-making process has been negligible while its functioning has encountered numerous problems; the implementation of the directive of free access to environmental information has more or less stalled and most of the EIAs simply pay lip service to the right of interested parties to participate.

Thirdly, institution- building has responded to the need to come to terms with developments taking place outside Spain: the CAMA was set up to keep pace with EU recommendations about the opening up of environmental politics to social participation, whereas the National Council of the Climate originated in obligations acquired after the signing of the Kyoto agreements. However, even if Spanish public authorities remain reluctant about letting social groups partake in environmental protection, some recent trends are easing the way towards the 'democratization' of this public policy. The European Commission, lacking the means to closely monitor the implementation of environmental directives in the member-states, has expressed, more or less openly, its willingness to cooperate with conservationists to control governments' performance. These groups have seized the opportunity and are providing Eurocrats with important information which domestic authorities would like to keep away from public scrutiny. Besides, national officials need increasingly to count on the expertise of interest groups in order to improve the Spanish position in relation to a growing number of issues and to facilitate putting directives into practice.

Lastly, the assumption of power by the conservative party, which publicly adheres to the idea of state withdrawal from certain areas so that private interests can play a larger role in the direction of some public issues, may contribute to the 'participatory mood'. Although this last factor is currently hypothetical, the other factors elaborated above may well lead to public authorities losing control of events. By contrast, if authorities persist in following the traditionally secretive approach, it will not be without political costs and policy failures.

REFERENCES

Aguilar Fernández, S. (1993) Corporatist and statist designs in environmental policy: The opposing roles of Germany and Spain in the community scenario, *Environmental Politics* **2**, 223-247.

Aguilar Fernández, S. (1994) Convergence in environmental policy? The resilience of national institutional designs in Spain and Germany, *Journal of Public Policy* **14**, 39-56.

Aguilar Fernández, S. (1996) The conflict surrounding the declaration of the regional park "Cuenca Alta de Manzanares": A Spanish case study, in M. Kousis, S.Aguilar and T. Fidelis, *Grassroots Environmental Action and Sustainable Development in Southern Europe. Final report submitted to EC, DG XII*, Final Report, European Commission, DG XII, Contract No. EV5V-CT94-0393.

Aguilar Fernández, S. (1997) *El Reto del Medio Ambiente: Conflictos e Intereses en la política medioambiental Europea*, Alianza Editorial, Madrid.

Aguilar Fernández, S. (1998) *Habitats directive?* Unpublished manuscript.

Aguilar Fernández, S., and Jiménez, M. (1998) *Exhaust emissions*, Unpublished manuscript.

Aguilar Fernández, S., and Subirats, J. (1998) Political conflict and industrial waste, in B. Dente and P. Fareri, *Deciding about Waste Facilities Siting*, Kluwer, Rotterdam, pp.69-98.

Alonso, L.E. (1991) Los nuevos movimientos sociales y el hecho diferencial español: una interpretación, in J. Vidal-Beneyto, *España a debate*, Tecnos, Madrid.

Beck, U. (1998) Poilitics of risk society, in J. Franklin, *The Politics of Risk Society*, Polity Press, Oxford, pp.9-22.

CAMA (1994) *Memoria del Consejo Asesor de Medio Ambiente*, Draft, Madrid.

Font, J. (1997) *El consells ciutadans: concepte y experience*, Unpublished manuscript.

Font, N. (1996) *La europeización de la política ambiental en España. Un estudio de implementación de la directiva de evaluación de impacto ambiental*, Ph.D. Thesis, Universidad Autónoma de Barcelona.

Font, N. (1998) Democracía y participacío ciutadana, *Politíques 22. Ed. Mediterrània.*

Foweraker, J. (1985) Franco's Corporatist Strategy and its Implications for Corporate Interest Representations in Spain Today., Essex Papers No. 24.

Giddens, A. (1998) Risk society: The context of British politics, in J. Franklin, *The Politics of Risk Society*, Polity Press, Oxford, pp.23-34.

Gunther, R. (1991) The Dinamics (sic) of Electoral Competition in a Modern society: Models of Spanish Voting Behavior, Working Papers No. 28 Barcelona, Institut de Ciències Polítiques i Socials.

Gunther, R. (1996) Spanish Public Policy: From Dictatorship to Democracy, Estudios Juan March, Centro Estudios Avanzados en Ciencias Sociales, Working Paper 1996/84 Madrid, Ediciones Peninsular.

Información de Medio Ambiente, (1996) , Bulletin No. 43 Madrid, MOPTMA-MIMAM.

Kousis, M., Aguilar Fernández, S., and Fidélis, T. (1996) *Grassroots Environmental Action and Sustainable Development in Southern European Union*, Final Report, European Commission, DG XII, Contract No. EV5V-CT94-0393.

Kraft, M.E. (1996) *Environmental Policy and Politics: Toward the Twenty First Century*, Harper Collins, New York, NY.

Linz, J.J. (1981) A century of politics and interests in Spain, in S. Berger, *Organizing Interests in Western Europe*, Cambridge University Press, Cambridge, MA, pp.8-21.

Martín Rebollo, L. (1984) Las relaciones entre las administraciones públicas y los administrados, in Juan J. Linz, *España: un presente para el futuro*, Instituto de Estudios Políticos, Madrid.

Melucci, A. (1988) Getting involved: Identity and mobilization in social movements, in B. Klandermans, H.-P. Kriesi and S. Tarrow, *From Structure to Action: Comparing Social Movement Research Across Cultures*, JAI Press, Greenwich, CT, pp.329-348.

Rose-Ackerman, S. (1995) *Controlling Environmental Policy. The Limits of Public Law in Germany and the United States*, Yale University Press, New Haven.

Schwerin, E.W. (1995) *Mediation, Citizen Empowerment, and Transformational Politics*, Praeger, Westport, CT.

Subirats, J. (1997) *Dossier: Els Consells Ciutadans*, Unpublished manuscript.

Valiente, C. (1997) *Políticas públicas de género en perspectiva comparada: la mujer trabajadora en Italia y España (1990-1996)*, UAM Ediciones, Madrid.

Chapter 11

Interaction Between State and Non-state Actors in the Implementation of the CAP Agri-environmental Measures

The Greek Experience

Leonidas Louloudis, Evi Arahoviti and Dimitris Á. Papadopoulos

1. INTRODUCTION

The increasing awareness of the need for a financially sounder and environmentally friendlier policy has inspired attempts from the 1980s onward to adjust the CAP[1]. The reform of the CAP (approved in 1992) marked a turning point, aimed primarily at restructuring agricultural markets. One of the central elements of the CAP reform was the encouragement of farmers to use less intensive production methods, thereby reducing their impact on the environment and preventing the generation of unwanted surpluses. As part of the CAP reform, the Union agreed on a set of complementary agri-environmental and afforestation measures. Regulation 2078/92 concerns agricultural production methods compatible with the requirements for the protection of the environment and the preservation of the countryside. The objective of the regulation is twofold: to combine beneficial effects on the environment with a reduction of agricultural production; and to contribute to

[1] In December 1968 the Commission produced its 'Memorandum on the Reform of Agriculture', which generally became known – after the name of the Commissioner for Agriculture – as the 'Mansholt Plan'. The Mansholt Plan broke new ground. Both the problem of over-supply and the need for basic adjustment were recognized. A long-term program – 'Agriculture 1980' – would contain a new approach to price policy, under which prices were again to play their true role in guiding production. Structural policy would aim to create 'modern production units' through selective investment aids. The Plan provoked violent opposition among farming circles throughout the Community. After much debate, the outcome was modest (Tracy 1997).

K. Eder and M. Kousis (eds.), Environmental Politics in Southern Europe, 277–298.

agricultural income diversification and rural development. To achieve such objectives, member-states have to draw up schemes. These schemes may provide aid for farmers who undertake practices that effect the environment and the countryside[2].

EU environmental policy is moving in parallel directions. Already the 'Groundwater Directive' of 1980 had aimed to protect groundwater from potential sources of pollution; the 'Drinking Water Directive', also of 1980, had established quality standards, including limits to nitrates content and residues from plant protection products. The 'Nitrate Directive' of 1991 aims to reduce water pollution caused by nitrates from agricultural sources. If these aims were strictly followed, they could require significant changes in production systems, especially in areas of intensive plant and livestock production such as the Thessaly plain or Amvrakikos bay (Central Greece). The 'Habitats Directive' of 1992 aims to protect or re-establish natural habitats for wild flora and fauna, and provides for the creation of a Europe-wide network of protected zones, known as 'NATURA 2000'.

The whole legal framework of EU environmental and agri-environmental policies, if implemented alongside the anticipated impacts of deregulation policies and biotechnological innovations, will change dramatically the profile and the wider perspectives of European agriculture in less than two decades (Pitman 1992). Thus there is an urgent need for all public, civic, social or private agencies involved with agricultural or environmental policies to quickly adapt to the new realities which the sector faces. This need is even more difficult to satisfy in Southern Europe, where the developmental role of the state, and particularly of public administration, is rather weak. In these areas, as has already been realized, "the implementation of environmental policy can be particularly problematic ... since environmental protection requires complex administrative and managerial development, co-ordination between numerous state and quasi-state agencies and the involvement of a large and complex set of actors .... Such co-ordination may be

[2] The schemes available are diverse, reflecting different regional conditions. Member-states can set up programs either at the national, regional or local level, depending on the degree of administrative decentralization as well as on the environmental and agricultural characteristics of the relevant areas and their specific needs. While member-states are obliged to set up schemes, farmers' participation is voluntary. The schemes allow for farmers to commit themselves for a minimum of 5 years (or 20 years). The cost of the program is shared with the member-states. Co-financing takes place on a 50% basis except for regions of Objective 1, where costs are co-financed at a rate of 75%. As regards the Community resources available, the total allocated within EAGGF Guarantee Section for co-financing of expenditures under this regulation in the period 1993-1997 is about 5 billion ECU's (EUR 15) (Cammarata 1997).

hampered because of inadequate administrative structures, institutional fragmentation, lack of resources and staff-power to commit to environmental management, or because actors do not treat environmental concerns as matters of priority" (Yearly, Baker and Milton 1994).

This paper attempts to give a first account of the way the Greek agricultural network moves towards the aforementioned adaptation to these new perspectives, focusing mainly on the relationship between public administration, particularly the Ministry of Agriculture (MoA), and certain environmental NGOs. Although the role of NGOs is crucial for the implementation of agri-environmental policy, the change of the socioeconomic framework in which they have operated during the last two decades should not be underestimated. More specifically, "the environment agenda no longer belongs to the environmental NGOs; the movement today is 'owned' by a multitude of different agents in society ... This is particularly the case with the business community. Since the late eighties, there has been a deeper and deeper engagement in the environment agenda by individual companies and organizations representing business interests in general" (Porritt 1997). If this statement holds for Northern Europe, it is even more valid for Greece. As will be argued below, from the 1970s onwards, the penetration and intervention of the private agrochemicals and farm machinery marketing sector in rural areas was decisive. Insofar as the agri-environmental measures become the potential axis of rural development, the management of natural resources gains new momentum. Accepting that "popular participation is a crucial component" of rural sustainable development, especially in the South, and as environmental demands "bring new social relationships into being, and with them new power relations" (Redclift 1993), this work ultimately attempts to explore the struggle between human agents over environmental resources in rural Greece after the first implementation of agri-environmental measures.

Data were derived by an elite semi-structured interview survey conducted in the framework of a wider project funded by the Commission and aiming towards a first evaluation of the implementation of 2078/92 in nine European countries[3]. In total, 20 semi-structured elite interviews (Moyser and Wagstaff 1987) were conducted between May 1997 and May 1998 with key-persons from a wide spectrum of public, non-governmental, and private agencies (Ministries, regional research institutes, agricultural professional

[3] FAIR1 Project (CT-95 0274) 'Implementation and Effectiveness of Agri-environmental Policy measures under EU Reg. 2078/92'. The Greek Research team was comprised of the authors and their colleagues N. Beopoulos, G. Vlahos, all working in the Agricultural Economics Dept, Agricultural University Athens.

organizations and cooperatives, private firms and consultants, NGOs, and farmers) in order to cover the views of the main actors involved directly or indirectly in the planning and implementation process of Reg. 2078/92.[4]

2. IMPLEMENTATION OF REG. 2078/92: THE ADMINISTRATIVE APPROACH

The implementation of Regulation 2078/92 in Greece makes provision for various agri-environmental schemes. Four of these schemes are national (horizontal), to be applied throughout Greek territory, and five are designed to be local (zonal) within certain designated zones. Only three aid schemes, approved by the Union, have been implemented since 1995. These three are: a) the horizontal scheme of "biological agriculture", b) the zonal scheme "for the reduction of nitrate-pollution related to farming in the plains of Thessaly" and c) the horizontal scheme providing "for the long-term set aside of agricultural land" that was approved by the Commission in May 1996 and whose implementation began the same year.

For the time being only two schemes, namely that of biological agriculture and that of the reduction of nitrate pollution in the plains of Central Greece are fully implemented in the sense that there are 'tangible' results; the one concerning the long term set-aside of agricultural land was still in the process of the evaluation of applications when this text was written. As far as the conversion to or continuation of the biological farming methods scheme is concerned, according to the results reported, the rather modest objectives concerning the area to be covered were fulfilled. Furthermore one could comment that a great part of the beneficiaries are farmers already implementing Regulation 2092/91 on biological production methods. The most important biological cultivation is that of olive trees for olive oil production, covering almost 60% of the overall area under this scheme. The scheme for the reduction of nitrate pollution caused by agriculture in the region of Thessaly (Prefectures of Larisa, Karditsa, Magnisia and Trikala) is a pilot scheme. The farmers' participation as judged by the areas covered by the program can be considered satisfactory. Nevertheless, the area under the

[4] The elite interviews were structured around the following five variables: Career and background of the respondent, relation of the organization in which the respondent is employed to the agri-environmental policy, relationships with other actors, organizations and agencies in the framework of 2078/92 planning and implementation process, accounts of the expectations derived from the compatibility of the Regulation, implementation, evaluation and suggestions for improvement of the Regulation.

targeted crop, i.e., cotton, is 1/8 of the regions' cotton cultivated land, which accounts for almost 50% of the available arable land in Thessaly and the major part of irrigated land.

The Directorate of Spatial Planning and Environmental Protection (DSPEP) of the Ministry of Agriculture (MoA) is responsible for the implementation of the regulation. The elaboration and design of all programs has been accomplished in what apparently could be defined as a typical top-down approach. The Greek administration had no previous experience in the planning and implementation of relevant schemes, which are quite complicated, as they should have clear environmental objectives, something so far almost totally ignored by Greek agricultural policy makers; at the same time they should be attractive to farmers strongly accustomed either to the productionist and/or the protectionist policies of the recent past. The fact that there was a need for the involvement of regional authorities in the process complicated the situation due to poor co-ordination between central decision making and the periphery. Thus, for problems already solved in other, experienced, administrative mechanisms, solutions should be reinvented quickly in the Greek case. Furthermore, one should add the reluctance or even resistance of the agricultural network (Collins and Louloudis 1995) regarding such measures. Nevertheless, although on a limited scale, there was collaboration between the central services and extension agronomists in the initial stages of framing the program. It should be noted here that after Greece's entry into the Union in 1981, the burden of administering the implementation of the CAP measures was automatically assigned to the extension service. Meanwhile, no major reorganization of the extension service has been undertaken. Hence, the mainly bureaucratic role of extension workers has been maintained, and even exacerbated; extensionists have been given further duties to control the implementation of the CAP measures. Thus, they are severely restricted in their capacity to provide advice and are able to assist information seeking individuals only in a rather fragmented, inadequate, and inefficient manner (Koutsouris and Papadopoulos 1998). Just before the adoption of the regulation, thanks to the initiative of the DSPEP, a scientific Committee was established. Apart from the representatives of the MoA, representatives of the faculties of the Agricultural University of Athens, the National Foundation for Agricultural Research (NFAR) and of PASEGES (Panhellenic Confederation of Agricultural Co-operatives Unions) comprised this committee which assessed the state of the environment in Greece as it is affected by modern farming practices. The hierarchical order of the national objectives of agri-environmental policy was ultimately decided at the ministerial level. The elaboration of the program was carried out, and continues to be carried out by the DSPEP in collaboration with the prefectural (NUTS III) departments of agriculture. During the elaboration of the programs, information meetings and consultations with NGOs (Greek Ornithological Society, WWF-Greek division) representatives of biological farming associations, as well as with PASEGES, were held in order for the Ministry to become familiar with their views. To what extent these interactions between the state agencies and the above mentioned NGOs were sufficiently organized and effective, we shall examine in the next chapter. What is clear so far, is that during the course of these interactions the main responsibility remained with the DSPEP.

Providing information to the prefectural departments for the approved measures involves issuing implementation circulars and regular information meetings. In addition, supplying information to farmers involves the regional agencies, farmers' professional associations and co-operative organizations, local authorities and the mass media. According to the official view of PASEGES, the role of the organization in the dissemination of information about the regulation could have been more effective if the MoA was more collaborative in its design and implementation. In the absence of this collaboration, the role the private sector has played in the actual implementation procedure was of critical importance. Thus, to a significant degree, private agronomists and companies substituted for the state extension service and the co-operative or agricultural professional organizations as far as information diffusion and consultancy services were concerned. In order to explain this, besides considering the reluctance of the state and regional administration, one should consider the innovative character of the measures and the established role of the private sector in the areas of their implementation. In one case (Larisa) the private sector, represented mainly by a well established local firm providing agricultural inputs (fertilizers, pesticides) to farmers, went so far as to create an ad hoc NGO for the participants in the nitrate reducing program under the name 'Friends of agricultural practices compatible with environmental protection' in order better to promote their commercial product (crystalline fertilizers) – an inspired but not unexpected initiative, as the amount of agri-environmental subsidies at a local level was significant[5].

[5] The contracts signed by the beneficiaries, apart from the obligations concerning agricultural methods and practices, included additional terms i.e., the elaboration of an Environmental Management Plan by an agronomist to be submitted with the application form and an obligation to keep books on activities and observations related to the program under contraction. These terms, of course, imply an increased transaction cost for the farmer, which is, in the case of the Environmental Management Plan, a paid cost since, as far as the other terms are concerned, they are incorporated in the overall cost of consultancy. Procedures designed to control the actual conformity of farmers with the terms

3. INTERACTION OF STATE AND NON-STATE AGENCIES: THE ACTOR'S VIEW

3.1 Interaction with NGOs

As we have already said, in Greece prior to 1992, no special provisions had been proposed for the protection of the environment from agricultural activities, with the exception of certain regulations concerning the hygiene standards that agricultural installations are expected to meet, or some general and non-systematic advisory measures, such as appeals for the rational use of fertilizers and pesticides, taken by the Department of Agricultural Extension of the MoA. Moreover, Greece, like other Southern member-states of the EU and unlike Northern partners, such as England and Denmark, missed the opportunity offered by article 19 of Reg. EU/797/85 (currently 2328/91) concerning the modernization of agricultural structures. This article highlighted the 1985 shift in the CAP's attitude towards the environment and included provisions for financial support of all environmentally friendly practices. However, since the late 1980s, agriculture has gradually been associated with nature and conservation, and it currently connotes a production activity that causes pollution through the intensive use of chemical inputs. Regarding this shift of public opinion towards a recognition of the adverse effects which modern agriculture has on the environment, one should not under-estimate, among other factors, the positive role of Greek environmental NGOs. These changes in attitudes were incited by special attention paid to wetland protection.[6] Consequently, the EU financed a substantial number of projects related to wetland management. These studies not only revealed the relationship between modern agricultural practices and natural (wetland) degradation but also strengthened the political role of NGOs as potential consultants of the state on environmental issues. Several traditionally scientific organizations, such as the Ornithological Society and many new NGOs, rallied to the protection of wetlands. For the first time, agricultural practices were extensively scrutinized, and the resulting

---

are performed by specialized agencies, in one case (nitrate monitoring in Thessaly) state run, while in the other (biological agriculture) privately run. Both types of agencies are paid by the beneficiary, thus increasing the eventually paid administrative cost of the measures.

6 In fact, the country's most important wetlands were protected by both the Ramsar Convention (signed in 1973) and the EC Directive with respect to the protection of birds (Dir. 79/409 and the Greek government's Ministerial Resolution no. 9655/1985).

criticisms reached the mass media. Hence, the need for wetland protection in Greece made the wider public acknowledge environmental damage caused by agricultural practices, and sparked off a series of debates concerning existing agricultural development policies (Beopoulos and Louloudis 1997).

In any case, Reg. 2078/92 marked a significant change in the communication between the state and NGOs for two reasons. Firstly, by focusing on the relationship between agriculture and environment, it engaged the MoA to think positively about agricultural practices that reduce pollution and improve the conservation of natural resources, thus abandoning for the first time its productionist policies, although only to the smallest possible degree. Secondly, it gave opportunities to NGOs to broaden their spectrum of thinking by getting involved not only with the impacts of agriculture on the environment, but also with agricultural practices themselves, and especially those that may be compatible to environmental protection and conservation. Thus, Reg. 2078/92 launched a new extremely interesting and promising dialogue between the MoA and NGOs, though, as was to be expected, this dialogue was not without problems of communication, to say nothing of those of collaboration.

The MoA was not prepared to deal with an agri-environmental policy in 1992. In discussions in Brussels prior to voting for Regulation 2078/92, the two Directorates of the MoA maintained that Planning and Agricultural Structures were more relevant and that the DSPEP were trying to avoid this responsibility. Soon after the regulation came into force, in July 1992, DSPEP undertook responsibility for its implementation in Greece. Unfortunately, this delay was not without cost. Greece was absent from the critical negotiations during the preparation of the regulation in Brussels. Nevertheless, from this point on, the requirements of elaborating the implementation of regulation motivated novel procedures of collaboration inside the ministry. In an attempt to divide the country up into eco-geographical zones and to designate environmentally sensitive areas within those zones, information was requested from the Agriculture Directorates in the 56 prefectures of the country, which usually perform their bureaucratic, routine jobs indifferently. Surprisingly, however, 35 of them sent back an extremely interesting volume of data about cultivated varieties, endangered species, agricultural landscapes etc., found in their areas. It is a pity that this valuable material was soon forgotten by the central headquarters of the MoA. Some contacts were established between the DSPEP and the Bureau of Biological Products as well as the Directorate of Planning of the Ministry of Aegean, which is responsible for the small islands of the Aegean. Resistance also appeared – not unexpectedly – since, even inside the DSPEP, the ideology of productionism and unrestricted growth is dominant. In addition, within the MoA, a lobby supporting animal production was against the program of the long term set-aside, which, it was though, would compete with the production sector in certain islands of the Aegean. What was really in danger was the local system of organized interests around the appropriation of subsidies. The same type of bureaucratic resistance combined with conflicts among vested local interests explains partly why in the region of Thessaly only 2 out of 4 prefectures adopted and promoted the program of nitrate reduction.

The regulation provided an opportunity for the DSPEP to develop new forms of collaboration with other state agencies: the Directorate of Environmental Planning of the Ministry for Environment, Spatial Planning and Public Works (ESPPW), the Directorate of Planning of the Ministry of the Aegean, NFAR, the Agricultural University of Athens, and the Universities of Salonica and Patras. All of these interactions, with the exception of that with the MoESPPW, finally formed an informal network of collaboration. Collaboration with the MoESPPW and to some extent with NFAR had an institutional character. A representative of the MoESPPW participates in the committee for the evaluation of the long-term set-aside program. This collaboration originates from the common responsibility of both ministries for the program Natura 2000 (implementation of Directive 92/43). The Greek Center for Biotopes-Wetlands (GCBW), an NGO founded in 1991 as a branch of The Goulandris Museum of Natural History, initially co-funded by WWF-Greece, undertook the co-ordination and control of selection and registration of areas comprising the national list of habitats. This list was finalized in December 1995 under the supervision of the two ministries[7]. Thus, although worthy of note, it is not inexplicable why, soon after the acceptance of the long term set-aside program by the Commission, the MoA officially invited the MoESPPW to participate from the outset in the evaluation process of the management plans for the eligible areas of the program.

NGOs participated neither officially nor directly in the preparatory committee, established by the MoA, for the elaboration of regulation implementation. Nevertheless, the consultation of the Greek Ornithological Society (GOS) and WWF-Greece was queried, yet with no particular results. Both parties, state agencies and NGOs, seem to disagree on the reasons for this ineffectiveness. The MoA considers NGOs unprepared, for the moment,

[7] Since the country was divided into 3 geographical areas, the responsibility for the registration of habitats was assigned, correspondingly, to research teams from the Departments of Biology of Athens, Salonica and Patras Universities. The range of the work done can be judged by the size of the area covered which encompass almost 18% of the total national land.

to offer realistic solutions to technical problems related to the implementation of regulation, for example, the concrete agricultural practices compatible with the management of bird fauna from a conservationist's perspective. GOS argues that, in 1992, they were informed about the regulation through their contacts with Bird Life International. In the following year a report was submitted to the MoA about regulation planning and an informative trip to France by a small team (consisting of ministerial staff, NGO members, and farmers) was organized with the assistance of IEEP-London. Since then, collaboration has stopped, because the Committee established by the MoA to monitor progress on regulation implementation – and in which a representative of GOS has been formally invited to participate – is still inert. So far the organization is kept informed about the regulation through its representative in Brussels (which participates in meetings with Bird Life International, European Environmental Bureau, WWF International). This lack of communication can be partially explained by the fact that GOS is mainly interested in the long term set-aside program and less in the priorities set by the MoA in favor of nitrate-pollution reduction and biological agriculture. Nevertheless, GOS was happily surprised by the initiative of the MoA to include all areas of Natura 2000 in the program of the long term set-aside, thus encompassing areas important for the conservation of bird fauna – although its view is that priority could have been given to areas protected by the Ramsar Convention. At the local level and within the framework of its environmental conservation projects, GOS encourages farmers to participate in the scheme by providing information about the program and assistance in preparing environmental management plans. It actually uses the program as a 'compensation tool': "we are very interested in applying the scheme within the specific protection areas for birds, and so far this is the only aid offered for this purpose". The example of the relationship between the MoA and the GOS (one of the oldest NGOs and also one of the most recognized by the state) is characteristic of the transitory type of communication which agri-environmental regulation inaugurated between the two agencies. Within this institutional framework, both sides recognize the need for collaboration; nevertheless, it does not produce concrete results, since the two partners are still reluctant to accept the other party's efficiency and/or availability. The interactions of WWF-Greece and the MoA, though still in the initial stages, appear to be more or less the same as the ones already discussed (i.e., between the MoA and GOS). WWF-Greece became interested in agri-environmental measures through its collaboration with WWF International, when the latter was asked by the EU to comment on AGENDA 2000 and the CAP. This collaboration will result in proposals to be discussed with the MoA. In this way, WWF-Greece intends to exert pressure on the MoA and consequently to promote its own interests in the area of agri-environmental measures. We should note, however, that these interests have not yet been fully articulated, as the organization intends to use the program of biological agriculture within its wider framework of nature protection and conservation programs.

The obstacles in the path of a working collaboration between the state and NGOs are also apparent in the case of the DIO. The DIO was founded in January 1993 as a non-profit organization officially recognized by the MoA as an agency for the inspection and certification of biological production. In this capacity, it participates formally in meetings organized by the Biological Products Bureau held in the headquarters of the MoA. Consequently, it is not an NGO in the formal sense of the term but, as its founding members were involved in the ecological movement of the 1980s, they often organize seminars and conferences, as well as publishing books and magazines on ecological issues outside their main interest in biological agriculture (for example, on genetically modified organisms, air-spraying of pesticides, etc). For the same reason they maintain constant and working relations with other NGOs like the Society for the Protection of the Prespes Lakes, WWF-Greece, and the Greenpeace-Greek Office. The DIO has participated since 1994 in meetings organized by the DSPEP concerning the implementation of the biological agriculture scheme under Reg. 2078/92. During these early meetings they disagreed with several issues proposed by the administration. Amongst them were the total budget allocated to biological agriculture, the segmentation of biological agriculture into zones, and the estimated cost of production for biological products for which, as the DIO admits, neither they nor the MoA had anything more than scanty empirical approximations. Furthermore, they offered their services as consultants in preparing the documents an applicant should submit to state authorities in order to be considered eligible for participation in the biological agriculture scheme of Reg. 2078/92. According to their opinion, the MoA did not follow their advice, but instead adopted a much more bureaucratic procedure. Today the DIO participates alongside GOS in the Monitoring Committee which, as mentioned above, is still inert. Nevertheless, the DIO is ready to accept that after 1993-1994 the MoA's attitude towards biological agriculture has been more positive.

Yet what the DIO fears most is that, although their role is to control biological agriculture production, they are gradually becoming part of the bureaucratic mechanism of the state and instruments of its own rationale, which is to achieve the highest possible uptake of the 2078/92 available funds. It seems worthwhile to mention at this point the organization's views on Reg. 2078/92 in general. The DIO appears to encounter the Regulation as if it refers almost exclusively to the biological agriculture program. Thus its assessment of the recent shift of the CAP towards agricultural practices compatible with the protection of the environment, concerns mainly the implementation of biological farming practices. Furthermore, the expected outcomes from the implementation of the Regulation account mainly to the development of biological agriculture at the European level. In addition, the suggestions for improvement point to the creation of one administrative center for all biological agriculture policies in order to overcome the contradictory and fragmentary approaches that still prevail in the MoA. Although this view is to be expected from an organization such as the DIO, which is so strongly interested in the development of biological agriculture, one should underline that this is a rather narrow view on the broader trends in the 'greening' of the CAP.

As the CAP and the EU environmental policy "are influencing social perceptions and discussions about countryside and rural problems", it is interesting to follow the discourse of the environmental NGOs on the rural environment and the countryside, and to see how they perceive the notion of rurality (Billaud et al. 1997). It seems that NGOs, with the exception of the DIO, conceive of the rural environment and countryside without reference to the role of farmers as a traditional social group for the development of rural society. It appears from their discourse on conservation strategies that the conception of farmers as both food producers and landscape guardians has almost entirely disappeared. Instead, it is the environmental dimension of rurality as 'free nature' that is emphasized. A clear indication of this trend comes from a GOS interviewee: "We think that with this aid offered under the long-term set-aside scheme, farmers could change their farms into biotopes. They could at least stop farming and start doing things with positive effects on the environment". The case of the DIO is of course different since its role of inspecting and certifying biological agricultural production entails the social dimension of rurality. Finally, WWF's attitude towards the farming community has not yet been formulated, since the organization has only very recently started to be interested in the biological agriculture scheme of Reg. 2078/92. Within this framework it has established a very small pilot project on biological agriculture within its wider project in the national park of Dadia, in Northern Greece.

There is no doubt that agri-environmental policy has so far had a positive impact from an administrative point of view. Two ministries, the MoA and the MoESPPW, which were rarely collaborative, found themselves in a position of intense communication because of the objectives and the content of Reg. 2078/92 and Dir. 92/43. Today this collaboration tends to extend to new policy areas which concern, for example, desertification, biodiversity, and the nitrate directive (91/676). Directly or indirectly, the same policy imposed the need for the interaction of state agencies with NGOs, research institutes, university departments, and private firms in a formal/informal network of communication and collaboration. This progress, though, should not be overestimated by attributing to state agencies a readiness to respond to changes in policies inaugurated by the reform of the CAP, thus ignoring the role certain individuals have taken in creating an agri-environmental policy network. Greece is a small country with a large and inefficient public sector, but a long tradition in the utilization of interpersonal relations. Only a few competent civil servants, especially in the DSPEP, are familiar with individuals from universities, research institutes and NGOs or the private sector who have some interest and experience in agri-environmental issues. To some extent these interpersonal relations compensate successfully for the many qualitative and quantitative weaknesses of the public sector, yet only to some extent. The lack of a national strategy and plan for the integration of agri-environmental policy into other existing rural development policies is more than obvious. So far, with the exception of one meeting for the discussion of biological agriculture at ministerial level, the political leadership of the MoA simply follows passively without intervening in the decision-making process of planning and implementing Reg. 2078/92.

Interactions between state agencies and NGOs are characterized by some mobility, but without concrete results or the elimination of the traditional mistrust both parties harbor for each other. This lack of a working relationship increases the isolation of the public sector from social actors and the society of farmers at large, because the co-operative movement and the professional organizations of farmers are not informed, thus aggravating their rather hostile attitude towards ongoing changes of the CAP since 1992. Spain, on the other hand, has a successful working example in that respect. The program 'Estepas Cerealistas' of Castilla-Leon aims at combining the protection of flora and wildlife with the extensive farming systems of this area. The program, administered by the regional authority of Castilla-Leon, is the outcome of collaboration between two social organizations. The first is the Sociedad Española de Ornitologia (SEO), a member of Birdlife International Association. The second is the farmers' union, Union de Pequeños Agricultores (UPA). On the one hand, the SEO takes care of issues such as biodiversity, anti-erosion, and non-abandonment of agricultural activities. On the other hand, the UPA secures a viable income for small and medium farmers and enhances its political influence in the area. From the perspective of the administration, this program is conceived as a partial solution to the problem of integrating a less competitive agriculture into a more market-oriented policy regime (Laddys Rueda Catry and Arribas Macho 1997).

Nevertheless, in the Greek case, it should not be underestimated that the objectives and content of the agri-environmental measures exercised pressure on the MoA to recognize the need for collaboration with NGOs and other social agents, even though this already existed in the form of interpersonal, informal channels of communication. The MoESPPW has been more open in this respect, but both ministries are still far from an institutionalized, formal collaboration with environmental NGOs and other social agents.

Finally, the Greek experience confirms that NGOs, initially weak in peripheral areas, have been strengthened in two ways as European integration is deepened. First, European integration has brought these areas to the attention of (internationally) established campaign groups and, second, it has given them new legal instruments to apply (Yearly, Baker and Milton 1994). However, it also indicates that, as the strengthening of NGOs in peripheral areas originates mainly from Brussels' political initiatives and funding and not from the increase of their popular support, they are in danger of becoming more of an extension of state bureaucracy rather than a political movement with its own independent ideology and voice.

3.2 Interaction with private agencies

In contrast to the NGOs rather marginal involvement, the role of the private sector in the promotion of the regulation was decisive in both case-study areas (Larisa - Mani). Before the implementation of the regulation in Larisa the aforementioned private company provided drop irrigation systems to a small group of farmers (owning 1000 ha in total) and prepared the field for the application of the suggested sophisticated fertilization techniques. During 1995 the company provided the new crystalline fertilizers to this team of farmers, thus initiating the implementation of the regulation in an informal manner. The agronomist owner of the company was a member of 'Eurogreen' (European consulting agency) and, through that network, he was informed that the implementation of the regulation could start in 1995, and not one year later. The same company established a close, albeit informal collaboration with the prefectural Directorate of Agriculture (DoA) in order to organize and facilitate the heavy and complicated administrative work. In parallel, during the preparation phase, the owner of the company had regular contacts with the central headquarters of the MoA, acting as an informal consultant. As a result, by the time the Commission gave the green light for the implementation of the specific program (reduction of nitrate pollution), this company was the only organization well-prepared for its implementation. Other agencies, and particularly representatives of agricultural cooperative Unions, argued that the program should start in 1996, so that everyone could compete with each other more fairly. On the other hand, local political pressures originating in the deepening crisis of the cotton production sector pushed the MoA to direct part of the Reg. 2078/92 budget to the area as a response to these demands.[8] Well positioned to take advantage of these developments by the end of 1997, the specific private company managed to establish contracts covering 10.000 ha of the program area's 26.000 ha total, leaving the rest of it to numerous other private firms and agronomists.

In the Mani case (biological agriculture oriented in olive oil cultivation), another private agency started to organize a network for the production, marketing and export of biological olive oil almost 10 years before the implementation of the Reg. 2078/92 program. The transition from conventional to biological agriculture was not a difficult process since, in certain communities, the traditional olive oil growing process was not far from the standards set by biological agriculture regulation. Before the Reg. 2078/92 started, this firm secured significantly higher prices for the biological producers (30-40% above the market prices). With its implementation the firm could pay lower prices to them in relation to the prices paid before. As a result, it benefited indirectly from the distributed subsidies. Similar to the Larissa case, this company established firm liaisons with a few agronomists of the prefectural DoA, especially the ones in charge of administrative work and control of the insect *Dacus oleae*. However, it could not establish the same kind of relations with the central headquarters of the MoA, which was not as flexible during the program preparation and design stage. The most important problem concerned the design of the Environmental Management Plan, which proved to be a very complicated process. Finally, in order to facilitate the implementation of the program, this private agency performed an administrative role, apart from its marketing and advisory function. More specifically, its technical personnel keeps the books concerning activities and observations related to the program, since it is difficult for farmers to do this themselves.

Subsequently, at least as the cases of Larisa and Mani illustrate, the private sector proved to be much more effective in its collaboration with the

[8] The cultivation of cotton had reached 440,000 ha in 1995, in other words, over one-third of the irrigated lands. As an indirect result, Thessaly, the principal agricultural region in central Greece and the center of cotton cultivation, became the 'heart' of the farmers' movement nationwide. When the government refused to draw on national resources to compensate for the falling price of cotton caused by overproduction and the application of the co-responsibility levy set by CAP, strong popular opposition arose against it.

public agencies than the NGOs. Its contribution, especially in terms of facilitating the administrative process as well as establishing a pre-existing network of producers/candidate participants, was crucial. Of course, the main drive behind this involvement was the simple fact that the companies benefited a lot – directly or indirectly – from the subsidies distributed via Reg. 2078.

4. DISCUSSION AND CONCLUDING REMARKS

Post-1992 agri-environmental policy, if combined with the environmental policy of the EU and the strategic perspectives summarized as Agenda 2000, may result in a breakthrough for the future of European agriculture. New emphasis is given to the environment of the agricultural sector and more generally to the concept of a sustainable countryside. These initiatives require a re-evaluation of the objectives and instruments by which the CAP has been developed since the 1960s. The dominant ideology of productionism, protectionism, and one-dimensional emphasis to the rural sector at the expense of social and environmental issues should be abandoned. In this respect, the concept of an integrated rural development strategy, which was introduced at a conference in Cork, Ireland (November 1996), enters the discussion with the aim of raising the multi-functional role that many farmers can play in generally managing the countryside in the context of their rural, rather than their solely agricultural activity.[9] This indicates that progress must be made towards environmentally sustainable production and consumption. It is true that relevant EU action was restricted to the development of various measures and programs, the sum of which gradually formed the structural dimension of the CAP. To this extent, although the reformed structural policy has stressed rural development by introducing the accompanying measures, it is argued that, in the years to come, there must be an optimal mobilization of synergies in order to make progress in an integrated rural policy.

Greece, a Southern European country which continues to have a large agricultural sector, favorable eco-geographical conditions, and great potential for sustainable rural tourism development, should seize the benefits of

[9] Although the farm Ministers did not accept the inclusion of the Cork Declaration into the Summit conclusions at that time (Dublin, December 1996), the concept of overall sustainable development was ultimately included in the Amsterdam Treaty (June 1997) as one of EU's objectives from now on.

the new rural development policy[10]. But this new policy framework for the countryside cannot be implemented by the existing agricultural policy network. New actors are urgently needed – actors capable of understanding and deepening politically, to say nothing of elaborating technically the novel aspects of the CAP. Environmental NGOs are the proper social agents to accomplish this task if they are given enough room further to develop their ideas and expertise. In this regard, prospects are not so gloomy as they used to be not long ago. After a first difficult period of adaptation (1981-1993), dominated more by the Greek government's resistance to, rather than compliance with the obligations of EC membership, today the majority of political parties support the European integration of the country. Yet, the environment is a policy sector where, for many historical, socio-economic and cultural reasons, the harmonization to European norms still meets with resistance. Even if, in due time, this resistance weakens, exploitation of the new policy instruments and financial means requires the fulfillment of three political prerequisites: the modernization of public administration, the autonomous and independent political expression of the farming professional/co-operative movement, and the organization and mobilization of civil society. In Greece, all these prerequisites for the implementation of a sound agri-environmental policy are more or less non-existent.

The objectives and content of the EU agri-environmental policy constitute an institutional framework favorable for an opening of public administration to environmental NGOs and other social agents. Planning and implementation of Reg. 2078/92 was impinged upon to some degree by the narrow time limits available for spending the budget allocated for this purpose. Our survey showed that, under the prevailing ideology of productionism, what mostly contributed to this course of action was the centralized bureaucratic structure of public administration which, in addition, suffers from inter- and intra-ministerial fragmentation and lack of coordination. Nevertheless, the initial collaboration of the MoA central services with the agricultural extension service at prefectural level in order to collect agronomic and environmental data nationwide, proved their competence when they were asked not simply to distribute and control subsidies and other EU provisions to farmers. At the same time, the status of DSPEP, a department

[10] The Luxembourg European Summit (12-13 December 1997) reached a final decision on the whole issue of Agenda 2000. Concerning the CAP, the Summit concluded that "the process of reform begun in 1992 should be continued, deepened, adapted and completed, extending it to Mediterranean production. The reform should lead to economically sound, viable solutions which are socially acceptable and make it possible to ensure fair income, to strike a fair balance between production sectors, producers and regions and to avoid distortion of competition" (Pezaros 1998).

of rather marginal importance in the framework of a 'production-oriented ministry', was upgraded, though its main problem of limited personnel and lack of expertise remains and is becoming more acute.

Furthermore, collaboration of the two ministries (MoA and MoESPPW) both in the implementation of Dir. 92/43 and later in the integration of data derived by the elaboration of the same directive toward the implementation of Reg. 2078/92, had a twofold effect. First, to create a new common policy ground, known ever since as agri-environmental. Second, to bring one ministry (MoA) less open to other social movements – except the farming co-operatives – into contact with another ministry (MoESPPW) more open to social agents, such as the environmental NGOs (e.g. GOS, GCBW, Arctouros-society for the protection of bear, Societies for the Protection of Monk Seal and *Caretta caretta* Turtle, Greek Society for the Protection Of Nature). We should note that the involvement of Greek NGOs in the elaboration and implementation of a sound environmental policy under the umbrella of an agri-environmental institutional framework is still in its infancy. After the collapse of the Green Party in 1990, the existing or newly formed NGOs focused their attention primarily to the protection of wild life and wetlands threatened by modern agriculture. Consequently they have tended to ignore the relationship between the natural environment and agriculture. Moreover, in Greece, they are not ready to associate their conservationist objectives with the role of farmers as a dominant social component of rurality. Thus, their experience as well as their views about the rural environment and countryside are restricted compared to the new role which the post-1992 agri-environmental policy envisions for them. While some of the aforementioned NGOs were informed from the very beginning about the regulation, ultimately they were not invited to participate in the ministerial Committee designing its implementation. Nowadays, the relationship between the MoA and NGOs has taken on a more formal character as the MoA invited NGOs along with other representatives from universities, research institutes and cooperatives' confederation to participate in a committee to monitor the progress of regulation – a committee which, a year and a half after its appointment, has never worked. The MoA, although more collaborative now with NGOs than it used to be, still considers them as lacking the expertise necessary to make an essential contribution to regulation implementation. Notwithstanding the validity of these claims, the responsibilities of the MoA remain intact. Even if an informal network of agri-environmental policy has already been established through the initiatives and interpersonal links of certain officers of the MoA with members of academia, NGOs, and private consultants, this does not adequately address the current technical or, more importantly, strategic needs of the expanding field of rural policies. A lot more has to be done. DSPEP personnel should be doubled at the very least. Collaboration with extension workers at prefectural level should be strengthened, thus obliterating its current bureaucratic character. The committee monitoring of regulation progress should be convened as soon as possible. Last but not least, the political leadership of the MoA should intervene actively to shape national expectations of agri-environmental policies. Four years after the Greek state's adoption of Reg. 2078/92, it is only now that the responsible ministry has begun to consider – although still not to act toward – the establishment of a committee for the harmonization of agri-environmental and other related and/or contradicting agricultural/structural policies. Typically Greek though this may be, success or failure of the regulation implementation still greatly depends on individual initiatives, political influence, and interpersonal links of certain active staff members of the DSPEP. Consequently, long term questions of strategy and targeting of agri-environmental policies are still not even being posed by the government and responsible state agencies. Essentially, the problem is political in nature. The Greek state is still reluctant to accept environmental NGOs as equal partners in shaping policies of major economic importance, while the political weakness of these organizations makes them unable to press for their views to be heard.[11]

This lack of long term considerations might have been less important if agri-environmental policies were not such totally new policy area and not such a self-evidently positive instrument for the sustainable development of the EU's Southern member-states. For one, when deconstructing the explanation and logical thrust of Reg. 2078/92, the ideal argument supporting it (i.e., the shift from reducing overproduction to promoting environmentally sound uses of natural resources and from decreasing the intensity of production to promoting of biological agriculture) reveals the preponderance of 'Northern problems' and of the Northern, rich countries of the EU (Billaud

[11] This is not another Greek peculiarity. Even in the UK, the predominant political structure has prevented the emergence of a strong Green Party, which in many other EU countries has actively lobbied for improved rural conservation through AEP (e.g. Germany). Thus, in the UK, groups such as RSBP, the Wildlife Trusts or the Council for the Protection of Rural England have assumed an important lobbying role with regard to agri-environmental issues. As with Regulation 797/85, article 19, the formulation of AE schemes under Regulation 2078/92 is the responsibility of the central government, with advice and technical support coming from the statutory agencies. So the green political voice has been relegated to environmental NGOs, as well as government agencies such as English Nature or the Countryside Commission. Although these groups have some influence over government AEP, they are nonetheless politically marginalized and may lack the crucial political credentials at times of important decision-making (Hart and Wilson 1998).

et al. 1997). This critique interprets the concluding remarks of a recent comparative research project on the implementation of the regulation in Germany, Portugal, and Spain. Apart from the differences of implementation found throughout Germany, this research shows two things. First, it shows that agri-environmental policies (AEP) are an appropriate basis for a redefinition of agricultural modernization paths in that – from the perspective of managing and conserving rural landscapes – they revalue, in positive terms, those traditional forms of production which were formerly regarded as technically and economically inefficient. Second, it shows that there are differences in the strategy adopted by the two Southern member-states. The Portuguese Ministry of Agriculture has interpreted the AEP by considering multiple aspects of rural society, including social, demographic, economic and environmental problems. The AEP were introduced into rural society as an attempt to re-address problems associated with traditional agriculture. The primary focus was not reconversion to extensification, but rather the maintenance of existing traditional, extensive forms of production. The bulk of AEP has been targeted toward the conservation of traditional agricultural systems by, in a rather diffuse manner, indirectly subsidizing either the preservation of land use or the existing state pensions. In Spain, although AEP do not respond to the main environmental problems plaguing Spanish agriculture, that is, erosion and drought, and although their integration with other policies of rural development is poor, they have stimulated an interesting debate within the agricultural policy network, especially within the farmers' co-operative movement, on the relationship between agriculture and environment. This debate has adopted the concept of sustainability as a central reference point (Garrido/Moyano 1996, cited by Bruckmeier, Patricia and La Calle Dominguez 1997).

Targeting is a fundamental feature of any public policy. Increasing public expenditure on AE schemes will bring with it ever closer scrutiny of their environmental effectiveness and value for money. Many of the environmental problems associated with modern intensive agriculture – such as soil erosion and nitrate pollution – lend themselves to area-based targeting, requiring for their solution often quite drastic and long term shifts in land use within precisely defined vulnerable locations. An effective program of habitat reconstruction and restoration would also require careful targeting. And yet, there is still little public debate about, much less research into the different targeting strategies that might be available (Potter, Cook and Norman 1993). If in Europe this debate is 'still little', in Greece the attention of the agricultural policy network is drawn exclusively to the preservation of protectionist policies which traditionally supported the intensive farming systems on irrigated plains and which neglected mountain and marginal areas (Louloudis and Maraveyas 1997).

On the other hand, private interests, especially at the local level, were ready to exploit a new source of subsidies and EU transactions. State policy promoted a bundle of measures (e.g. reduction of nitrate pollution, biological agriculture, long-term set-aside), while ignoring other more acute environmental problems (e.g. erosion) in the face of pressure from political circumstances (the reduction of cotton subsidies in Thessaly) or vested interests at the local level (the use of crystalline fertilizers in Thessaly, biological olive-oil production and exports in S. Peloponnisos). Representatives of the private sector performed, in some cases, the role of a driving force behind the regulation's initiation and implementation. They collaborated with the central and regional headquarters of the MoA in order to get things done and to overcome certain bureaucratic obstacles. In this way they proved to be one step ahead of other actors and agencies. Of course, both the bureaucratization of the Greek public extension service and the remarkable development over the past three decades of the private agrochemical sector, contributed greatly to this active private involvement in the programs' design and implementation stages. It is evident that the agri-environmental subsidies 'package' represents a new arena for private companies to reorganize themselves and establish alternative expansion strategies. These strategies fit well with the public administration's pressing interest to achieve high rates of absorption in the relevant funds.

Finally, it is of paramount importance that the implementation of agri-environmental policy be the result of collaboration and synergy between the agricultural policy network, the environmental NGOs, and the farmers themselves. The aforementioned example of Castilla-Leon illustrates the potential of such an approach. Things cannot not be so readily complementary anywhere else, but still, this example is of some value, since it shows how different agencies, departing from their own perspectives and interests but sharing a common understanding of new agri-environmental policy, are able to collaborate creatively.

REFERENCES

Beopoulos, N., and Louloudis, L. (1997) 'Farmers' acceptance of agri-environmental policy measures: A survey of Greece, *South European Society & Politics* **2**, 118-137.

Billaud, J.P., Bruckmeier, K., Patricio, T., and Pinton, F. (1997) Social construction of the rural environment. Europe and discourses of sustainability in France, Germany and Portugal, in H. Haan, B. Casimis and M. Redclift, *Sustainable Rural Development*, Ashgate, Aldershot, pp.9-34.

Bruckmeier, K., Patricia, P., and La Calle Dominguez, J. (1997) *A New North-South Development Axis with the CAP Reform?* Paper presented at the XVII Congress of the European Society for Rural Sociology "Local Response to Global Integration. Towards a New Era of Rural Restructuring", 25-29 August, Chania (Crete), Greece.

Cammarata, A. (1997) Agriculture and Environment, CAP Working Notes, Special Issue Brussels, European Commission.

Collins, N., and Louloudis, L. (1995) Protecting the protected: The Greek agricultural policy network, *Journal of European Public Policy* **2**, 95-114.

Hart, K., and Wilson, G. (1998) *Agri-environmental Policy in the UK: From Policy Shaper to Policy Receiver*, Unpublished manuscript.

Koutsouris, A., and Papadopoulos, D. (1998) Extension functions and farmers' attitudes in Greece, in N.G. Roling and M.A.E. Wagemakers, *Facilitating Sustainable Agriculture*, Cambridge University Press, Cambridge, MA, pp.88-99.

Laddys Rueda Catry, C., and Arribas Macho, M. (1997) *The Role of Spanish Farmers' Unions in the Implementation of Agri-environmental Programs: The Case of "Estepas Cerealistas"*, Paper presented at the ESRS Congress, 25-29 August, Chania (Crete), Greece.

Louloudis, L., and Maraveyas, N. (1997) Farmers and agricultural policy in Greece since the accession to EU, *Sociologia Ruralis* **37**, 270-286.

Moyser, G., and Wagstaff, M. (1987) Studying elites: Theoretical and methodological issues, in G. Moyser and M. Wagstaff, *Research Methods for Elite Studies*, Allan & Unwin, London, pp.1-24.

Pezaros, P. (1998) Agenda 2000: Reforming the common agricultural policy further, *Eipascope* **1**, 22-29.

Pitman, J. (1992) Changes in crop productivity and water quality in the United Kingdom, in K. Hoggart, *Agricultural Change, Environment and Economy*, Mansell, London, pp.89-122.

Porritt, J. (1997) Environmental politics: The old and the new, in M. Jacobs, *Greening the Millennium? The New Politics of Environment*, Blackwell Publishers, Oxford, pp.62-73.

Potter, C., Cook, H., and Norman, C. (1993) The targeting of rural environmental policies: An assessment of agri-environmental schemes in the UK, *Journal of Environmental Planning and Management* **36**, 199-216.

Redclift, M. (1993) Sustainable development: Concepts, contradictions and conflicts, in P. Allen, *Food for the Future. Conditions and Contradictions of Sustainability*, John Wiley & Sons, New York, NY, pp.169-192.

Tracy, M. (1997) Agricultural Policy in the European Union and Other Market Economies, APS - Agricultural Policy Studies Brussels, AGRA FOCUS.

Yearly, S., Baker, S., and Milton, K. (1994) Environmental policy and peripheral regions of the European Union: An Introduction, in S. Baker, K. Milton and S. Yearly, *Protecting the Periphery: Environmental Policy in Peripheral Regions of the European Union*, Frank Cass and Co, Essex, pp.1-21.

Chapter 12

Images of Sustainable Development in Italy[1]

Dissenting Voices of Environmentalists, Business and Civil Servants

Mario Diani

1. PRELIMINARY REMARKS

The emergence of new institutional arrangements requires new cultural models and discursive practices. This applies regardless of the particular view of institutions we subscribe to. Whether we refer to them in a procedural and organizational sense, or, more loosely, as taken for granted patterns of social interaction, we cannot conceive of institutions without reference to peculiar cultural models (Powell and DiMaggio 1991). It is through cultural production that emerging institutions consolidate by developing specific identities, securing legitimacy, and being perceived as universal (Lanzalaco 1995: 61-65). In this chapter I draw upon this general principle to analyse the institutional impact of the idea of sustainable development (henceforth, SD) in Italy in the 1990s. I discuss whether this concept has at all affected institutional innovation in environmental policy networks, by providing a common cultural frame and thus allowing previously distant actors, with conflicting stakes in environmental policy, to develop some degree of mutual understanding and some procedural consensus on how to address environmental issues.

[1] An earlier version of this piece (Diani 1996) was produced for Institute IARD in Milan as part of a comparative project on 'Environmental Sustainability and Institutional Innovation', funded by DGXII of the European Commission (contract EV5V-CT94-0389), and coordinated by Patrick O'Mahony of University College Cork. I'm grateful to Gianpietro Gobo and Anna Lisa Tota for their collaboration in the conduction of interviews to senior civil servants.

K. Eder and M. Kousis (eds.), Environmental Politics in Southern Europe, 299–320.

First, I illustrate how the idea of SD was elaborated in different ways by different types of actors (NGO leaders, civil servants, and business representatives) in the mid-1990s.[2] In the second part of the paper, I examine whether the emergence of SD as a potentially unifying interpretative frame actually facilitated new institutional arrangements, driven by the principles of discursive democracy in the second half of the decade (O'Mahoney and Skillington 1996; Ruzza in this volume). Finally, I relate briefly my findings in relation to broader questions about the institutionalization of environmental politics in Southern Europe. The bulk of empirical evidence comes from in-depth interviews, conducted between the Fall of 1993 and the Spring of 1994 in Milan and Rome (see Appendix), and from written documents (Sassoon and Rapisarda Sassoon 1992; Conte and Melandri 1993). I also draw upon recent investigations of specific environmental policy networks (Giuliani 1998; Tebaldi 1998).

2. CONCEPTIONS OF SUSTAINABLE DEVELOPMENT AMONG ENVIRONMENTAL POLICY ACTORS

2.1 Environmentalist views: Legambiente's 'scientific environmentalism'[3]

Environmental actors tended to define SD along the lines proposed by the Brundtland Commission, according to which a) natural resources should not be exploited at a pace faster than the time required for their reconstitution; b) emphasis should be placed upon recycling, and priority assigned to the use of renewable materials and energy sources; c) emissions of polluting agents and waste disposal should not exceed the environment's capacity to absorb them (Alberti and Parker 1993). They approached environmental

[2] The contested and ultimately ambiguous nature of the concept of SD has been repeatedly underlined (Baker et al. 1997; O'Riordan and Voisey 1997). The Italian case is broadly consistent with this picture.

[3] This analysis is based mainly on documents and statements by leaders of Legambiente, the major political ecology organization in the country, whose strategies of action are, according to its members, very close to those of Friends of Earth International (Interview 4; Diani 1995: chapter 2; Donati 1996). Accordingly, it should be received as an account of the view of SD held by the Italian movement's most prominent organization, rather than a reflection of the position of a presumably homogeneous movement.

action from a strongly scientific and public interest perspective, arguing that environmental problems should be regarded as technical as much as political problems, and definitely not as partisan issues. Reasonable and acceptable solutions should be acceptable to all those, including business actors, who are far-seeing enough to envisage the general pay-off that would result from the large scale adoption of environmentally sounder technologies (Interview 4; Conte and Melandri 1993, 1994).

In contrast to the industrial perspective on SD, advocating a more efficient use of conventional energy sources (see 2.2 below), stronger attention was paid to alternative energy sources (hydroelectric, aeolian, etc.). Energy consumption, in particular of non-renewable sources, was still growing at unacceptable rates (Interview 5). Accordingly, Italian environmentalists pushed for the introduction of a 'carbon tax' along the lines followed in Northern European countries, an option which faced severe criticism from the industrial ranks, including supporters of SD (Sassoon and Rapisarda Sassoon 1992).

Altogether, however, the symbols and the arguments supporting these proposals largely differed from the anti-industrial, counter-cultural jargon, traditionally popular among left-wing ecology groups (Richardson 1997). The roots of the Italian crisis were identified in "the simultaneous absence of an efficient welfare state and of credible, transparent mechanisms of market competition .... [the existence] of a protected industry [that has not relied] upon entrepreneurial skills but rather on systematic political exchange with policy-makers" (Legambiente 1994: 2). References to economic inefficiency and the ensuing risks of marginalization for the Italian economy were indeed frequent, to the point of using the metaphor of 'azienda Italia' (the Italy company), a "condensing symbol" (Snow and Benford 1988) traditionally more popular among economists and business operators than among environmental activists (Legambiente 1994: 1).

This approach struggled not surprisingly to identify the threshold, beyond which the social costs of natural protection policies became intolerable; or, in other words, the extent to which the consumption of non-renewable sources could be reduced in developing countries, in light of the (real or presumed) economic and social advantages it otherwise entailed. In spite of the substantial differences between the 'new' politics of the 1980s and 1990s and previous examples of oppositional political action, ecology activists with a background in either social movements or traditional left-wing organizations still were particularly concerned with issues of equity and global social justice. The notion of SD was indeed directly related to the persisting, and even deepening, power inequalities between the Northern and the Southern hemisphere (Melandri 1993). SD was seen as a chance to respond to massive migration movements from the poorer to the wealthier areas of the planet, and to the rising gap in income and consumption styles between the former and the latter. Environmental defense strategies should therefore incorporate a broader set of social and political problems. These included the reconversion of former Communist countries to market economies; the need to cope with the intolerable constraints public debt posed on developing countries, strengthening their subordination to Western countries.

References to unemployment issues also highlighted the connection between SD and social justice (Conte and Melandri 1994; Legambiente 1994). Environmentalists challenged the common wisdom relating technological innovation to a necessary reduction of the labor force – an argument which has been routinely raised against environmental concerns (Pridham and Konstadakopulos 1997, with specific reference to Southern Europe). New employment could indeed be created if the basic guidelines of employment policies, uniformly adopted by otherwise ideologically different governments, were abandoned. Resources should be subtracted from expensive and environmentally damaging public works like motorways, tourist harbors, etc. More money should rather be invested in the maintenance of cities' architectural and urban quality, the expansion of public transport, the management of natural areas, etc. Research and innovation on environment friendly goods and technologies should also be promoted, and massive reduction of working hours be considered.

While all these proposals were consistent with an SD perspective, environmental leaders differed in their explicit recognition of the importance and centrality of the concept. Former anti-nuclear campaigners feared that the concept of SD was more in tune with industrial perspectives on the environment than previous political ecology approaches. In particular, they warned that excessive emphasis on issues like global warming – at the core of any SD strategy – might provide nuclear power hard-liners with the opportunity to revitalize their arguments, and try to reverse the decisions taken in the aftermath of the 1987 referenda (Diani 1994b). Nor did the campaigners involved[4] regard as adequate the amount of organizational resources actually devoted by Legambiente to campaigns most explicitly related to Rio 1992 deliberations, such as the one on global warming (Interview 4; Interview 6).

[4] This might however depend on the fact that, following shortage of funds, the amount of resources groups like Legambiente pour into specific initiatives is largely dependent upon adequate private sponsorships. Campaigns targeting more concrete issues like the quality of Italian beaches were not surprisingly more likely to attract business support than those on broader, more immaterial problems like global warming (Interview 4).

2.2 Business views: Sustainable market opportunities

The business approach to SD took a quite different, if not wholly incompatible, path. Some of its most outspoken advocates viewed it as "the achievement of the firm's goals in such a way that the production process is shortened, the use of energy and raw materials – and not just of labor force and capital – are kept to a minimum, and wastes of any kind are cut down as much as possible" (Sassoon and Rapisarda Sassoon 1992: 25). For individual firms, SD was, in the last analysis, "a chance to use the environmental question in order to increase their market competitiveness. I do not see an ethical approach to this problem on the part of the business world, although there is a line of thinking which frames the question in terms of social responsibility .... there are always economic reasons lying behind" (Interview 13). Global SD was regarded as unlikely, since current differences between the North and the South of the planet tended – if anything – to be deepening (Interview 13).

Conditions for SD in Italy were increasingly favorable. The end of ideological conflict allowed actors, previously on different sides, to join forces on behalf of shared goals, such as environmental protection – although attitudes towards ecology movements were still largely hostile and suspicious (see below). Environmentally sounder technologies were central to this view of SD, as they allowed firms to increase the efficiency of production processes, and therefore to enjoy competitive advantages in both national and international markets. Advanced technologies also required less energy and raw materials than outdated ones (Interview 13). In order to achieve these goals, however, substantial innovation was needed, especially in the relationship between public authorities and the business world. As public regulation had largely been conceived of in restrictive terms, based on 'command and control' approaches, and on a misrepresentation of the firm merely as a source of pollution, excessive constraints had been imposed on economic activities (Sassoon and Rapisarda Sassoon 1992: 55).

This approach had been modified to some extent only with the EU Fifth Action Program. A modern notion of environmental policy should be inspired by the following basic principles: a) modifying the relationship between public agencies in charge of environmental controls and the 'controlled' economic subjects; b) modifying the relationship between environmental regulations and market competition principles (Sassoon and Rapisarda Sassoon 1992: 64); c) replacing the 'Polluter Pays Principle' with a 'prevention pays principle' (Sassoon and Rapisarda Sassoon 1992: 19). Requirements for a new balance between regulation and deregulation included a) simpler Italian, but also European, legislation; b) a smaller number of institutional bodies dealing with permits and authorizations related to the environmental implications of any business activity; c) clearer environmental competencies between different public agencies; d) better coordination between ordinary regulation and new tools of environmental control, like eco-auditing, eco-labeling, etc. (Sassoon and Rapisarda Sassoon 1992).

The popularity of such an approach among ordinary Italian firms was, however, unclear. While the most impressive actions to cope with SD ideas has been undertaken by the biggest, usually multinational, firms (see details below), Italian industry largely consists of small and medium firms, with stricter margins for technological innovation, and greater exposure to market competition (Donati 1994). This did not encourage the diffusion of new ideas in a milieu which many observers regard as fairly narrow-minded and resistant to innovation (Interview 3; Diani 1994a; Donati 1994; Pridham and Konstadakopulos 1997: 135). Survey data still identified the need to comply with environmental regulations as the major reason behind firms' attempts to improve their environmental performance. There was little appreciation of the possibility of combining technological innovation and better environmental performances with competitive advantages.[5] Some interviewees indeed explicitly imputed Italian entrepreneurs' failure to secure market niches in recently developed areas, like automotive anti-pollution devices, waste disposal, etc., to their indifference to global changes in the environmental field (Interview 3). Firms' internal structure also cast doubts on the real impact of SD notions in the business community. In most firms environmental problems were still the responsibility of personnel and industrial relations branches, i.e., they were still cognitively associated with workers' health and working condition problems in the first place (Donati 1994: 27).

Advocates of a business approach to SD placed a strong emphasis on competitive advantages in order to render the idea appealing to a largely unreceptive audience. At the same time, though, they were most careful at preempting charges of insufficient social responsibility in favor of pure deregulation. Awareness of the social role of business had steadily increased throughout the 1980s in the Italian business community (Sassoon and Rapisarda Sassoon 1992: 63). This secured the effective combination of economic growth and environmental protection. Italian proponents of green industry redefined producer's responsibility by drawing upon the idea of

[5] See Dionisio (1994). These findings were part of a project on the diffusion of eco-auditing and eco-management procedures among 6,000 firms in five European countries. Unfortunately, the low rate of response (about 13%) and the differences between national sub-samples rendered any comparative evidence seriously flawed.

'extended producer liability'[6]. It was no longer restricted to the quality of the goods produced, but encompassed both the environmental impact of the production process and the problems related to the disposal of goods when their life-cycle had come to an end (Pavesi 1994).

2.3 Civil servants' views

Some civil servants held rather hostile views of SD. As a senior officer in the Industry Department put it: "Well, it's a word I don't like... anyway, by SD I mean as much development as possible, trying of course to render it compatible with environmental needs... I don't like it because it suggests that there is some environmental barrier, which must not be crossed... I emphasize the notion of the development, in as much as it is development which allows to improve environmental standards, not vice versa" (Interview 9). "Environmental degradation is negatively correlated to levels of development, Third World countries are the ones with the worst environmental record... only the environmentalists fail to understand that Malthusian notions of limits to growth are actually against the environment... I call this environmental fundamentalism" (Interview 9).

Other positions were, however, quite close to the more standard notion put forward in the Brundtland Report. One senior officer in the Environment Department identified for example SD as "a type of social and economic growth which is compatible with the amount of actually available environmental resources, and with the need to reproduce the eco-systems" (Interview 10). Another related it to pre-industrial ages, when economic growth was slow enough not to threaten natural resources, and viewed it as an attempt to keep the pace of economic growth and the destruction of natural resources under control; SD also aimed "... at redressing current imbalances between the different socio-political regions of the world" (Interview 11). Conceptions of growth where GNP was the only indicator of welfare and development were also criticized. The costs of environmental damages generated by natural disasters such as floods, landslides, etc. should not be included in the GNP. By current standards, what should be treated as sheer losses were paradoxically included among gains (Interview 11).

In contrast to business, technological innovation did not play a key role in civil servants' approach to SD. Some respondents ranked it behind changes in public opinion – possibly the most important precondition for better environmental preservation – and economic operators' responsibility

6 Although the source cited here refers to 'liability', it more plausibly means 'responsibility' in the broadest, non legalistic sense of the term.

(Interview 9). Technological change was considered useless in the absence of restrictions to sources of pollution: "If I improve equipment to reduce car pollution, but the number of cars increases, then in terms of global pollution I have not achieved any improvement" (Interview 12). On this ground, even programs like Agenda 21 were very generic and void of empirical content (Interview 12).

There was a tendency to emphasize producers' rising responsibility and to expect them to take care of the whole product life-cycle. Interestingly, this need not necessarily to happen through legislation. Rather, greater emphasis should be placed on market mechanisms, which were more easily perceived by those directly in charge of production processes. "Thinking of waste disposal, legislative regulation is hardly feasible, as it is the responsibility of those who have no direct control of production processes. Instead, market mechanisms are in principle more effective because they render it immediately clear to producers where potential gains are" (Interview 11). In any case, the usefulness of the responsibility principle for better environmental preservation was doubtful. Its fair application might have encouraged proper behavior from both producers and consumers; otherwise, they might have been tempted to escape the costs (Interview 11). General ideas about private property should also change somewhat: "Producers should realize that they will increasingly get involved in new arrangements, where they will remain owners of their products, and consumers will pay no more to secure property, but rights of use... it will be like some sort of generalized leasing scheme." (Interview 11)

Views about the involvement of public opinion were also diversified. Some feared that extensive applications of the subsidiarity principle on the local level might lead to any public project being stopped by citizens' NIMBY local opposition, whatever its specific merits: "I can't see how one could further extend citizens' powers in our country... any local council can call referenda to stop any project, one can't go any further" (Interview 9). Other senior bureaucrats held more positive views of citizens' involvement. For instance, senior staff of the Environmental Department had facilitated the involvement of over 300 local action groups, independent of city councils, in consultations regarding the environmental impact of planned high-speed trains in Northern Italy (Interview 10).

There were none the less a few points of convergence. In particular, subsidiarity was regarded as a potential danger, as it offered national governments the arguments to challenge European directives. Several governments, including Italy, were often reluctant to implement European legislation in their own countries (Giuliani 1992; Pridham and Konstadakopulos 1997). In this respect, according to an Industry Department official, "subsidiarity has fuelled inter-state controversies, at least in the short term... any time one state does not like any specific directive, it invokes the subsidiarity principle to slow down its implementation..." (Interview 9). It was also noted – this time by a member of the Environment department – that "a great deal of Italian environmental legislation has been pushed forward by European laws... we hope the Union does not relax their monitoring over national policies" (Interview 10).

2.4 Shared and conflicting frames

Representations of SD differed in their definition of both principles of 'responsibility' and 'subsidiarity'. The environmentalists emphasized polluters' direct and specific responsibilities for the impact of their activities, i.e., a notion of responsibility strongly related to a view of economic actors as sources of environmental damage. In contrast, business advocates of SD rejected the linkage between responsibility and 'polluter pays principles', relating the former to a closer, more systematic monitoring of the whole life cycle of goods on the part of producers. This should be driven by producers' commitment to the broader collective welfare. Civil servants targeted most powerful social actors – i.e., business – as those with greatest responsibility for environmental protection. At the same time, they also expressed sympathy for a reduction of their own direct control and regulatory capacity, in favor of self-regulatory mechanisms.

As for subsidiarity, environmentalists related it primarily to the inclusion of ordinary people in the policy process, especially though, not exclusively, at the local level. At issue was not the role of regulation *per se*, but rather the influence that concerned citizens should be able to exert upon it. The same notion was also framed, though more indirectly, in business discourse. Sometimes subsidiarity consisted of the reduction of controls by public bureaucracies and environmental protection agencies on the production processes, and therefore, of the greater importance assigned to the sense of responsibility of individual producers. At other times, citizens' participation was equated with consumers' behavior and, as such, it was seen as a major source of both challenges and opportunities for the business community. However, an explicit political role for citizens was feared rather than required, because the spread of irrational, anti-industrial feelings within public opinion was precisely behind many of the problems currently experienced by firms. Civil servants were actually the most suspicious towards subsidiarity, either because in their views it was already very high (Interview 9), or because they feared it might reduce the stimulating role of EU regulation.

These different framing approaches were not totally incompatible, though. For instance, the need to monitor the whole product life cycle was underlined by business operators, ecologists, and civil servants alike (Bianchi 1993: 136). All parties also advocated the growth of environmental consciousness among the general public, albeit the reasons behind such hopes were, not surprisingly, rather different. Another broad convergence could be found in the mainly technical way of framing environmental problems. Far more so than in the past, Legambiente documents reflected their efforts to legitimize their claims by sophisticated technical and economic reasoning (Conte and Melandri 1993, 1994; Legambiente 1994). This did not prevent strong disagreements on specific problems, as the carbon tax controversy demonstrated. Still, greater homogenization than in the past could be found in the definition of key problems.

Differences were stronger in the most explicitly 'political' aspects of frames; aspects, for example, which were expected to provide a rationale for the adoption of the SD perspective by overcoming of traditional barriers. In particular, a preliminary analysis suggests that business representatives' (but also staff in the Industry Department) were indeed more explicitly critical of ecologists than the other way around.[7] This might have been due to the fact that in Italy the industrial approach to SD was still largely unpopular among its potential audience. As a consequence, its supporters may have felt pressured to convince their fellow industrialists that their approach had nothing to do with the environmentalists' positions: "Earlier proponents of a green approach to business activity risked being misinterpreted as an environmental 'fifth column' by their fellow industrialists. Therefore, they had – and somehow still have – to make it very clear that their position differed drastically from that of ideological and political ecologists." (Interview 13)

3. EMERGING ARRANGEMENTS AND SUSTAINABLE DEVELOPMENT

The potential for the development of more cooperative and less confrontational relationships between environmentalists, industrialists, and government has been quite good throughout the 1990s. Environmentalist organizations have consolidated their position of relative prominence and influence

[7] Contrast for instance Sassoon and Rapisarda Sassoon (1992) and Conte and Melandri (1993, 1994).

in the polity. At the national level, the Greens' inclusion in the government and the appointment of a Green MP to the Ministry of the Environment in 1996 has encouraged institutional innovation and, at the very least, enlarged opportunities for informal consultation (Giuliani 1998; Giuliani, Rinkevicius and Sverrisson 1998). At the same time, though, preliminary data on environmental protests between 1988 and 1997 suggest that the capacity of the environmental movement to promote protest actions of national relevance – if not protest actions in general – has gradually decreased over the 1990s. Environmental NGOs appear increasingly oriented towards pressure through a combination of mailing campaigns and petitions, and scientifically informed pressure in the form of technical reports or counter-plans (Donati 1996; Diani and Forno 1999). A more complex relationship between environmentalists and other actors has apparently developed, where dialogue and cooperation on some issues parallel conflicts and disagreements on others.

Business' propensity to establish more dialogue-oriented relationships with both government and environmentalists may have been encouraged by at least three changes: first, the increasing pressure coming from environmental legislation, especially from the EU; second, the growing awareness of the relationship between rising ecological standards and market dynamics, and therefore of the opportunity the former present for both technological innovation and the acquisition of new market niches; third, the acceptance of environmental groups as a force to be reckoned with. At the very least, environmental groups may display strong local opposition to specific industrial projects, or they may damage the public image of goods they regard as dangerous for the environment and the consumers alike.[8] Some members of the business world indeed acknowledged that without pressure from environmental groups business operators would have probably been far slower at seizing the opportunities created by the rise of environmental consciousness: "Today, many environmentalists think of their role as a stimulus for industry to adopt stricter environmental standards... while earlier environmentalists regarded industry as inherently dangerous." (Interview 13)

At the time of the interviews, some civil servants were still very skeptical of this supposedly developing dialogue with the environmentalists. Although their national leaders were given some credit for their more moderate orientations, ecology groups were largely seen as obstacles to any policy implementation, given their indiscriminate opposition to any infrastructure, including those supposed to tackle environmental problems such as water depurators or waste incinerators (Interview 9). According to these critics,

[8] On the relationship between business, consumerism, and environmentalists see Donati (2000, especially chs.4-5).

this approach testified to a dangerous lack of global strategies, and to the ecologists' tendency to capitalize on people's irrationality (Interview 12). Others, however, differentiated between national organizations and grass-roots groups, which the former managed to control and steer only to a limited extent (Interview 13).

Nor was the relationship between civil servants and business unproblematic. Business interests had solid links with the Industry or the Budget Department. Both had a strong say – possibly a stronger say than the Environment Department – in environmental policies and were usually regarded as friendlier to business' arguments. In contrast, the environmental ministry was often perceived by business as a source of further complication and bureaucratic obstacles, rather than as a step towards the rationalization of environmental policy (Donati 1994). It was also often regarded as the ecologists' ministry.[9] Yet, in spite of this perception, the Environment Department's positions on a number of issues were quite close to those of business. At the time of the interviews there was, for example, convergence on the rejection of carbon tax proposals, or on the view that further steps towards stricter environmental standards (e.g. on the issue of CO2 emissions) should only follow general agreements at the international level (Sassoon and Rapisarda Sassoon 1992: 125-126; Interview 5). We also found broad orientations within the Environment Department, arguing for greater interaction and more formal consultation with business interests, while public bodies still retained the ultimate power of decision-making (Interview 10).

Overall, despite persistent mistrust and a lack of communication, the climate seemed to be broadly conducive to the development of new relationships between the different actors with strong stakes in environmental policy. Evidence of a change can actually be found in a variety of domains. First, environmentalists have been increasingly involved in committee work at the national level. Formal procedures for the consideration of their position had been missing during the whole period of committee work on the implementation of Rio deliberations. Members of the major environmental associations sat in some committees (e.g. in the Commissione Ambiente Globale) as individual experts rather than as representatives of public interests; other times, they were not included in any working group (e.g. in the case of the preparatory work for the ratification of the Rio convention on global warming). Although informal consultation was viewed as routine practice by both the environmental ministry and officials from ecology

9 See for example Interviews 3 and 9. That environment and industry departments are strongholds of the respective interests is however a world-wide perception (Lowe and Goyder 1983) rather than an Italian peculiarity.

NGOs (Interviews 4, 5, and 10), this depended on personal relationships rather than public recognition of the importance of certain interests and points of view for the policy process.[10]

A certain degree of tension actually existed between politicians, scientific experts working within public agencies like ENEA, and representatives of ecology NGOs. Sometimes, as in the preparatory work for the Rio 1992 conference, relationships between politicians (especially the Minister) and NGOs were particularly fruitful, yet with some dissatisfaction on the part of in-house bureaucrats, who were frustrated at not being assigned the prominent role they would deserve (Interviews 6 and 7). At other times, as on the occasion of the implementation of Agenda 21, interactions between civil servants of different public agencies were stricter than between the Environment Department and the ecologists, with the former being given the final say and NGOs being just part of the consultation process.[11]

More recently, however, there have been several signs of growing involvement on the part of environmental NGOs. At the national level, the National Environmental Agency was finally established in 1997 under the leadership of a prominent environmental activist. Environmental groups have also been involved in more systematic consultations with the Minister of the Environment on key policy areas like plastics waste (Tebaldi 1998). Environmentalist research centers also represent a more systematic source of up-to-date statistical information on the environment than do the official bodies.[12] Local initiatives to implement Agenda 21 have been far from systematic, but where they have taken place, they have seen a significant presence of members and organizations from the environmental sector. Members of major groups like Legambiente and WWF have played active roles in projects promoted by city councils, like Milan, Bologna, Rovigo, Livorno, or Venice; research centers close to environmental NGOs, like Istituto Ambiente Italia, have also been involved (Giuliani 1998; Andringa et al. 1998: 86-89; Di Giulio and Pareglio 1998).

Interaction between public agencies has also improved. Most tensions actually developed with those actors who "operate in sectors let's say op-

[10] In line with the country's tradition in new policy areas Lewanski (1997), Pridham and Konstadakopulos (1997), Giuliani (1998) and Giuliani, Rinkevicius and Sverrisson (1998).

[11] While we find informal consultation in both cases, differences emerge for instance in the visibility assigned to internal experts and NGOs on the occasion of public meetings, international conferences, etc.

[12] While Legambiente has delivered a yearly report on the state of the environment since 1989 (Melandri 1989), the Minister of the Environment has delivered only three (in 1989 1992 and 1996) and ISTAT, the national institute of statistics, four (Giuliani 1998: 151).

posed to our own [environmental defense], i.e., dealing with the transformation of territory [e.g. the Public Works Department] or with industrial policy" (Interview 10). The situation has, however, improved to some extent: "I must acknowledge that when our department [environment] opposed some projects by ENEL [the Electric Power agency] which had a big environmental impact, they took our remarks seriously, a few years ago there would have been total opposition." (Interview 10) Relationships are easier where an objective convergence of interests can be found. One example comes from the cooperation between the Environment and the Transport departments, both of which encourage the spread of public transportation (Interviews 10 and 12). More recently, the Minister of the Environment established a task force on waste management which brought together environmentalists, bureaucrats and business leaders and which eventually led to the creation of a National Waste Committee in 1997, coordinated by the Environment Department, but in which the Department of Industry also plays a major role (Tebaldi 1998).

Multiple partnerships have often developed in areas of industrial activity with high environmental impact. By the early 1990s agreements had already been reached between public institutions and business, whereby firms in a given sector committed themselves to stricter environmental procedures in exchange for lighter bureaucratic requirements, financial or fiscal incentives, or practical support from the public administration.[13] The strategy of establishing consortia in areas like plastic waste (Tebaldi 1998) or glass recycling (Boari and Odorici 1996) has also expanded. Partnerships should contribute toward overcoming one of the traditional constraints on technological innovation in Italy, i.e., the small size of most firms. Consortia supported by public institutions should make funds and expertise more easily available to small businesses, unable to afford substantial investments on their own. The territorial concentration of activities in specific areas across Italy should also help (Interview 13). Advocates of consortia have to overcome the skepticism regarding the public administration's capacity effectively to promote new agreements and arrangements. With a few remarkable exceptions, public officers tend to be perceived as passive actors in this policy network, where most initiatives come either from environmentalists

[13] For example, the Industry Department had got involved in several agreements of this nature with firms in the oil sector (Interview 9); or, the Lombardy regional council had negotiated a scheme with producers of printer toners, according to which the latter undertook the task of collecting and recycling otherwise polluting wastes, while public institutions would launch promotional campaigns with large-scale hardware clients (Interview 13).

or business representatives (Interview 13). However, when they work, consortia contribute to the spread of trust and thus to the integration of broader environmental policy networks. This has happened, for example, in Milan where a scientist and leading figure of Legambiente was co-opted as an individual expert into the local government – at the time led by the Northern League. This prompted the creation of a consortium for managing waste disposal and, as a result, the amount of recycled waste in the city quickly rose from 5% to 30%. On that occasion, an agreement was reached between the local agency for waste disposal, paper producers, and the environmentalists, in which the ecologists took responsibility for promoting selective waste disposal among citizens (Interview 13; Giuliani, Rinkevicius and Sverrisson 1998: 29).

This has not been the only example of collaboration between ecologists and business. Workshops, round-tables, and public meetings have been organized and often promoted by foundations like Fondazione Mattei,[14] Istituto per l'Ambiente,[15] Fondazione Lombardia Ambiente,[16] research bodies like the National Research Council (CNR), or public agencies like ENEA-National Agency for Alternative Energy (Interviews 3 and 13). Cooperation with ecology NGOs has also grown substantially on specific issues. The most common form of cooperation is the sponsorship of specific events and campaigns; more rarely, cooperation develops on projects where firms use their specific technical competence. At other times, however, even some observers within the business world are suspicious of the way industry is actually managing these experiences of cooperation. "Often, big firms look for cooperation with the environmentalists just to improve their public image... they try to appear as environmentally and socially responsible, without necessarily being so" (Interview 13).

Open conflicts between environmentalists and organizations representing business have only occasionally developed in recent years, for example with Agrofarma (chemicals) on the occasion of the referendum against the use of pesticides in agriculture, or with Assoplast (the plastic producers' association) during campaigns against plastic bags (Donati 1994: 52ff). Overall, however, consultation between business and environmentalism is still only occasional. Some observers (e.g. Interview 3) feel that there is still strong misunderstanding and suspicion on both sides. This is both an outcome and a precondition of the lack of systematic opportunities for dialogue apart from

[14] A body funded by public firm ENI which includes members of the scientific advisory board of Legambiente among its senior researchers.
[15] Environmental foundation sponsored by business associations.
[16] A foundation mainly working for the Lombardy regional council.

public debates in which everyone stands by their official position. The current opportunities for substantive exchange are still rare, moreover, they are restricted to an intellectual elite on both sides. There is nothing comparable to the levels of consultation between industrialists and trade unions (Interview 13).

4. SUSTAINABLE DEVELOPMENT AND THE INSTITUTIONALIZATION OF ENVIRONMENTAL POLICY IN A SOUTHERN EUROPEAN COUNTRY

I would like to conclude by discussing the strength and nature of the relationship between the emergent idea of SD and new institutional arrangements, as well as the existence of any Southern European peculiarities in terms of the institutionalization of environmental politics. Regarding the first question, the strong version of the neo-institutionalist hypothesis – i.e., that Habermasian discursive democracy might result from the spread of a common cultural framework, and that such a framework may be provided by the idea of SD – is simply impossible to test with this evidence (as well as, I'm inclined to think, with any type of evidence). Several other factors, which relate only partially to the emergence of the concept of SD, have also contributed to growing communication and partnerships in the field of environmental policy. These include the de-radicalization of the movement, the strengthening of environmental agencies and departments, the emergence of 'environmentalists' within business, and the growth of international pressure, mainly through EU legislation. European legislation and policy in general has indeed represented an enormous opportunity for modernizers within business and for the scientific, policy-oriented people within major environmental NGOs – though by the 1980s the later were already dominant within Italian environmentalism, well before SD was perceived to be a potentially unifying concept (Diani 1995).

SD has surely constituted a common label for a set of otherwise quite heterogeneous national and international policies, and has probably offered some common ground for dialogue between different actors. However, one might wonder if other key ideas have not played a similar, and possibly even greater, role than SD. In particular the idea of 'scientific environmentalism' seems to have gained substantial popularity. Many scientific actors have been prepared to serve as intermediaries between other policy actors, from foundations (e.g. the Fondazione Mattei[17] and Fondazione Lombardia per l'Ambiente mentioned above) to research institutes (Istituto di Ricerche Ambiente Italia or Istituto per l'Ambiente) and universities (Giuliani 1998: 162; Tebaldi 1998). In these cases reference to science and technical knowledge, rather than the idea of SD, emerges as the main condensing symbol.

Focus on SD may well have rendered it easier for environmentalists to engage in dialogue with business, but the latter have proved far more resistant to SD – somewhat ironically, given that SD has often been identified as business's environmental ideology (Donati 2000, chapter 5; Richardson 1997; Schnaiberg 1997). Environmentalists have apparently been more prepared to compromise and come closer to business than the other way around. At the very minimum, the process of mutual recognition around a shared definition of the issues at stake has not been symmetric.

The hypothesis that SD is the new common frame of reference for environmental partnerships faces another objection, namely, that mutual recognition does not depend exclusively on shared cultural principles; rather, it also depends on broader trust as well as on prospective partners' perception of a relatively balanced distribution of power and influence in the specific policy network. Accordingly, dialogue might be more difficult than the spread of SD ideas would suggest because of the feelings of powerlessness, and the ensuing uncertainty of status, experienced by the key contenders. This applies not only to environmentalists, but also – more surprisingly – to business. The latter feel cornered by the ecologists' capacity to turn ordinary people's irrationality into a strongly anti-industrialist interpretation of environmental issues (Donati 1994, 2000).

This has resulted in widespread feelings of isolation among the business community, especially in light of policy-makers' sensitivity to public opinion. Industry was condemned to play the role of scapegoat, following the masses' reluctance to modify the real causes of environmental degradation, in particular, their own life-styles (Donati 1994: 101-102), and politicians' refusal to hamper their electoral chances by taking a clear stance against environmental populism. This situation may at least partially account for Business' half-hearted support of SD ideas and therefore for their difficulty in overcoming traditional misunderstandings, despite the availability of new common frames of reference.

One might also wonder whether the Italian experience reflects any Southern European peculiarity. There is no clear answer to this question. From the point of view of the environmental movement the Italian situation

[17] Fondazione Mattei is also central in transnational innovative networks like the Greening of Industry Network (Diani and Sverrisson 1999).

shows – if anything – more proximity to Northern European countries than to other Southern European ones: rates of associational membership and propensity to engage in protest may be lower than those recorded in the countries where environmentalism is at its strongest, but not impressively so.[18] Likewise, the interviews reported here suggest a convergence between business and environmentalist views which, at least according to recent work on waste management policies, is easier to find in a Northern European country like Britain (Rootes and Whale 1998) than in a Southern European one like Spain (Laraña and Pascual 1998).

Moving to environmental policy styles, the answer is at the same time easy and complex. Italian environmental policy is regularly characterized in terms of late and uneven development, scarce coordination, excessive formal legal production, and uncertain implementation capacity (Bulsei 1990; Lewanski 1990, 1997; Giuliani 1998); a policy in which innovation is scattered, restricted to specific cases, and largely dependent on the roles of specific entrepreneurs. The overall reception of SD ideas and policy measures seems to shows no exception to this profile. Recent analyses of environmental policy – and in particular SD policy – in Portugal and Greece are also broadly consistent with this picture (Ribeiro and Rodrigues 1997; Fousekis and Lekakis 1997).

All this resonates well with the conventional portrait of environmental policy – and indeed of any policy – in Mediterranean countries as characterized by 'ineffectiveness', 'political corruption', 'administrative lethargy' and 'defective coordination' by an overdeveloped state (Pridham and Konstadakopulos 1997: 127). By this token, a 'Southern European dimension' is definitely there. And yet, while the picture is quite compatible with the stereotype, even a quick look at Northern European experiences in SD and Agenda 21 implementation suggests a more complex profile. For one, response to Agenda 21 has been highest in the UK, a country whose credentials in terms of effective environmental protection are far from immaculate (O'Riordan and Voisey 1997). On the other hand, SD implementation in countries which are actually leaders in terms of environmental policy innovation shows a pattern surprisingly close to the one we have just described. In Germany, SD policy has been quite episodic and has received no system-

[18] For example, while membership is significantly lower in Italy than in the Netherlands, differences to UK and Germany are now much closer than it used to be the case; propensity to protest is lower than in West Germany, but well in line with other Northern European countries (Biorcio 1999). Differences are instead pronounced to Southern European countries such as Spain or Greece (van der Hejden, Koopmans and Giugni 1992; Jiménez 1999).

atic attention from a state preoccupied with the problems of reunification (Beuermann and Burdick 1997). Even in Norway, commitment to SD has been variable: though it may have profited from an established institutional platform, as a specific issue it has had shifting fortunes on the political agenda (Sverdrup 1997). The same applies to Sweden (Andringa et al. 1998).

In sum, while the adoption of SD strategies may take different forms in different countries, there is only partial evidence of the existence of a peculiar "Southern European syndrome" (La Spina and Sciortino 1993). Nor there is any particular support for the view of SD as an overarching master frame, creating common mutual recognition and trust where before there was none. Other related approaches, such as scientific environmentalism in Italy or established institutional procedures, as in Norway, have been as effective as SD at developing dialogue between competing actors in environmental policy networks.

APPENDIX

List of interviews:

1. Researcher, Istituto Ambiente Italia, Milan
2. Researcher, European environmental task force, Brussels
3. Environmental economist, Enrico Mattei Foundation, Milan
4. Project leader, Legambiente, Rome
5. Senior civil servant, Environment Department, Rome
6. Environmental economist, Legambiente, Rome
7. Researcher, ENEA (National Agency for Alternative Energy), Rome
8. Senior researcher, Lombardia Risorse, Milan
9. Senior civil servant, Industry Department, Rome
10. Senior civil servant, Environment Department, Rome
11. Senior researcher, ENEA, Rome
12. Senior civil servant, Transport Department, Rome
13. Journalist, L'Impresa Ambiente, Milan.
14. Member of the Presidential Board, Confindustria (Italian Industry Association)

REFERENCES

Alberti, M., and Parker, J. (1993) Gli indicatori di sostenibilità ambientale, in G. Conte and G. Melandri, *Ambiente Italia '93*, Koinè Edizioni, Rome, pp.61-81.

Andringa, J., Giuliani, M., van Zwanenberg, P., and Ring, M. (1998) Participation by mandate: Reflections on local Agenda 21, in A. Jamison, *Technology Policy Meets the Public*, Aalborg University Press, Aalborg, pp.81-108.

Baker, S., Kousis, M., Richardson, D., and Young, S. (1997) Introduction: The theory and practice of sustainable development, in S. Baker, M. Kousis, D. Richardson and S. Young, *The Politics of Sustainable Development: Theory, Policy and Practice within the European Union*, Routledge, London, pp.1-42.

Beuermann, C., and Burdick, B. (1997) The sustainability transition in Germany, *Environmental Politics* **6**, 83-107.

Bianchi, D. (1993) Il ciclo dei consumi e dei rifiuti, in G. Conte and G. Melandri, *Ambiente Italia '93*, Koinè Edizioni, Rome, pp.127-140.

Biorcio, R. (1999) *Environmental Protest and Ecological Culture: A Comparative Analysis*, Paper for the ECPR Joint Sessions, Mannheim, 26-31 March, 1999.

Boari, C., and Odorici, V. (1996) *Pressioni istituzionali, P.M.I. e relazioni interorganizzative nel settore del riciclo del vetro*, Working Paper, Department of Business Economics, University of Bologna.

Bulsei, G.L. (1990) *Le Politiche Ambientali*, Rosenberg e Sellier, Turin.

Conte, G., and Melandri, G. (eds.) (1993) *Ambiente Italia '93*, Koinè Edizioni, Rome.

Conte, G., and Melandri, G. (eds.) (1994) *Ambiente Italia '94*, Koinè Edizioni, Rome.

Di Giulio, E., and Pareglio, S. (1998) Sostenibilita' urbano e azione locale, *Equilibri: Rivista per lo Sviluppo Sostenibile* **2**, 25-45.

Diani, M. (1994a) *Framing and Communicating Environmental Issues in Italy: The Role of Politcians and Civil Servants*, Project Report, European University Institute, Florence.

Diani, M. (1994b) The conflict over nuclear energy in Italy, in H. Flam, *States and Anti-Nuclear Movements*, Edinburgh University Press, Edinburgh, pp.201-231.

Diani, M. (1995) *Green Networks. A Structural Analysis of the Italian Environmental Movement*, Edinburgh University Press, Edinburgh.

Diani, M. (1996) *Environmental Sustainability and Institutional Innovation in Italy*, Research Report, IARD, Milan.

Diani, M., and Forno, F. (1999) *The Evolution of Environmental Protest in Italy 1988-1997*, Paper for the EPCR Joint Sessions, Mannheim, 26-31 March, 1999.

Diani, M., and Sverrisson, A. (1999) *The Greening of Industry Network. Transnational Linkages and Environmental Science and Technology Policy*, Research Report, Bocconi University-POLEIS, Milan/ University of Stockholm.

Dionisio, S. (1994) Ecogestione: aziende italiane al banco di prova, *L'impresa ambiente* **6**, 30-34.

Donati, P.R. (1994) *Framing and Communicating the Environmental Issues: Industry and the Environment in Italy*, Project Report, European University Institute, Florence.

Donati, P.R. (1996) Building a unified movement, in L. Kriesberg, *Research in Social Movements, Conflict and Change*, JAI Press, Greenwich, CT, pp.125-157.

Donati, P.R. (2000) *The Industrial Class in Post-Industrial Age: Corporate Political Strategies in the Environmental Issue*, Ph.D. Thesis, European University Institute, Florence.

Fousekis, P., and Lekakis, J. (1997) Greece's institutional response to sustainability, *Environmental Politics* **6**, 131-152.

Giuliani, M. (1992) Il Processo Decisionale Italiano e Politiche Communitarie, *Polis* **6**, 307-342.

Giuliani, M. (1998) Environmental S&T policy in one country: The public-policy interface in Italy, in A. Jamison, *Technology Policy Meets the Public*, Aalborg University Press, Aalborg, pp.147-172.

Giuliani, M., Rinkevicius, L., and Sverrisson, A. (1998) Making participation happen: The importance of policy entrepreneurs, in A. Jamison, *Technology Policy Meets the Public*, Aalborg University Press, Aalborg, pp.19-48.

Jiménez, M. (1999) Consolidation through institutionalisation? Dilemmas of the Spanish environmental movement in the 1990s, *Environmental Politics* **8**, 149-171.

La Spina, A., and Sciortino, G. (1993) Common agenda, Southern rules: European integration and environmental change in the Mediterranean states, in J.D. Liefferink, P.D. Lowe and A.P.J. Moll, *European Integration and Environmental Policy*, Belhaven Press, London, pp.216-234.

Lanzalaco, L. (1995) *Istituzioni, Organizzazioni, Potere*, Nuova Italia Scientifica, Rome.

Laraña, E., and Pascual, E. (1998) *The Socio-Pragmatics of Two Contending Discourses*, Paper for the XVI World Congress of Sociology, Montreal.

Legambiente (ed.) (1994) *Ambiente, lavoro, futuro. Idee, scenari, proposte per la riconversione ecologica dell'economia*, Unpublished Research Report, Rome.

Lewanski, R. (1990) La Politica Ambientale, in B. Dente, *Le Politiche Publiche in Italia*, Il Mulino, Bologna, pp.281-314.

Lewanski, R. (1997) *Governare l'Ambiente*, Il Mulino, Bologna.

Lowe, P.D., and Goyder, J.M. (1983) *Environmental Groups in Politics*, Allen & Unwin, London.

Melandri, G. (ed.) (1989) *Ambiente Italia*, Isedi, Turin.

Melandri, G. (1993) Il ruolo dell'Europa dopo UNCED, in G. Conte and G. Melandri, *Ambiente Italia '93*, Koinè Edizioni, Rome, pp.17-27.

O'Mahoney, P., and Skillington, T. (1996) Sustainable development as an organising principle for discursive democracy? *Sustainable Development* **4**, 42-51.

O'Riordan, T., and Voisey, H. (1997) Governing institutions for sustainable development: the United Kingdom's national level approach, *Environmental Politics* **6**, 24-53.

Pavesi, F. (1994) Imprese in corsa per lo sviluppo sostenibile, *L'impresa ambiente* **6**, 20-28.

Powell, W.W., and DiMaggio, P.J. (eds.) (1991) *The New Institutionalism in Organizational Analysis*, University of Chicago Press, Chicago, IL.

Pridham, G., and Konstadakopulos, D. (1997) Sustainable development in Southern Europe? Interactions between European, national and sub-national levels, in S. Baker, M.Kousis, D.Richardson and S. Young, *The Politics of Sustainable Development: Theory, Policy and Practice within the European Union*, Routledge, London, pp.127-151.

Ribeiro, T.G., and Rodrigues, V.J. (1997) The evolution of sustainable development in Portugal, *Environmental Politics* **6**, 108-130.

Richardson, D. (1997) The politics of sustainable development, in S. Baker, M. Kousis, D. Richardson and S. Young, *The Politics of Sustainable Development: Theory, Policy and Practice within the European Union*, Routledge, London, pp.43-60.

Rootes, C.A., and Whale, R. (1998) *Competing Discourses in the Debate over Waste Disposal Facilities*, Research Report, University of Kent at Canterbury.

Sassoon, E., and Rapisarda Sassoon, C. (1992) *Management dell'ambiente*, Edizioni del Sol-24 Ore, Milan.

Schnaiberg, A. (1997) Sustainable development and the treadmill of production, in S. Baker, M. Kousis, D. Richardson and S. Young, *The Politics of Sustainable Development: Theory, Policy and Practice within the European Union*, Routledge, London, pp.72-88.

Snow, D.A., and Benford, R.D. (1988) Ideology, frame resonance, and participant mobilization, in B. Klandermans, H.-P. Kriesi and S. Tarrow, *From Structure to Action:*

Comparing Social Movement Research Across Cultures, JAI Press, Greenwich, CT, pp.197-217.

Sverdrup, L.A. (1997) Norway's institutional response to sustainable development, *Environmental Politics* **6**, 54-82.

Tebaldi, M. (1998) *Environmental Policies and Brokerage Roles in Italy. The Plastic Recycling Policy*, Research Report, Bocconi University-POLEIS, Milan.

van der Hejden, H.-A., Koopmans, R., and Giugni, M. (1992) The West-European environmental movement, in L. Kriesberg, *Research in Social Movements, Conflict and Change*, JAI Press, Greenwich, CT, pp.1-40.

Chapter 13

From Common Interest To Partnership[1]

Changes in the Italian Environmentalist Discourse

Anna Triandafyllidou

1. A COMMON INTEREST FOR ENVIRONMENTAL PRESERVATION

The idea of a common, collective interest in environmental preservation lies at the origins of the environmental movement and is often proposed as the rationale behind environmental policy. As a matter of fact, the environment is by definition considered as a collective good: it cannot be divided up in an equal manner, neither between groups of people, e.g. states, nations or regions, nor between individuals. Nonetheless, until recently, the environment remained under the jurisdiction of the separate nation-states. It was a collective, national good. The effects of air and water pollution, soil degradation or the depletion of resources were geographically limited. Over-exploiting or careless modes of production compromised mainly the local resources without endangering the existence and/or survival of neighboring countries. However, the development of ever more intensive modes of pro-

1 The author would like to thank Patrizia Bigotti for her invaluable help in the collection of the data for the 1995-97 period and Klaus Eder for his constructive comments on earlier drafts of the paper. Data concerning the 1987-91 and 1992-94 periods were collected during the author's employment as research fellow under the auspices of two international research projects funded by DG XII of the European Commission ('The Making of an Issue: A Comparative Study of Cultures of Environmentalism in Europe,' 1992-94; 'Environmental Sustainability and Institutional Innovation', 1993-95).

K. Eder and M. Kousis (eds.), Environmental Politics in Southern Europe, 321–342.

duction and, more particularly, the discovery of nuclear power has changed dramatically the landscape.

Nowadays, risks deriving from environmental pollution cannot be confined within the territory of a continent, much less of a specific country, as accidents such as that of Chernobyl have shown. Thus, the concept of a common, collective interest in environmental protection has acquired a universalistic overtone. The interest and responsibility for environmental preservation embraces not merely a national government or a people but humanity as a whole.

Apart from being a non-divisible good, the environment is also a pool of limited resources. A zero-sum balance governs the use of the natural capital: if one country abuses its resources, other countries are also affected by the damage caused. Hence, the defense of the environment is a matter of common interest.

2. THE SUSTAINABLE DEVELOPMENT APPROACH

Environmental policy had become an issue in many European countries soon after the end of World War II.[2] Initially, environmentalist movements were concerned with specific problems such as air pollution in urban areas or river water pollution. However, soon it became clear that an efficient and effective environmental policy had to address the overall relationship between ecology and economy. As a matter of fact, the former was seen as a burden to the latter, albeit an unavoidable one.

The concept of sustainable development (SD) was introduced as early as 1980 (IUCN 1980). However, it acquired prominence only in the late 1980s when the WCED report (1987) was launched. The new approach maintained that environmental protection was not contrary to economic growth. Instead, the preservation of natural resources started to be seen as a lever for further and durable socio-economic development (Pearce, Markandya and Barbier 1989; Turner 1993). In political and policy terms, the new approach aimed at striking a balance between "the anti-growth groups and the development-at-all-costs advocates" (O'Riordan 1988: 36).

[2] For a discussion of the historical roots and evolution of the concept of sustainable development see Donati (1994: 1-24). Other works to be consulted include: Archibugi and Nijkamp (1989), Barbier (1993), Cleveland (1981), Ghai and Vivian (1992), Redclift (1988, 1992), Simonis (1990), and Turner (1993).

The idea of a common collective interest in environmental preservation is inherent in the SD approach. This interest is defined synchronically, i.e., with regard to the human population in the world today: "sustainable development requires meeting the basic needs of all and extending to all the opportunity to fulfill their aspirations for a better life" (WCED 1987: 8), and diachronically, with reference to the future generations: "to ensure that it [development] meets the needs of the present without compromising the ability of future generations to meet their own needs" (WCED 1987). In other words, from a SD perspective, the common interest is enlarged to include humanity as a whole in the present as well as the future.

Moreover, the SD approach introduces the notion of 'social partnership'. According to this view, not only governments, industry or environmental organizations but also the public, consumers, the media, financial institutions and other social agents are 'partners' in a sustainable development model and, therefore, have their own share of responsibility as well as an obligation to participate in the struggle for environmental preservation (European Foundation 1992). Thus, the common collective interest in environmental protection is interpreted as a right and also as a duty to participate actively in policy-making and implementation.

In conclusion, the SD model bears with it a new conception of the common collective interest in the environment. Firstly, this interest is no longer understood in national or regional terms but with reference to humanity as a whole, including present and future generations. Secondly, by affirming the common interest, it also introduces a common responsibility and a set of tasks to be undertaken by all, individuals and groups alike.

3. FROM COMMON INTEREST TO PARTNERSHIP: THE AIMS OF THIS STUDY

The notions of a common interest in the environment and the idea of partnership outlined above are but rough and simplistic summaries of some aspects of the contemporary environmentalist debate, which includes divergent and conflictual interpretations of the relevant concepts. Not only are these concepts contested, but also the overall sustainability approach remains, to a large extent, debatable (Triandafyllidou and Fotiou 1998). Moreover, the interpretation of the ideas involved in a common interest or partnership approach and their adoption as policy principles depends upon the specificities of the national context within which policy decisions are formulated. This context is defined by external parameters, such as international regulations and political pressures to conform with a given policy paradigm, and internal factors including power relations between industry and environmental groups as well as the national political culture (Triandafyllidou 1996) and dominant policy practices in a given country. However, it is worth noting that at the European level, SD tends to become a more general cultural framework within which European integration will take place (Triandafyllidou 1998). In this context, new terms and policy options promoted by the European Commission will interact with, influence, and be transformed by national understandings of politics and practices of policy-making. Clearly, the political culture of each nation-state and the prevailing understandings of citizen-state relations and notions of the public good will mediate social partnership definitions.

My aim in the following sections is to explore the ways in which national specificities, and in particular the national political culture, may interfere with the definition of a common collective interest in the environment and the notion of a social partnership in environmental preservation. More particularly, I will highlight the ways in which interpretations of the common interest or partnership principles are conditioned by nationally specific discursive frames concerning policy-making and implementation in Italy. Furthermore, I will investigate the extent to which different conceptualizations of the collective interest in environmental protection contribute to the formulation of different policy options. Last but not least, by means of analysing empirical material, namely newspaper articles, EU documentation and interviews with EU officials, I shall seek to achieve a better understanding of the very notions of common interest and partnership in relation to the environment.

4. RESEARCH DESIGN AND METHODOLOGY

The aims of the study will be achieved through a comparative analysis of the Italian press discourse regarding the environment and the European environmental policy discourse. The study will be divided into three chronological phases. The first period covers five years starting in 1987, when the WCED report launched the concept of sustainable development as the guideline for environmental policy. During this period, the SD model gained prominence internationally, not merely as a concept but also as a policy approach. Nonetheless, in Italy the environmentalist discourse continued to concentrate on specific issues such as nuclear radiation, air or water pollution, or the problem of urban waste (Donati and Diani 1995: 6-11; Fiore

1991; Triandafyllidou 1995). The Italian press discourse on environmentalism will be analysed during this period with the aim of highlighting the ways in which a common collective interest in the environment is conceptualized. The relationship between the definition of the public interest and the policy measures proposed will also be examined.

The second period under examination encompasses the years between 1992-1994, during which the Fifth Environmental Policy Programme and a number of other important policy documents (The Community Framework for Sustainable Mobility and Agenda 21) were issued by the European Commission and other international organizations. During this period, SD became the dominant policy paradigm, at least within the European Union, and was incorporated into Community legislation through the Fifth Programme. It is therefore assumed that it started influencing the environmentalist debate as well as national policy decisions in the member-states. My aim here will be to highlight the different elements involved in the notion of partnership through an analysis of EU documents and interviews with the social, economic and political actors participating in the policy-making process. The study will also seek to show the differences between the notion of a common interest presented in the Italian press in the first period and the idea of partnership put forward by the Commission in the early 1990s.

The last period comprises the years 1995-1997. Although the meaning of environmental sustainability remains contested among social actors (Triandafyllidou and Fotiou 1998) during this period, SD may be considered the dominant policy approach within the EU that conditions environmentalist discourse and guides policy-decisions both at the European and the national level. By means of studying the Italian press during this period, I will seek to assess the extent to which in the media discourse on the environment the notion of partnership has replaced earlier ideas of a common collective interest. The study will also assess the extent to which interpretations of these two notions are influenced by the Italian political culture.

The main reason for choosing Italy as a case study is its contradictory character. On the one hand, Italy seems to be the 'greenest' member-state and, on the other hand, it fails to comply with the most fundamental environmental provisions. Italian environmental legislation has grown dramatically during the last decades but legal provisions and policy measures are rarely implemented, either because they are too strict or because of the Byzantine character of the national administration (Donati and Diani 1995: 129). Besides, the Italian public has often reacted in contradictory ways, showing either excessive or too little sensitivity with regard to environmental problems. Thus, the Chernobyl accident and the nuclear issue in general has been the main issue around which environmentalist mobilization took place, while, for instance, relatively less attention was paid to domestic tragedies such as that of the Seveso accident (Donati and Diani 1995: 12). Indeed, the main nation-wide protest campaigns were organized against nuclear power and attracted widespread popular participation (Barone 1984; Del Carria 1986; De Meo and Giovannini 1985). Moreover, Green lists increased their shares in local and national elections and environmentalist associations multiplied their members from the late 1970s onwards (Fiore 1991). Thus, Italy provides fruitful ground for examining the relationship between the concept of a common collective interest in the environment and the idea of partnership introduced by the SD approach.

Moreover, the particularities of the Italian political culture, namely the mistrust that characterizes the relations between citizens and the state and the growing dissatisfaction of Italians, as the discrepancy between their expectations of public management and perceived policy outcomes has become wider (Triandafyllidou 1996: 373-374), offer interesting ground for testing my hypothesis that general policy concepts may be 'adapted' to national realities. It is indeed worth studying the interpretation of concepts such as 'public interest' and 'social partnership', which appeal to solidarity, trust, and the public good, in a context of mistrust and discontent, where public officials are perceived as people making disproportionately large personal gains by means of abusing their public position (Mannheimer and Sani 1987: 16).

With regard to the press discourse, I will analyse the articles dealing with environmental issues published in the two largest Italian dailies[3], *Corriere della Sera* (CdS) and *La Repubblica* (LaR), during the last week of April[4] in the period between 1987-91 and 1995-97. The EU policy discourse will be studied through the official documents issued by the European Commission

[3] LaR is a large progressive daily newspaper published in tabloid form. It has a specific content feature; the material published in it is divided into two sections. The general section deals to domestic and international news and general commentaries on politics, the arts, sports or other social issues. The second section deals with news of local relevance and varies according to the place of publication of the newspaper, namely Naples, Rome, Florence, Bologna, Milan, Turin or Genoa. LaR belongs to the De Benedetti-L'Espresso media group. CdS, on the other hand, is the Milan newspaper par excellence. It is published in the same city, but also prints a Rome edition with a short local section, and has a large national circulation. CdS belongs to the Fiat-Rizzoli industrial group and situated politically on the center-right. Regardless of their party affiliations, which have varied during the past decade, both newspapers may be considered mainstream and politically moderate.

[4] This week (22-29 of April) was deliberately chosen as a sampling period because it coincides with the Chernobyl accident anniversary and thus ensures a relatively large representation of environmental issues in the news reports.

during the period 1992-94 concerning the environment and the sustainable development and also through interviews conducted in 1994 with EU officials, NGOs, and industry associations' representatives.

Newspaper articles as well as interview texts have been coded using a common set of coding categories inserted into a database management system (Foxpro2 software) application. The coding scheme included a set of identifying categories (newspaper name; date; page; type of article; title; and subtitle for newspaper articles; and name of interviewee; position; date; name of interviewer for interview texts); six issue categories (general environmental issues; global effects on climate and nature; chemical effects/radiation; flora and fauna effects; effects on public life; other)[5] and two coding categories (the first concerning the appeal to the collective interest for environmental protection and the second regarding the social partnership notion).

The common collective interest concept has been operationalized as a coding category using the following indices: a) references to the common interest of a collectivity, including references to the local community, the nation or humanity as a whole; b) definition of the environment as a public good which cannot be divided among groups or individuals; c) references to a common, collective right to enjoy nature and its resources; and d) references to a common responsibility towards environmental preservation. The partnership concept has been operationalized through the identification of its main components: a) the notion of a common responsibility towards the environment and of burden-sharing with regard to environmental costs; b) a call for individual or collective participation in policy-making and implementation; and c) references to a common collective interest (or to competing collective interests) regarding environmental protection. It is obvious from this operationalization that there is at least partial overlap between the two concepts. In fact, my aim here is to explore further the commonalities and differences between them.

Word strings and/or short text passages including catch-phrases, rhetoric devices, or metaphors, which highlight the particular conceptualizations of the common interest or the partnership notions, have been inserted into the database. Two strategies have been followed for the analysis of the texts. First, the different versions of the concepts presented in the texts were systematized into categories so as to highlight the different interpretations of each concept. Secondly, the links between elements of the text and more general cultural categories have been investigated in order to check their cultural resonance. This type of analysis may be called semiotic-structuralist (Hijmans 1996) to the extent that structural or semiotic elements of the texts are used to make inferences about the context in which they are produced.

[5] The full version of the coding scheme may be found in the appendix.

5. NOTIONS OF 'COMMON INTEREST' IN THE ENVIRONMENT IN THE ITALIAN PRESS

The database regarding the first period of the analysis includes 325 articles published in LaR and CdS during the sampling week in the period 1987-91. The idea of a common collective interest in the environment was mentioned in 61 of these.

A plurality of views is expressed in the Italian press discourse with regard to the collective interest for environmental protection. The reader is often reminded that by helping protect and preserve natural resources – by means of avoiding excessive waste (LaR #228)[6], saving energy (CdS #163) or recycling (CdS #170), for instance – they defend their own collective interest. The same happens when they use their bicycles instead of driving their cars (CdS #131) or they promote the creation of public gardens in the city center (LaR #258; CdS #122). Nonetheless, this appeal to the common interest is framed in terms of penalties and interdictions. Rather than evoking civic consciousness among the public, this type of discourse mobilizes the fear of sanctions. This characteristic can be attributed to the particular nature of the Italian civic culture which is marked by reciprocal mistrust between the citizens and the state and discontent of the former towards the latter (Calvi 1987; Fabris 1977; La Palombara 1965; Sani 1980). Fines or other types of sanction are seen as the only way to impose a certain policy measure on citizens. Given that citizens are convinced that the public officials and the political elite look after their own private gain rather than the public interest, they would tend to protect their individual interest rather than what is defined as the common collective good.

Within this version of the common collective interest, a hint is dropped concerning the importance of voluntary participation. However, references to the need for popular mobilization or the right, as well as the duty of citizens to be informed and to contribute to environmental policy, remain indirect: the law and order approach prevails over a participatory view of policy-making. Nonetheless, the seeds of the partnership notion may be identified in some of the articles of this period (LaR #210; LaR #277; CdS #101)

Moreover, the protection of the environment is seen as a means to preserve the natural capital of the country, the common national heritage of its people, which is a source of material profits and contributes to a better quality of life (LaR #230; LaR #349; CdS #122). Even though this version defines the environment as a national good, ecological initiatives are also interpreted with ref-

[6] See list of articles cited in the appendix.

erence to the common, global good. The international character of environmentalism is emphasized (LaR #222; LaR #214; CdS #121; CdS #119).

The notion of a common interest in environmental protection is, however, called into question in some articles where competing collective interests are identified around an ecological issue. Thus, where state authorities close down an industry that does not respect environmental regulations, the collective interest of the local population for a healthier environment is contrasted to the common interest of the workers that will lose their jobs (LaR #304, LaR #340). Moreover, a NIMBY approach of local inhabitants regarding transport infrastructure is contrasted to the collective benefit to be drawn by the whole region, or even a wider area, from better transport services (LaR #352). In such cases, the traditional dilemma between ecology and economy is reproduced and the two are viewed as mutually exclusive. Environmental protection is thus considered to take place at the expense of economic development.

Nonetheless, the wider collective good may also coincide with the strictly local interests: this is often the case when environmental preservation is a precondition for tourism. Thus, in the Ticino valley for instance, local dockyard owners agreed on stricter control of their production activity in order to prevent pollution of the river waters and beaches (LaR #320). Besides, competing collective interests may be reconciled through improvement of the environmental policy projects, which take into account social, economic and ecological concerns (CdS #111, LaR #289).

In conclusion, appeals to the common collective interest are intertwined with a regulatory framework, where emphasis is put on restrictions, controls, and penalties. However, the notion of common responsibility begins emerging in the Italian press discourse and the need for individual and collective contributions to environmental protection is pointed out. Nonetheless, the idea of social and political actors as partners-stakeholders is still absent from the press discourse. During the period 1987-1991, the focus of the discourse is on the conversion from "man (sic) as exploiter-thief of natural resources" to "man (sic) as their guardian and manager". And it is also emphasized that "no one can 'take issue' with a topic of such wide implications" (CdS #174).

6. PARTNERSHIP IN THE EUROPEAN POLICY DISCOURSE

The second section of the study investigates the definitions of partnership inherent in the EU environmental policy discourse. Twenty-one official policy documents were selected from the database system[7] of the European Commission library and the documentation center of DG XI. The selection of the material was primarily based on the author of the texts[8], namely the documents were issued either directly by one of the European institutions[9] or by expert teams or organizations delegated by the Commission or the European Parliament (EP). A large number of the texts collected were issued as reports of the Commission or the EP or as articles or dossiers in one of the official EC publications.[10] All of these documents may thus be considered to express, as closely as possible, the official views of the EU or, at least, of its main political institutions.

The documentation database has been complemented by a set of eight interviews[11] conducted with economic actors, environmental organizations and Commission officials in 1994. Interviews were conducted on the basis of a common though flexible scheme: essential points were discussed in all interviews, however the order and the phrasing of the questions did not always follow the same sequence. Interviewees were encouraged to express freely their opinions, i.e., the opinion of the organization they represented, with regard to specific concepts or policies. These main points included 'sustainable development', 'subsidiarity with regard to SD policy', 'voluntary approach/ dialogue', 'responsibility/polluter pays principle', 'technological vs. market

7 Legal documents have been retrieved from the database CELEX while other official documents and publications of the EC have been collected from the databases SCAD and ECLAS. For further information, see also EUR-OP (1995).

8 The complete list of the documents has been constructed after multiple bibliographical researches in terms of 'author' (=the Commission, the EP, the Social and Economic Committee or the Council), 'time period' (1992-1994), 'document type' (=UE, i.e., official publication of the EU) and 'subject' (=sustainability, environmental policy & sustainable development (SD), SD, environment & development). Furthermore, two complementary lists concerning, on the one hand, environmental policy in general and, on the other hand, development (economic or social) as a generic term, have been constructed. These additional research strategies aimed at ensuring that all environmental policy documents which referred to partnership and/or sustainable development were included in the database.

9 Namely, the Commission, the Council, the European Parliament (EP) and the Economic and Social Committee (ESC).

10 Texts that offer different versions of the same document, the initial and the revised version of a Council resolution on sustainable mobility, for instance, or a report issued by the European Parliament and the Opinion (Avis) given by the relevant Parliamentary Commission, have both been included in the material. This research strategy may sometimes have led to partly overlapping documents, however, it was deemed necessary to ensure completeness of the database.

11 The names of the interviewees are withheld for reasons of confidentiality. The full interview texts may be made available upon request to the author and after consultation with the funding institutions.

instruments for the achievement of a sustainable society', 'the role of the public', and 'economic growth and sustainability'.

The main feature that characterizes definitions of partnership, both in policy documents and interviews, is citizen participation and shared responsibility (#16; #21, par.7 and 8; #10, par.5; #8, par.1; #19, par.1; #17, par.1; #1, par.4; #2, par.4).[12] In other words, partnership does not simply imply a common collective interest, but it is a principle that should guide policy-making: Citizens are seen as stakeholders with regard to the common, natural capital and are invited to participate in its management:

> the extension of the definition of interests in sustainable development from those of government and the social partners (industry employers, managers, workers) to include other stakeholders (e.g. the public, consumers, the banks and financial institutions) ..., collective acceptance of responsibility, and willingness to participate in the process of change. (#16, par.1)

> members of the public and citizens' organizations should contribute to the planning and execution of policies and projects directly affecting their quality of life and their surroundings. (#19)

The idea of sharing responsibilities and working together is emphasized by the interviewees as well (Ind1; Ind2; EU3)[13] who recognize the necessity for collaborating and assuming a share in the task of protecting the environment.

In conformity with the above basic definition, partnership is related to the 'polluter pays principle.' The sharing of responsibility is thus expressed in concrete financial terms (#6, par.3; #5, par.1; #15, par.1). It is also linked with the principle of equity: environmental costs should be carried by those who reap the relevant benefits (#5, par.2; #3, par.3). This principle applies to individual consumers, to business, and also to the EU as a transnational agent. Similarly, assistance is foreseen for those who cannot pay their part in environmental preservation. Thus, welfare benefits are envisaged for the less favored sections of the population to assist them with increasing transport fees due to the internalization of environmental costs (#5, par.1).

Moreover, partnership is seen as a guideline for international policy within and also beyond the EU. The global nature of environmental problems and the necessity of transnational co-operation is emphasized in most of the docu-

[12] The documents cited may be identified by their record number (#..). Par. stands for paragraph.

[13] Abbreviations used: Ind = Industry association representative; EU = EU official; NGO = Non-governmental organization representative.

ments analysed (#21, par. 1 and 2; #18, par.4; #6, par.4, 6, 9 and 14; #8, par.2, 3 and 6; #2, par.2 and 9). Concerns are expressed with regard to the assistance that should be offered by the EU to third countries in developing their economies without destroying their environmental resources. Hence, partnership is in conformity with a view of the collective interest in global terms.

Nonetheless, the contradictions inherent in the idea of a common collective interest found in the Italian press discourse, namely the presence of competing collective interests with regard to environmental protection, also make up part of the partnership debate. Thus, according to the Commission, partnership involves not only a common interest in the preservation of nature, but also the accommodation of divergent regional interests, so that local or regional development is not compromised by international concerns. More particularly, attention will be paid to ensuring that the interests of the less favored regions are integrated into the environmental policy framework so that they manage to develop rapidly and without compromising their natural resources (#18, par.2; #6, par.10; #10, par.3 and 4; #4, par.2; #8, par.6; #2, par.7). In this respect, the principle of partnership is seen as a lever for development because, on the one hand, it ensures a better quality of life through the protection of the environment and, on the other hand, it aims at integrating ecology with economy, thus contributing to growth. At the same time, this does not mean that environmental quality in the EU should be entirely compromised in order to protect interests which, although collective, are strictly local (#6, par.8)

Even though the policy guidelines derived from the principle of partnership are clear, the issue of accommodating competing collective interests is far from being resolved. Thus, the debate concerning job creation vs. environmental protection remains open (#4, par.2; #11 par.1). However, EU documents point to the long-term implications of pollution and argue that deterioration of the environmental conditions compromises in the long run the health of the workforce as well as the resources available for production (#3, par.2). Thus, the idea that ecology and economy are complementary – which lies at the heart of the SD approach – is emphasized. Similarly, reductions in purchasing power are seen as short-term costs that contribute to long-term benefits, i.e., durable growth, improved environmental quality, and long-term prosperity for all (#3, par. 4).

The fact that partnership presupposes a common interest where, in reality, there exists a variety of divergent collective interests, is a matter of concern for the industry representatives and EU officials interviewed. Business associations maintain that an effective partnership regime can only be realized through direct negotiation between the parties involved (a local community and an industry, for instance) (Ind1; Ind2). Thus, from their point of view, which is also supported by some EU administrators (EU2; EU3), partnership implies a voluntary approach instead of a regulatory framework. The market economy will provide the incentives for environment-friendly production techniques. This is clearly contrary to the NGOs' point of view. For them, environmental policy involves a regulatory regime, which is the only way to ensure that the public interest for a clean environment is taken into account (NGO1).

In conclusion, partnership may be defined as an extended version of the common collective interest idea, incorporating the principle of responsibility sharing and participation. It may also, however, be interpreted as a totally new approach to policy-making. My findings suggest that the latter definition prevails in the EU environmental policy discourse. The principle of partnership implies a right and a duty to participate in policy decisions. It thus ensures that the interests of all social actors are taken into account and competing collective interests are negotiated with the aim of achieving maximum benefit for all parties. It also entails the adoption of a new policy approach: the classic regulatory framework where abuses are prevented through sanctions, is replaced by the voluntary action model, according to which environmental preservation should be left to the initiative of social actors. Their actions are eventually conditioned by market mechanisms, which ensure that the maximum possible benefit is achieved. Of course, the prevalence of a voluntary approach over the regulatory framework in environmental policy is, for the moment, debatable. However, the notion of voluntary participation is intrinsic to the partnership concept.

7. EUROPEAN INFLUENCES AND ITALIAN POLITICAL CULTURE: ENVIRONMENTAL DISCOURSE IN 1995-97

Two important changes have been identified in the Italian press discourse on the environment in the period 1995-97. On the one hand, the idea of partnership as sharing of environmental costs is mentioned in the news reports. On the other hand, however, the notion of a common collective interest in environmental protection to be defended through a regulatory policy approach persists together with some typical features of Italian political culture, namely distrust in the authorities, mismanagement and personalization of environmental issues (Triandafyllidou 1996).

Out of a total of 56 articles covering the period 1995-97, 11 contained references to a common interest in environmental preservation and 11 included notions of burden sharing of environmental costs. Thus, there is a clear shift towards a new conception of environmental policy, which sees different social and political actors as co-participants in environmental protection. Nonetheless, quite surprisingly, all the articles, including notions of partnership in this respect, referred to international issues, the risks and damages related to the Chernobyl accident aftermath, in particular, and cost sharing between the government of Ukraine, the Soviet Union, and 'the West' (LaR #R11, CdS #R14, #R15, #R23). In this respect, notions of individual participation and burden sharing include acts of solidarity or courage of specific individuals or families towards the victims of Chernobyl (CdS #R15, LaR #R28). More rarely, notions of individual or collective participation referred generally to the preservation of the global natural heritage (CdS #R7) and to environmental policy decisions (CdS #R1).

A more market-oriented view of partnership, however, has been identified with regard to non-chemically-treated food. Consumer responsibility and the impact of consumer preferences on environmental and agriculture policy through market mechanisms is emphasized in two articles published by LaR in 1996 (#R41, #R42). Attention is paid to the power of individual consumers to influence the behavior of business and politicians. Consumers, industry and the state are seen as partners-stakeholders, even though they often have conflicting interests in the preservation of natural resources, the protection of animals, and the production of organic food.

Overall, although the idea of burden-sharing and collective or individual participation permeates the Italian press discourse, it is to a large extent confined to international or global issues. The concept of partnership and of a market-oriented approach to environmental policy that prevails in the EU documentation and interviews remains marginal in the Italian context. National environmental problems remain related to the approach identified in the late 1980s (see section 5 above), based on a law-and-order policy framework and characterized by the mistrust towards public authorities.

In fact, reports on environmental disasters abroad also adopt a 'traditional' common interest approach where the 'goodies' and the 'baddies' are identified and mistrust towards governmental authorities, in this case the Ukrainian and Russian administration, prevails (LaR #R13, #R35). The conception of nature as a common public good and the need for international collaboration for protecting the natural capital is emphasized in some articles (CdS #R16, LaR #R30). However, there is no mention of the need for active participation at the individual or collective level, nor is the environmental preservation logic related to market mechanisms.

There are four articles (CdS #R26, #R47, #R52, LaR #R56) in the 1995-97 sample, which may be deemed to reflect the Italian-style regulatory approach. Environmental resources are defined as collective goods that ought to be protected. However, their preservation or abuse is related to competing individual or collective interests. These are not to be regulated through market mechanisms but they are to be held in check by interdictions, fines, and a strict regulatory framework. Nonetheless, the implementation of such a framework is hindered by the inefficiency of the administration and the clientelistic networks and individual interests that interfere with the state apparatus. Thus, problems and solutions concerning environmental issues of local relevance, such as the management of water resources, noise reduction, and the protection of natural reserves, depend upon government regulations and the imposition of fines against the perpetrators. Moreover, in this respect, the relationship between the state and the citizens or the economic actors (including workers and industry) is characterized by discontent and mistrust, features typical of the Italian political culture (Almond and Verba 1980). The indifference of politicians and their focus on narrow party interests is emphasized, in conformity with the prevalent cultural norms (Calvi 1987; Fabris 1977; Sani 1980)

There is only one article published in LaR in 1997 (#R54) in which the role of market mechanisms in environmental policy is discussed: this regards the recycling of old car materials. Nonetheless, even in that one instance in which the role of the market in Italian environmental policy is recognized, the mistrust towards governmental measures prevails: "the script [of the ecological demolition of old cars] is very much Italian-style". There is disagreement about the economic feasibility of the operation and opinions are divided between small business and big industry. Regardless of the usefulness of the specific measure, this article is the only one that introduces a market logic in the debate over a national environmental policy measure.

8. CONCLUSIONS

The aim of this paper has been to discuss critically the notions of partnership and of a common collective interest in environmental preservation. More specifically, my aim has been to highlight the differences between the two concepts and also to investigate the extent to which the introduction of the concept of social partnership into the European environmental policy debate is reflected in the Italian press discourse. Secondly, the relationship between the two concepts and specific policy options has been examined. Moreover, the study has aimed at revealing the ways in which the national political culture may influence the framing of environmental policy concepts and measures.

The idea of a common collective interest in the environment is prevalent in the Italian press discourse in the period 1987-91, despite the fact that the SD model had already gained international recognition. The common interest in the environment is defined in regional, national and international terms and is related to a growing emphasis on human responsibility for the depletion of natural resources. Furthermore, the existence of competing collective interests with regard to environmental preservation and their possible accommodation is discussed. The traditional economy vs. ecology framework seems prevalent in the Italian press discourse in this period. Thus, a regulatory policy model is proposed. Instead of being encouraged to comply voluntarily with environmental measures, citizens are threatened with sanctions. The prevalence of this approach in the Italian press is in line with the overwhelming mistrust, which characterizes the relationship between the citizens and the state in Italy.

The analysis of the EU documentation and interviews shows that the notion of partnership is radically different from the common collective interest perspective. Partnership implies a participatory approach, where costs and benefits are shared among individuals and among collectivities, nations or international bodies. This implies a principle of intra- and international equity, which in fact lies at the heart of the SD model, as well as a principle of justice: those who reap the benefits should pay the costs. Moreover, the notion of partnership opens the door to an alternative policy regime, where environmental protection is regulated by market mechanisms. To put it simply, the market is expected to offer the incentives for adopting environment-friendly patterns of production and consumption. This voluntary action approach is viewed as the most suitable solution for the accommodation of competing collective interests

The analysis of the Italian press during the period between 1995 and 1997 confirms my hypotheses. On the one hand, it shows that European policy discourse and decisions are reflected on the public discourse at the national level: the notion of partnership has thus become part of the Italian debate in recent years even though the term in itself is not used by the press. The necessity of sharing environmental costs as well as the importance of individual and/or collective participation in environmental policy is emphasized. Moreover, the role of consumers in influencing public policy through market mechanisms and the interdependence of business, administration, and consumers is pointed out. Thus, even though the term of partnership in itself is not present in the Italian press discourse, elements of the SD model and the partnership-voluntary approach to environmental policy-making and implementation have permeated the press' discursive *repertoire*.

On the other hand, the persistence of the 'old' regulatory policy approach and the prevalence of notions of public mismanagement, mistrust towards politicians, implementation of policy through the threat of sanctions, and the interference of clientelistic networks in the negotiation of competing collective interests, show that the environmentalist discourse is heavily influenced by Italian political culture. As a matter of fact, local or national issues are predominantly discussed within such a law-and-order-plus-distrust perspective, while the burden-sharing approach is confined mainly to international issues and, more particularly, to nuclear disasters.

This study shows that new concepts and policy approaches introduced at the Brussels level cannot but influence the terms of the national press discourse. Quite surprisingly, this is not done at the level of slogans but rather in relation to accounts of problems and solutions proposed. At the same time, old models and deeply rooted notions of the national political culture, including particular understandings of policy-making and implementation (clientelism, mismanagement and mistrust), persist in the environmental discourse. The presence of both elements in the press reports suggests that this may be a transitional period: as European integration grows, national elements may gradually be replaced by common European tendencies. Nonetheless, this has still to be proven. It is also worth noting that the national political culture does not function as a distorting lens that leads to euphemistic notions of partnership, which would disguise a clientelistic framework: it rather seems to hamper the introduction of the new perspective altogether.

The findings of this study provide for a useful starting point. Future research could examine the media discourse in other member-states with a political culture similar to that of Italy, e.g. Greece, Portugal or Spain, so as to assess the extent to which national patterns persist there too. The identification of such patterns would be useful for the design and implementation of EU environmental policy in the future.

APPENDIX I – CODING SCHEME

A. Identifiers for newspaper articles
A.1 Newspaper name (2 col.) - [NAME]
1 Corriere della Sera (CdS)
2 La Repubblica (LaR)
A.2 Date (8 col.) - [DATE]
A.3 Page (3 col.) - [PAGE]
A.4 Type of article (1 col.) - [ARTITYPE]
1 news reports
2 dossier

3 commentary/editorial
4 interview
5 letter
6 other
A.5 Title (MEMO variable) - [TITLE]
A.6 Subtitle (MEMO variable) - [SUBTITLE]
Identifiers for interviews
A.7 Name of interviewee (INT_EE)
A.8 Position within organization (POSIT)
A.9 Date (DATE)
A.10 Name of interviewer (INT_ER)
B. Environmental Issues
B.1 Generic environmental problems/issues [PROBGEN]
1 air pollution
2 water pollution
3 soil pollution
4 depletion of resources
5 general pollution
B.2 Global effects on climate and nature - [PROBSP1]
1 global warming
2 ozone hole
3 deforestation/tropical forests
4 acid rain/radioactive clouds
5 radon
6 combination of two or more of the above
B.3 Chemical effects - [PROBSP2]
1 radiation
2 waste/chemical pollution
3 soil erosion/contamination
B.4 Flora and fauna effects - [PROBSP3]
1 landscape destruction
2 extinction of animal species
3 1 & 2
B.5 Effects on everyday life - [PROBSP4]
1 water shortage
2 food contamination
3 domestic waste
4 health
5 urban green
6 noise
7 combination of two or more of the above
B.6 Other (specify) - [PROBSP5]
C Appeal to the common collective interest
D Appeal/reference to the idea of partnership

APPENDIX II – LIST OF NEWSPAPER ARTICLES

CORRIERE DELLA SERA (CdS)

Rec	Date	Title
#101	27.4.90	Fiumi, mari e laghi sotto controllo: esami e sentenza due volte al mese
#111	25.4.90	Depuratore a Pero. La licenza c'è, i fondi non ancora
#119	22.4.90	Ecologia e industria, un giorno di pace
#121	22.4.90	Salviamo la Terra dal degrado ambientale e dalla retorica
#122	29.4.88	QT8, guerra legale contro il Comune
#131	27.4.88	Il Comune rilancia le bici gialle e offrira a nolo anche ciclomotori
#163	26.4.87	Polistirene espanso: un isolamento che migliora la qualità della vita
#170	26.4.87	Alimentata da tutti noi la crescita della 'montagna' di rifiuti solidi
#174	26.4.87	Uomo 'sfruttatore-rapinatore' delle risorse che la natura gli ha messo a dis..
#R1	22.4.95	Per diffendere la terra
#R7	26.4.95	Ti ramazzo Central Park
#R14	24.4.96	Incendio a Cernobyl, terrore radioattivo
#R15	24.4.96	Le testimonianze. Vivere nella zona con lo spettro atomico dietro la porta
#R16	24.4.96	Da Ginevra il monito: "E' come una bomba a tempo"
#R23	26.4.96	Cernobyl, allarme continuo. Un errore umano provoca una fuga radioatti..
#R26	28.4.96	Prigioneri dei pioppi. Agricoltori contro l'obbligo di sostituire gli alberi..
#R47	23.4.97	La giunta perde la guerra dell'acqua
#R52	28.4.97	Guerra all'inquinamento acustico. Milanesi assordati proposta la mappa ..

LA REPUBBLICA (LaR)

Rec	Date	Title
#210	28.4.87	La linea verde dei comunisti "Vogliamo salvare i parchi"
#214	28.4.87	Una festa contro il nucleare
#222	26.4.87	L'Europa riflette sulla 'lezione' di Cernobyl...
#228	25.4.87	La Regione dica basta agli imballi inquinanti
#230	25.4.87	Un grande parco per il pianeta Etna
#258	24.4.88	Parcheggi sotto il verde. No degli ecologisti
#277	27.4.90	'Cara città meriti zero in condotta'
#289	25.4.90	Quell'inceneritore spento riaccende le vecchie paure. Potrà funzionare se ...
#304	25.4.89	Si riaccende la guerra dell'Acna
#320	27.4.89	Un pass per le barche oppure addio Ticino
#340	28.4.89	Il Piemonte a Roma chiede la chiusura dell'Acna di Cengio
#349	27.4.91	Un Belpaese di plastica
#352	26.4.91	Bloccate le ruspe a Segrate. Stop ai lavori della dogana
#R11	27.4.95	Il mostro Cernobyl uccide ancora. Tre milioni di Ucraini a rischio
#R13	28.4.95	Ottomila metri di fuoco. Esplode un gasdotto, panico in Russia
#R28	22.4.96	Mai più Cernobyl
#R30	23.4.96	Capocotta, il WWF chiede garanzie
#R35	26.4.96	Quelle mutazioni rimarranno per sempre
#R41	29.4.96	La muca pazza ed. il virus del business
#R42	29.4.96	Coltivazioni troppo "al verde".
#R54	25.4.97	I demolitori contro Corso Marconi
#R56	29.4.97	Task force antirumore arrivati quaranta fax

APPENDIX III – POLICY DOCUMENTS

#1 Council (1994), Projet de résolution du Conseil Environnement sur le Livre Blanc: Croissance, Competitivité et Emploi, 7252/94 restreint ENV 121. Brussels: EC, le 31 mai 1994.

#2 European Commission (1994), Examen interimaire de la mise en oeuvre du programme communautaire de politique et de l'action en matière d'environnement et de développement durable: "Vers un développement durable", Communication de la Commission, Bruxelles: EC, COM 94/0453 final.

#3 European Commission (1994), Economic growth and the environment: some implications for economic policy-making. Communication to the European Parliament and Council, Brussels: EC, COM 94/0465 final.

#4 Economic and Social Committee (1993), Masucci Ettore, Rapporteur. Co-rapporteurs: Mr. Bleser, Mr van der Decken and Mr. Gafo Fernandez. Information report of the section for external relations, trade and development policy on EU relations with the associated countries of Central and Eastern Europe, focusing in particular on common problems in the fields of energy, environment, transport and telecommunications. Brussels: ESC, CES 862/93 final, EU4/07516.

#5 European Commission (1993), Le développement futur de la politique commune des transports: construction d'un cadre communautaire garant d'une mobilité durable. Communication de la Commission. Document établi sur la base du document COM(92)494 final. Luxembourg: CE, Bulletin des CE, Suppl. 1993/03, EU8/04688a.

#6 European Commission, Task Force on the Environment & the Internal Market (1993), La dimension environnementale de "1992": Rapport de la Task Force L'Environnement et le Marché Intérieur. Bonn: Economica Verl., EU8/05027.

#7 European Parliament (1992), Verhagen Maxime, Session Documents, Report of the Committee on Development and Co-operation on the Community's environmental policy in relation to the developing countries. 28 January 1992, PE 154.190/fin.

#8 European Commission (1992), Congrès. Rio de Janeiro. 1992. Rapport à la Conférence des Nations unies sur l'environnement et le développement. Rio de Janeiro, juin 1992. Bruxelles: CE, SEC/1991/2448 final.

#9 European Commission (1992), Green paper on the impact of transport on the environment. A Community strategy for "sustainable mobility". Brussels: EC. COM/92/00046.

#10 European Commission (1992), Le développement futur de la politique commune des transports: construction d'un cadre communautaire garant d'une mobilité durable. Bruxelles: CE, COM/92/00494 final.

#11 European Commission (1993), Croissance competitivité et emploi. Les défis et les pistes pour entrer dans le XXI siècle. Livre Blanc. Bulletin des Communautés Européennes, Supplément 6/93. Luxembourg: CE.

#12 European Commission (1992), Vers un développement soutenable. Programme communautaire de politique et de l'action pour l'environnement et le développement durable et respectueux de l'environnement, vol. II, Bruxelles: CE, COM/92/00023 final

#13 Council (1992), Council Resolution of 3 December 1992 concerning the relationship between industrial competitiveness and environmental protection, 92/c 331/03.

#14 European Commission (1994), Commission Communication in accordance with Council Regulation (EEC) no 1973/92 of 21 May 1992 establishing a financial instrument for the environment (LIFE), relating to priority actions to be implemented in 1995, 94/c 139/03.

#15 Council (1993), Réglement (CEE) 1836/93 du Conseil du 29.6.1993 permettant la participation volontaire des entreprises du secteur industriel à un système communautaire de management environnemental et d'audit, JOL 1993, 168, p.1.

#16 European Foundation for the Improvement of Living and Working Conditions (1992), European roundtable on industry, social dialogue and sustainability, A general report, Rome.

#17 Council (1994), Council Resolution of 6 May 1994 on a Community strategy, for integrated coastal-zone management (94/c 135/02)

#18 European Commission (1992), Environment and development. *Courier (Africa, Carribean, Pacific, European Community)*, 133, pp.43-101.

#19 European Parliament (1994), Ruiz-Gimenez Aguilar Guadalupe, Rapporteur. Report of the Committee on the Environment, Public Health and Consumer Protection on monitoring by the European Community of the implementation of Agenda 21 of the UN Conference on Environment and Development (UNCED), 3 January 1994, PE 205.636 fin., (94/00001 PE)

#20 European Parliament (1992), Van Dijk Nel, Rapporteur, Report of the Committee on Transport and Tourism on the Green paper on the impact of transport on the environment. A Community strategy for 'sustainable mobility', 17 July 1992, PE 200.978 fin, (92/0256 EP)

#21 European Commission (1992) Vers un développement soutenable. Programme communautaire de politique et de l'action pour l'environnement et le développement durable et respectueux de l'environnement, Bruxelles: CE, COM/92/00023 final. (The same text appears also under the headings: EC. Commission (1993-) Vers un développement soutenable. Programme communautaire de politique et de l'action pour l'environnement et le développement durable et respectueux de l'environnement. Luxembourg: CE, EU4/06867a; Résolution du Conseil et des représentants des gouvernements des états-membres, réunis au sein du Conseil du 01/02/1993, conçernant un programme communautaire de politique et d'action en matière d'environnement et de développement durable. JOC 1993, 138, p.1.)

REFERENCES

Almond, G., and Verba, S. (eds.) (1980) *The Civic Culture Revisited*, Little Brown, Boston.

Archibugi, F., and Nijkamp, P. (eds.) (1989) *Economy and Ecology: Towards Sustainable Development*, Kluwer Academic Publishers, Dordrecht.

Barbier, E.B. (ed.) (1993) *Economics and Ecology. New Frontiers and Sustainable Development*, Chapman & Hall, London.

Barone, C. (1984) Ecologia: quali conflitti per quali attori, in A. Melucci, *Altri codici*, Il Mulino, Bologna, pp.175-222.

Calvi, G. (ed.) (1987) *Indagine sociale italiana*, Angeli, Milano.

Cleveland, H. (ed.) (1981) *The Management of Sustainable Growth*, Pergamon Press, New York, NY.

De Meo, M., and Giovannini, F. (eds.) (1985) *L'onda verde*, Alfamedia, Roma.

Del Carria, R. (1986) *Il potere diffuso: il verdi in Italia*, Edizioni del Movimiento non Violento, Verona.

Donati, P.R. (1994) *Sustainable Development and the Commission of the EU after the Fifth Action Programme*, Project Report on 'Environmental Sustainability and Institutional Innovation', European University Institute, Florence.

Donati, P.R., and Diani, M. (1995) *Framing and Communicating Environmental Issues: The Italian Case; Part II*, Project Report, European University Institute, Florence.

European Foundation for the Improvement of Living and Working Conditions (1992) Executive Summary, European Roundtable on Industry, Social Dialogue and Sustainability, Rome.

Fabris, G. (1977) *Il comportamento politico degli italiani*, Angeli, Milano.

Fiore, C. (ed.) (1991) *L'arcipelago verde. Geografia e prospettive dei movimenti ecologisti*, Valecchi, Firenze.

Ghai, D., and Vivian, J.M. (eds.) (1992) *Grassroots Environmental Action*, Routledge, London.

Hijmans, E. (1996) The logic of qualitative media content analysis: A typology, *Communications* **21**, 93-108.

IUCN (International Union for the Conservation of Nature) (1980) *World Conservation Strategy: Living Resources Conservation for a Sustainable Development*, IUCN - UNEP - WWF, Gland, Switzerland.

La Palombara, J. (1965) Italy: Alienation, fragmentation, isolation, in L.W. Pye and S. Verba, *Political Culture and Political Development*, Princeton University Press, Princeton, NJ, pp.282-322S.

Mannheimer, R., and Sani, G. (1987) *Il mercato elettorale. L'identikit dell'elettore italiano*, Il Mulino, Bologna.

O'Riordan, T. (1988) The politics of sustainability, in R.K. Turner, *Sustainable Environmental Management. Principles and Practices*, Westview Press, Boulder, CO, pp.29-50.

Pearce, D., Markandya, A., and Barbier, E.B. (1989) *Blueprint for a Green Economy*, Earthscan, London.

Redclift, M. (1988) Sustainable development and the market: A framework for analysis, *Futures* December 1988, 635-649.

Redclift, M. (1992) *Sustainable Development: Exploring the Contradictions*, Routledge, London.

Sani, G. (1980) The political culture of Italy: Continuity and change, in G. Almond and S. Verba, *The Civic Culture Revisited*, Little Brown, Boston, pp.273-324.

Simonis, U.E. (1990) *Beyond Growth. Elements of Sustainable Development*, Sigma, Berlin.

Triandafyllidou, A. (1995) *Framing and Communicating Environmental Issues: The Italian Case; Part I*, Project report, European University Institute, Florence.

Triandafyllidou, A. (1996) 'Green' corruption in the Italian press: Does political culture matter? *European Journal of Communication* **11**, 371-391.

Triandafyllidou, A. (1998) *Environmental Sustainability: A Cultural Framework for European Integration*, Paper presented at ISA, World Congress of Sociology, Montreal, Canada, 26 July - 1 August, 1998.

Triandafyllidou, A., and Fotiou, A. (1998) Sustainability and modernity in the European Union: A frame analysis of the policy-making process, *Sociological Research Online* **3**, www.socresonline.org.uk/3/1/2.html.

Turner, R.K. (ed.) (1993) *Sustainable Environmental Economics and Management. Principles and Practices*, Westview Press, Boulder, CO.

WCED (World Commission on Environment and Development) (1987) *Our Common Future*, Oxford University Press, Oxford.

Chapter 14

Non-Governmental Groups and the State

Environmental Politics in Portugal

Joaquim Gil Nave

1. INTRODUCTION

This article focuses on the political role of environmental collective action in Portugal. After a short historical-sociological overview of the political context within which the existing environmental group 'industry' emerged, particular emphasis is given to the formation of a new wave of environmental groups after the mid-1980s. Newly emerging environmental associations of this period blocked a green party strategy for the environmental movement. As a consequence, 'old' political ecology and protest action groups that had formed a fairly successful anti-nuclear movement in the late 1970s and early 1980s declined. Options for further mobilization of and by environmental associations were then shaped by changes in the political opportunity structure induced by the adhesion to the EC after 1986.

Our main argument is that in this process a relevant role must be attributed to state action toward environmental collective action. From the mid- to late-1980s, governmental actors initiated the construction of an environmental policy domain envisaging and encouraging the participation of environmental non-governmental organizations. This design of an environmental policy domain gradually changed toward an institutionalized participation of environmental groups. An institutional framework developed after 1987 on the basis of a Basic Law for the Environment and laws defining the status of environmental non-governmental organizations as 'public interest groups'. This made participation of groups in the policy-making arena a strategic option for environmental collective action. These new political conditions

K. Eder and M. Kousis (eds.), Environmental Politics in Southern Europe, 343–364.

for collective action and participation of groups in environmental politics in Portugal strengthened their role in agenda-setting processes of the environmental policy domain.

The analysis of phenomenology and resource mobilization of environmental groups shows how both the knowledge resources and the mobilization and organizational structures of the environmental movement 'industry' in Portugal have improved substantially since the 1980s, as has the contribution of environmental collective action toward shaping public opinion on environmental issues.

2. THE EMERGENCE OF ENVIRONMENTALISM IN PORTUGAL: THE POLITICAL OPPORTUNITY STRUCTURE

In the transition period to democracy, patterns of state action, institutional designs of policy-making, democratic practices and routines, and patterns of political culture have changed (Barreto 1994; Braga da Cruz 1995; Bruneau and Macleod 1986). The strengthening of authority and of the role of the state in this period restored state centralism and the self-sufficiency of policy and decision making by state actors (Barreto 1994). Aside from the effects on the polity, this also had particular influence on the shape of environmental politics. It influenced not only the emergence, forms, and impact of environmental collective action in this society, but also determined particular institutional modes of policy-making in the environmental domain.

This led to a particular constellation in which the institutional design of environmental policy-making (Aguilar Fernández 1993) still prevailing in the aftermath of the adhesion to the European Community and was able to minimize the participation of non-state actors. It practically excluded environmental movement organizations and other non-state actors as well (industry actors and independent expertise not part of the state administration). Participation was reduced to discretionary, occasional, and mainly informal consultation initiatives by governmental actors. Although the adhesion to the European Community promoted some changes in the participation of non-state actors in the policy-making process, it did not change practices, patterns, and modes of decision-making within most policy domains. Furthermore, the environmental policy field was created within a policy-making context subordinated to principles of economic and social development which favored other policy priorities at the expense of environmental policy-making.

This may be seen as a long-term effect of the political culture and of the *locus* of state action. The foundation of the political regime on the principles of liberal, parliamentary, and representative democracy was achieved against the participatory claims of grassroots groups and social protest (Barreto 1994; Bruneau and Macleod 1986; Graham and Wheeler 1983; Maxwell 1995).[1] The political system in Portugal developed toward a form of constraining civil society, which throughout the post-coup period had often blocked state action and authority in many domains. In other words, the stabilization of the political system and the re-establishment of state authority and legitimacy through constitutional and electoral means were made 'against' civil society.

Three factors have determined the specific rise and development of environmental collective action in Portugal: (a) the structure of state action, (b) the functioning of the politico-administrative system, and (c) the model of the party system. These are the focal dimensions of the political opportunity structure for environmental collective action.

The party system has excluded non-party organizations from playing a role within the political system. This made party politics practically the sole axis of political representation and democracy. Consequently, the space available in the political system for autonomous interest articulation remained for a long time very limited. The centralism prevailing in the structure of the state and of the administrative system even made the parliament a subordinate element of political life (Barreto 1994; Vitorino 1995). This allowed the government a lot of autonomy in policy-making. The space left to political action for non-party groups was the public space that was created in the transition to democracy. It often functioned also as an extension of party politics, in that it was widely used by parties to influence public opinion and to extend the control over other sectors of collective action emerging from civil society (Barreto 1994; Bruneau and Macleod 1986; Matos 1991).

Moreover, widespread and firmly established traditional patterns of political culture, which worked against the participation of the citizenry and gave state actors the monopoly of dealing with the *res publica*, blocked the development and improvement of participatory institutions (Santos 1990, 1994). These patterns of political culture have their roots in a tradition of authoritarianism and centralism of the state, the inception of which is attributed to the long lasting non-democratic paternalistic regime (Opello 1983;

[1] Such grassroots protest actions were led essentially by the working class movement under the auspices, and often the leadership, of military leaders that staged the coup, the communists, and other non-parliamentary radical left groups.

Giner 1986). The struggles against left radicalism of the post-coup revolutionary times reinforced these patterns of political culture not only within state action, but also in the political institutions and in civil society. Political channels for the exercise of participatory democracy – which had existed since the advent of democracy in 1975 – were mainly offered to, and viewed by, the citizenry as a gift from the state. This allowed helped to enhance the legitimacy of the state and extend the influence of party politics (Barreto 1994; Santos 1990, 1994).

Local authors focusing on the emergence of new movement politics in Portugal also explained the presumed weakness or 'deficit' of new social movements by the negative effect of the political culture and political system on civil society (Freire 1985; Santos 1994). In particular, the approach of Santos (1994) regarding the emergence of new social movement politics in Portugal has the virtue of coping with the problem of the specific paths of non-central societies joining modernity. This author stresses the fatality of 'feed-back' between 'old' and 'new' political issues and practices, that is, between the issues and practices of mainstream party politics and representative democracy, and those of participatory democracy and new social movements. Following this view, 'materialist' claims – better wages, welfare, etc. – and representative democracy had supposedly taken primacy over 'post-materialist' ones – such as ecology, anti-nuclear, gender and racial equality, participatory politics (Santos 1994: 90, 229-231).

Thus, the negative political opportunity structure for new movement politics has its roots in the historical contingency of the political and social processes structuring the political system. There were no wide mass mobilization waves for new kinds of issues like there had been in movements which, by the mid-1980s, were distressing mainstream politics in other central European nations. The weak dynamic of civil society, as a counterpart of the centralism of the state and of the exclusiveness of party politics in the political system, lies at the origin of the subsequent lack of autonomy of new social movement cultures. This means that two important dimensions of the political opportunity structure for the emergence of new political issues, such as the environment, remained for a long time unfavorable even after the democratic transition period.

Thus, the *revolutionary* climate of the transition period to democracy seems to have been without consequences. After a half century of backwardness and authoritarian rule, it was not easy to compete with basic welfare, economic, and political citizenship issues in seeking the attention of the public and of leading political actors. The widely unorganized and disconnected forms of environmental collective action, and the climate of political and ideological dissent that extended from the *radical left* to *new issue* politics did not facilitate the search for an organizational project that could give support to building the environmental movement's collective identity.

This context explains why the ecological critique of the late 1970s and early 1980s was confined to the nuclear issue, which from 1975 to 1985 remained a hot issue on the public and the political agenda. The environmental movement of this period in Portugal gained some relief more as a resonance of the anti-nuclear wave that swept across advanced industrial democratic nations by the late-1970s, than as a social movement able to give organizational and ideological support to the emerging forms of grassroots protests over environmental issues. In practice, the inability of groups to provide a political ecology frame to grassroots protests over environmental issues, which spontaneously emerged at the local level, isolated these groups. They turned into groups which were mainly preoccupied with internal ideological debates, similar to those of *new left* groups on the ebb of the post-coup revolutionary period (Fernandes 1982; Lopes 1984; Matos 1995).

In spite of this, the powerful public energy planning sector did not succeed in imposing the nuclear power solution, which was firmly opposed by a weakly organized, though convincing, discourse-oriented action that strongly mobilized public opinion. It is true that the anti-nuclear opposition took advantage of cleavages over the issue among state and governmental actors on the one hand and parties on the other. Nevertheless, the nuclear power issue debates show that what was at stake was the legitimacy of decision-making processes centered on the mechanisms of representative democracy and administrative technocracy. This may be explained in two ways.

First of all, it follows the same pattern of politics which dominated the transition-to-democracy period, in which non-institutional forms of influencing decision-making had become an important means of political action due to the lack of legitimacy and authority of state power. Secondly, the institutionalization of democracy gradually gave rise to new forms of public debate. Knowledge resources and argumentative means, with which most sectors of the anti-nuclear opposition were well equipped, started to become more important than political and ideological convictions in influencing policy decision-making. State actors did not manage to overcome the growing anti-nuclear sentiments of public opinion emerging from debates over the nuclear power program, which had been proposed by energy policy-makers as early as the 1970s and anticipated by the National Energy Provision Plan of the early 1980s.

However, the favorable resolution of the nuclear power issue can hardly be seen as an achievement of the environmental movement, as is often done by the guardians of the movement's collective memory (Os Amigos da Terra

1989; Eloy 1996). Other powerful and highly influential actors emerged defending a non-nuclear solution for energy provision, and they also actively (and decisively) participated in mobilizations over the issue. This was the case, for instance, of an active group of university experts on energy led by a young engineer of the Technical University of Lisbon (Domingos 1978). Still, influential leaders of conventional parties and of several governments in the period, and an elite of intellectuals and media opinion-makers played a decisive role in the resolution of the issue. As a result, the decline of the nuclear issue on the public and political agenda by the mid-1980s was followed by environmental groups withdrawing into their ghetto within the small and mainly ideological field of *new issue* and *new left* politics (Os Amigos da Terra 1989; Fernandes 1982, 1983; Lopes 1984).

The emergence of environmentalism during the first 10 years of democracy can also be explained in terms of civil society's weakness in shaping the environmental policy domain. Until the early 1980s, the green monarchists dominated the environmental sector in several post-coup, provisional, and constitutional governments. The party leader made it the green branch of mainstream, conservative politics. But the party lacked enough popular electoral strength to prioritize environmental policy issues within state action. Due in part to having neglected potential political support from groups mobilizing on environmental issues, the Popular Monarchic Party green leadership in charge of the environmental portfolio left nothing more than a few – though vitally important – proposals for nature protection and territorial planning. These proposals were hardly implemented or applied afterwards. Ultimately, the task of building a truly environmental policy sector was left to future governments (Os Amigos da Terra 1989; Eloy 1996; Melo and Pimenta 1993). In spite of their genuine efforts, the monarchists played a misleading role as advocates of environmentalism inside the administration.

In any case, the ideological gap between the traditionalist, conservationist (and conservative) approach of the monarchists to environmental policy issues, and the radical political ecology view and 'new left' rhetoric that predominated in environmentalist groups did not favor the development of a collaborative climate that could lead to institutional participation of such groups in policy-making. Furthermore, from the late 1970s to the early 1980s, the growing electoral support and political impact of the German Greens, which remains the paradigmatic case, started to exert a strong attraction on left-wing activists and ecologists in Portugal. The debate over political participation of environmental groups therefore moved toward attempts to found a green party. However, after having seen how some conventional parties were eagerly and successfully appropriating their ideas over the protection of nature, environmental defense, and natural resource managing, and after recognizing that green issues made old heralds of mainstream politics popular – as was the case of the small Popular Monarchic Party and the simulacrum of a Green Party founded under the auspices of the Communist Party – most ecologists chose to remain outside the party game (Fernandes 1982, 1983; Lopes 1984; Matos 1995; Melo and Pimenta 1993).

There were two reasons for this choice. First, there was an unequivocally awkward relationship to party politics that the youngest ecologists of the mid- 1980s shared with their age group peers. There was no general refusal by young people to participate in the resolution of public affairs. On the contrary, they kept demanding more opportunities to participate in decision making over issues affecting their age group interests and were eager to join associations as a means of participating in many aspects of social, political, and cultural life. However, this age group distanced itself profoundly from party politics 'as usual'.[2]

One indication of this was that environmental associations of the early 1980's declared themselves to be a-political, meaning that they were not only independent of parties and open to all kinds of individual political choices, but essentially that they did not want to align their action with any kind of vested economic, ideological, or political interests. This option implied their desire to avoid entering the party-politics game.

The second reason has to do with a clear refusal on the part of the younger generation of environmentalists to adopt the verbal radicalism surrounding conceptions, proposals, and ideological orientations of the previous generation. The refusal was extended to the so-called 'newcomer' ecologists, that is, people that, before embracing ecological values, had previously joined radical left groups, which had started to decay by the late 1970s. This radical, left-wing past turned out to be rather bad for the verbal

[2] Almeida (1995) offers an insightful analysis of this tendency among young people in the early 1990s in Portugal. In spite of the fact that several surveys have confirmed this tendency to stay away from party-politics, empirical evidence also revealed that there was still a strong disposition toward other forms of collective action. This means that the lack of trust in political institutions did not necessarily imply that new (or even partially new) forms of collective action and political participation were received with the same indifference by the young citizenry. After the decay of grand ideologies and the current 'protestantization of Catholicism', groups and individual citizens tended to preserve the autonomy of their own ideological choices, frequently resorting to variegated combinations of different ideologies. Being right or left wing usually meant a complete and articulate package of values and behaviors concerning several aspects of social and individual life, stretching from political and ideological options, to ethics, religion, and way of life. But things are changing. Young people have started to search more eagerly for a diverse combinations of options in all dimensions of social life (Almeida 1995: 68-69).

and conceptual radicalism of these 'newcomer' ecologists, who continued to distance these groups from the 'real world'. This made them look like cohesive political and religious sects which had transformed what, in the beginning, had been an open critique of established models of society, into fundamentalist beliefs. The environmental association movement wanted to avoid the decline experienced by the radical left movement.

This refusal meant that environmental associations were also giving up important ideological, cultural, organizational, political, and identity resources, which these activists had accumulated in many years of non-parliamentary radical left politics. It also meant the rejection of ideological means and resources susceptible of promoting the development of strong symbolic mechanisms of group and movement identity.

After the anti-nuclear mobilizations, environmentalism attracted contradictory segments among the youth and among *new issue* militancy in search of organizational means that could further political and civic participation. On the one hand, the anti-industrialism of the ecological discourse of this period and its critique of capitalistic growth at the expense of natural resource depletion made it very attractive to militancy abandoning *radical left* groups when the revolutionary period faded away. On the other, environmentalism also emerged as a field of public participation for youth in search for an alternative to the 'politics as usual' of *radical left* groups and of the mainstream parties' youth organizations. Thus, two opposed and contradictory views of politics and ecology joined the movement from the early to mid-1980s.

At the risk of being reductionistic, one may say that the former saw ecology as the continuity of the *radical left,* anti-capitalistic project, while the latter thought it was essentially an apolitical *moral crusade* (Eder 1985) which took, as its organizational project, the form of an apolitical association. One view assumed the ecological ghetto to be an emancipatory vanguard, which in some way revived the Leninist organizational model which prevailed in *radical left* politics. The other drew upon the potential for resistance and rational contention in ecological discourse, and aimed not at conquering a new political territory, but at expanding the new *culture of nature* (Eder 1995, 1996) as a social practice. The former was willing to take advantage of the upsurge of new political opportunities opened up by the stabilization of democracy – in particular of facilities and resources for electoral campaigning, propaganda, and political participation (Fernandes 1982, 1983; Lopes 1984; Franco 1984; Matos 1995); the latter viewed communicative action and apprenticeship as the means of cultural and social change.

However, both trends of the ecological movement found it hard to escape from their own ghettoized propaganda and conservation activities after the nuclear power issue. It is common knowledge that a lot of national branches of the environmental movement abroad resorted to anti-nuclear battles to further a well-organized and influential political movement capable of profiting from political opportunities opened by respective national political structures and conjunctures. In the Portuguese case, on the contrary, it was more the swan song of radical political ecology groups than a new impulse for the movement. In the end, only the effects of the adhesion to the European Community by the mid-1980s would finally contribute decisively to enhancing an environmental policy sector within the state, which had started to allow the participation of the citizenry and of environmental groups in policy-making processes.

3. THE PARTICIPATION OF NON-GOVERNMENTAL GROUPS IN ENVIRONMENTAL POLICY-MAKING

In 1987, the parliament approved the first basic law for the environment prepared by the public administration in this policy domain. As a complement to this, a bill for regulating the public status of environmental associations and institutionalizing their participation in public environmental issue arenas was also voted upon.[3] This represented an opportunity for movement organizations to step out from their 'ghetto' and actively to participate in policy and political arenas where environmental issues are at stake.

The law provided environmental groups with the legal status to participate as 'social partners' at several levels of the policy-making process in the environmental policy domain. It entitled them to be consulted and informed about environmental policy initiatives and plans. Furthermore, environmental associations were allowed to promote judicial proceedings aiming to prevent environmental damages, although they did have to resort to a public prosecutor for court proceedings. They also gained access to data and information about administrative procedures concerning environmental issues at the central, regional, and local level of the administration. On the other hand, the law committed the state to provide environmental groups with technical and financial support in order for them to improve their technical

[3] This law was revised by the parliament in 1998. The revision included many demands made by environmental groups and resulted in a more efficient and an enlarged scope of rights for participation of groups in environmental politics and decision-making.

and scientific skills and strengthen their means of propaganda and environmental education. Finally, it created a legal framework for the participation of groups in several consultation bodies of the administration dealing with environmental protection, territorial planning, and development policy issues and plans (Lopes 1984).

The list of rights given to environmental associations by this law was considered by many local observers as very innovative and 'generous' in comparison to rights given by the state to civil society associations of other policy domains (Amaral 1994: 376). Unfortunately, the failure to implement many environmental laws (approved in a hurry to implement EC commitments) also resulted in the bill on the participation of associations in regulatory state bodies at the central, regional, and local level not being fully applied.

Environmentalists had placed high hopes in both new laws. However, the deficiencies of further regulations and implementation hindered the impact and achievements of participation given to environmental groups (Miranda 1994). Nevertheless, groups still had the opportunity to extend considerably their scope of action and issue mobilization.

Most groups active from the mid- to late-1980s were formed by just a few young people, and their constituency was small. Fundraising hardly went beyond constituency contributions and meager subsidies from local and central state agencies. Resources were very scarce, as were campaigning skills and knowledge resources for effectively taking on a relevant role in the public sphere of environmental politics. Thanks to means made available by EC programs and national state authorities, environmental associations were provided with financial support, organizational resources, and other kinds of incentives to sustain regular activity. Many groups that were just struggling for survival and were mainly oriented towards environmental education and recreational activities for their young constituency then had a decisive opportunity to consolidate organizational structures. They were now given more incentives to continue and strengthen their action (Pimenta and Melo 1993: 151). The moment was also favorable for the emergence of new organizations and the renewal of some older ones.

The most skilled gained access to promotional programs for the environment sponsored by the European Community/European Union (EC/EU), which provided funds that associations used for improving their expertise and technical skills. Thereby they were able to enlarge their scope of field research on nature conservation activities, which attracted many young people to associations. Yet, the establishment of relationships with the EC/EU environmental milieu also gave the opportunity to enlarge contacts and collaboration at the international level.

In the meantime, the government created a public agency that aimed at promoting environmental education and supporting autonomous environmental organizations. Among its prerogatives, the *Instituto Nacional do Ambiente-INAMB* (National Institute for the Environment) would provide financial and technical aid to environmental associations according to requirements fixed by the law for environmental associations. More importantly, the agency also promoted what was then considered a turning point in the organizational structure of the environmental movement: the creation of a representative 'peak' entity comprising a great number of associations.

Since then environmental associations have received important stimuli from state actors. A real change occurred when the most important associations finally began collaborating with each other. They first started cooperating in the European Environmental Bureau, and after campaigning together against the Spanish government's plans for building a nuclear-waste disposal near the Portuguese border, they joined important campaigns against reforestation with exotic fast-growing tree-species. Other common initiatives to put issues on the public, political, and environmental policy agenda, ranging from seminars, meetings, and conferences to street protest actions, press releases, and other forms of propaganda, have since become regular events of environmental politics.

By promoting environmental associations, the environmental policy sector of the administration was also enhancing political support from within civil society to empower the sector within the state. The autonomy of most environmental associations vis-à-vis party-politics was well suited to this strategy. Paradoxically, even the strong criticism of associations about the environmental disregard of development policies helped to empower the environmental administration within governmental policy-making structures.

This political strategy of a tacit alliance involving the environmental policy leadership of the government and most influential associations – the strategy was launched during the first government of premier Cavaco Silva in the mid-1980s by the state secretary for the environment and natural resources – is well exemplified by the mutual collaboration and agreement on issue-debates and mobilization in the late-1980s. However, it was viewed by many ecologists as an equivocal strategy and caused a lot of debates within the movement (Lemos 1988: 50). Some organizations and activists blamed others for being dependent on the state, for 'getting used to state subsidies, and having developed close links to environmental state agencies and the government', and for being 'under the illusion of participating in the sphere of political power', which in practice 'kept them docile and controlled'. In other words, they were blamed for being manipulated by a government that did not care much about the environment, in spite of the 'environmentally friendly' profile of the policy sector leadership (Rosa 1990: 61).

This new framework of interchange and collaboration between associations and state actors of the environmental sector raised intense debates within the movement. The contradictions of this framework affected primarily groups close to political ecology, which started to decline because of internal dissent caused in part by the ambiguity of the relationships established with state actors.[4] The initiative of state actors from the environmental policy sector in proposing to environmental movement organizations a new collaborative and institutional framework definitely changed the contents of debates among environmentalist leaders about the organizational and strategic options of the movement. A new generation of environmentalists, who became the core members and leaders of the most resourceful associations, considered the new institutional framework of collaboration with state actors as invigorating, and not endangering the independence and organizational autonomy of the associations. Others, however, saw it as a zero-sum game that forced associations to give up their independence and intransigence vis-à-vis the government. Undoubtedly, the sponsorship of the state had substantially hindered the success of green party political initiatives.

Debates of the mid-1980s over organizational forms to enhance the political and social impact of the participation of environmental groups in public life and in the polity completely lacked a systematic and consistent analysis of the structural-political conditions and constraints. Nevertheless, there had been some important and consensual areas of debate within the movement over the targets of environmental collective action. In particular, there was consensus over strategic issues that could expand the movement organizational field by mobilizing mobilization potentials, as previous protest waves around some specific issues (e.g. the nuclear and eucalyptus issues) had already done. In spite of this, none of the political and organizational strategies proposed for autonomous participation in party politics were convincing enough to be accepted by most ecological groups and environmental associations. As a consequence, more political-biased activists, that is, activists with a vocation for political entrepreneurship disappeared in niches or chose the self-complacency and isolationism of radical discourse and practically abandoned activism in organizations.

[4] The best example is perhaps the group *Os Amigos da Terra*. The group was one of the most influential by the mid-1980s, but started declining due in part to internal splits and disagreements about how to face-up to this new political and institutional framework of environmental politics.

However, in order to face a project of institutional participation not restricted to simple consultation and collaboration with state agencies and the administration of the environmental policy sector, it was necessary to have targets and strategic goals clearly defined and based upon a model of social and political participation of collective actors in a civil society. That is, an alternative model to the mode to participation of autonomous citizen groups and a strategy capable of extending the scope of political participation were needed. This had been lacking in the environmental association movement when it responded positively to the invitation by state actors to adhere to a new institutional and regulatory framework of participation and collaboration with the state. Involving environmental associations in state action was a wise and convincing strategy launched by the young state secretary of the environment in the first government of premier Cavaco Silva in the mid-1980s, when the basic law for the environment and the bill regulating public participation of environmental associations were approved.

By this time, the newly founded federative coordination of the movement was incipient and maladroit. Moreover, it was launched more because of direct solicitation by state actors than because of real demands coming out of associational life. Participation of groups could thus be used by state actors as a substitute for interest articulation and against the autonomy of social processes and social conflict. The philosophical contents and assumptions of new social movements were absent from everyday life, from organizational contents, and from the mobilization goals of the majority of this second wave of environmental associations.

In this way, the acceptance by Portuguese environmentalists of the idea that consistent and coherent regulatory action through state policies – for the sake of environmental defense and nature conservation – was perfectly compatible with current market mechanisms. Furthermore, the functioning models of representative liberal democracy – provided the state was aware of its regulatory role and strongly inclined to intervene for the sake of the environment – had come to pass without – and regardless of – any kind of internal debate over ecological and environmental philosophies. Only the idea of entering politics by doing it 'another way' seems to have been present. In the end, the collaboration with the state in nature conservation policy-initiatives took place on the basis of an underlying trade-off between the government and associations, the state offering resources for associations to survive and help the nation to improve its environmental performance.

A new generation of environmentalists found that direct political participation using the party model of organization and intervention as an exclusive tool was not the most adequate choice for the movement under these circumstances. Given the constraints of party and political systems to autonomous collective action, the political culture, and the lack of a role for civil society in the political system, the political party model option risked becoming an exclusive, and likely deceptive model of action. It threatened to squander mobilization and organizational resources in potentially deceptive electoral campaigns, dreaming of electoral support that hardly would reach the ones Green parties were obtaining in other European nations. Instead of wasting mobilization and organizational resources, these could be applied toward improving knowledge and scientific skills of associations, toward achieving the right of representing public opinion in institutional arenas of the administration with regulatory and policy-making functions (though not in executive and decision-making institutions), or toward participating in, or collaborating with, peripheral and sectoral state administration organizations, municipalities, and communities, where important decisions were made on environmental issues.

These are some of the paths of action and civic participation that environmental movement organizations have gone through in Portugal from the late 1980's onwards. Instead of heading for more politically-oriented organizational structures and mobilization actions that could link environmental mobilization to grassroots and social conflict, the tendency was to join state policy action initiatives and to improve the scientific skills of associations as action resources. This furthered legitimacy and improved the representation of environmental interests in the state administration.

Herein lie the origins of the preponderance of moderate environmentalist trends – as opposed to those of political ecology – in the Portuguese Green movement after the mid-1980s. These paths pioneered the transformation of a couple of environmental associations into very influential public interest groups. Acting either by means of direct pressure upon the state administration or by influencing public debates on development and environmental issues through media channels, in which groups managed to find and were regularly offered exceptional conditions of receptivity to echo their standpoints and opinions, they became contributors of the environmental policy-making domain.

4. PHENOMENOLOGY AND RESOURCE MOBILIZATION OF ENVIRONMENTAL GROUPS

Participation of environmental movement organizations in policy-making arenas, according to conditions established by the law for environmental associations, was by the late-1980s still beyond the capacity of many groups. Only a small set of more skilled associations, which were emerging as the 'vanguard' of the movement and receiving wide public recognition, were able to gather a network of activists apt to participate fully in the new arenas of environmental politics. They firstly established devices of collaboration to guarantee full representation of the movement in agencies and advisory bodies where associations had representatives, and gradually took the lead of environmental collective action. Direct participation in public debates, collaboration in several EU Programs, and a couple of issue campaigns allowed the development not just of a comprehensive framework of dialogue, exchange, and mutual collaboration, but also fruitful competition among them. This framework of exchange imposed a strategy for maximizing the impact of common and autonomous protest actions and profiting from conditions allowed by the law for environmental associations.

In spite of slight differences in interests and distinct organizational strategies that secured the autonomy of each group in terms of mobilization and conduct, the environmental campaigns of the late 1980s and early 1990s revealed a lot of organizational and ideological similarities among these leading groups. This led to tacit agreement on a sort of 'division of labor', that followed and strengthened the original vocation, and on issue specialization among associations. For instance, *GEOTA* placed particular emphasis on energy, industrial wastes, and territorial planning issues, while the group *Quercus* developed skills in fauna and natural habitats conservation, fresh water pollution, and spectacular protest actions over some momentous issues and events. The *LPN* remained faithful to its original, more reserved stand, privileging scientific field research far from the public eye and putting particular emphasis on natural flora conservation and sea pollution problems.

Since it resulted mainly from strategic options autonomously decided upon by each group, this 'division of labor' did not mean exclusiveness in selected fields. Particularly, the growth of *Quercus* by the early 1990s and its style of action would lead the group to enlarge its scope of issue interest. Nor was it based upon any kind of formal agreement undertaken by associations. But it contributed greatly in improving the allocation of funds as well as the technical and scientific performances of associations in public and expertise arenas of environmental politics. It also facilitated and enhanced the development of international connections, mainly with EU centers.

Resorting to all kinds of funding facilities increasingly made available by state agencies, EU programs, and private foundations, groups improved their professional skills, and the technical and scientific expertise of their active membership. Associations found that strict independence regarding political parties, economic interests, and the government could be maintained, provided they were rigorous about knowledge and information resources. Associations succeeded in defining their own field of action and autonomy by never discussing political issues or participating in other kinds of public debates than those strictly related to the environmental field. By assuming the public interest of environmental issues they made their independence from parties and the government compatible with collaboration with political actors. The rule was to make the most of participation channels for the sake of improving the environmental performance of the economy and society, while resolutely opposing any public or private agents evading environmental laws. This made it possible to both criticize and to collaborate with the administration. Complaining to EU authorities about irregularities committed by state actors regarding environmental directives became vital to protest actions.

Similar to cases of local protest movements led by well-established associations which argued on the basis of alternative, environmental-friendly technical solutions and not on the basis of individualistic neighborhood interests, such a position avoided to compromise with local NIMBY movements. By strictly defining environmental protection and nature conservation issues as public interest issues, environmental issues could be defended and communicated as non-negotiable by principle. Intransigence was counterbalanced by the disposition to collaborate and participate in a common effort to find the best solution. Efforts were also made to be present on every public and institutional arena where environmental policy issues were debated, to let the public know their stand on environmental issues. The improvement of professional skills and technical and scientific expertise also allowed associations to base their discourse, opinions, and proposals over concrete policy issues on scientific and technical data. Moreover, this small group of associations managed to appear in agreement over most important issues, which, more often than not, had previously divided them in debate. They rarely allowed the public, the media, or the government to glimpse any disagreements, although divergence and rivalries existed. In this way they gradually gained the recognition of the public, media, politicians, experts, firms, interest groups, local and central administrations, and central state authorities.

The media played a very important role in producing such a credible and positive image of environmental groups. When environmental events and issues started to become news items, the search for information often led journalists to look for environmental associations which were both well informed about developments in their field and which could deliver up-to-date technical knowledge of the issues. The relationship between the media and the environmental groups was characterized by a sort of tacit trade-off between the movement and the media, associations guaranteeing information and expertise, and the media giving wide visibility to their environmental discourse.[5]

The use of scientific knowledge and expertise made them also well accepted among experts. Associations started to send representatives to debates on environmental issues organized by universities and scientific research agencies. It was no surprise that in these arenas association members not only were zealous in mastering the highly demanding scientific skills necessary to enter the debates, but they were also at ease with the procedures and rituals of the milieu. In fact, many activists are graduates in, or were students of, the natural sciences, physics, chemistry, etc., and group leaders are often recruited among young university professors. Thus, when the public sphere of environmental politics was shaped essentially as an 'expertise domain' in the 1990s, associations showed themselves to be well prepared to act accordingly.

Moreover, they also pioneered the concept of sustainable development and the discourse of ecological modernization in Portugal, which by the late 1980s has become mainstream ecological discourse in the EU and in other international environmental regimes. While promoting sustainability as a master frame for politicizing the environmental issue, environmental associations were able to go beyond the simple intention of nature conservation. The scope of claims directed against the disregard of environmental principles in policy-making extended to the development and economic growth policies of the government. This also contributed to the further cultural impact of the environmental discourse. Particularly, it gave rise to a new phase of environmental politics by the mid-1990s, when environmental issues started being framed as a 'politics of modernization' affair. This means the inclusion of environmental issues in mainstream political discourse over social and economic modernization of the nation.

Thus, environmental politics turned into a predominantly cultural type of collective action which was effective in the non-political, i.e., more specialized public discourse. Green political organizations never succeeded in bringing environmental policy issues to the center stage of political action. Mainstream parties tended to block off environmental issues from the core

[5] Only after the mid-1980s, when private broadcasting fostered intense competition for audiences, did radio and TV channels, as well as daily and weekly periodicals, begin to pay regular attention to environmental issues and environmental protest events. Consequently, they became an important 'screen' on which associations could perform. By the early 1990s, environmental associations offered them an outstanding alternative source of news and expertise on environmental issues (Valente 1994).

of important political debates, and they delegated participation in environmental debates to a few specialists in the field. Apart from electoral campaigns and some debates over more popular issues, the environmental policy domain was practically left free of party rivalries. Governmental and state actors of this field found this small, though very active, sector of environmental associations playing the opposition role within the environmental policy domain.

Associations thus retained their independence from party politics, and the improvement of expertise and organizational resources was the keystone to their strategy. First of all, this strategy allowed for a public image of trustworthy volunteer groups with expertise in environmental issues. But the intransigence of party politics, green politics included, meant essentially two different things: that they would not get involved in political debates or issues that were not clearly environmentally related; and that they strictly limited their protest initiatives to environmental issues or other policy issues with visible impact upon the environment. Furthermore, acting for the sake of the public interest became a master-frame of environmental discourse and issue definition by the associations. Thereby, they not only distanced themselves from party politics, but also strengthened their intransigence against the 'politics of interests' that by the early 1990s started to intrude into the institutional arena of the state.

5. A THEORETICAL INTERPRETATION

The simultaneity of achieving democracy, modernity, and higher levels of affluence and development played a determinant role in shaping the careers and organizational patterns of environmentalism in Portugal as both a social movement and a new ethics towards the natural environment. A first consequence of this was the political deficit of the social movement. Other local authors called it a 'deficit of social movement' *tout court*. However, in spite of the low degree of mobilization potentials of many environmental issues raised by environmental groups in the period between the advent of democracy and the stabilization of the nation as a full-fledged member of more advanced industrial democracies, the impact of protest actions around some of them – e.g., the nuclear power and reforestation with fast growing species issues – showed that we need a more fine-tuned assertion about environmental collective action in Portugal.

Changes in agenda setting, shifts in public issue attention and in state policy orientations pushed by the adhesion to the European Community fuelled the growth of other influential organizations within the environmental movement. The role of the state in shaping what I call the second wave of environmentalism is also important. Institutional arrangements deriving from the new legal framework that state actors 'offered' to environmental groups contributed substantially in shaping the pattern of a second wave of environmentalism in Portugal. The organizational model, patterns of action, phenomenology, and political orientation of this second wave were sharply different from those of the previous period. This model is better described by the concept of "cognitive praxis" (Eyerman and Jamison 1991; Jamison, Eyerman and Cramer 1990; Jamison 1996) and "cultural pressure groups" (Statham 1995; Eder 1996). By investing in knowledge resources and discourse, environmental associations emerged as intermediaries of scientific knowledge for practical purposes, which made them valuable contributors to the *construction of nature* as a social learning process (Eder 1996).

Indeed, it was the wise commitment of a small group of associations – which included *Quercus*, *GEOTA*, *LPN* and a few others acting mainly at the regional and local level – to invest in knowledge and communication resources that allowed them to maintain autonomy from state actors and party politics, and for them to strengthen their stand both in environmental policy-making arenas and in public *fora* where the definition and resolution of environmental issues were at stake. While knowledge resources facilitated the entry of associations into policy-making arenas, which otherwise were essentially reserved for administration expertise, communication in the public sphere disclosed the complexity of the environmental domain and made it accessible to the public.

Environmental politics in Portugal may be understood as a social learning process in which both democracy and the relationship of society *versus* nature were discussed, questioned, and gradually learned. The increase in ecological communication which Portuguese society has been experiencing since the late 1980s may not be just an effect of 'outside' factors alone, such as the international wave of environmental concern and the role of European Union environmental policy. The EU factor is one that acts primarily through institutional and state channels, particularly when it does not find an autonomous and active civil society.

Portuguese environmentalism was shaped as a moderate, apolitical, associational, non-mass movement, and has been much closer to soft versions of the ecological modernization discourse than to radical trends of political ecology, which since the mid-1980s virtually disappeared from the scene due to their inability to react to state action initiatives. This may be seen as an effect of a conjunction of factors, which include historical contextual aspects, political opportunity structure ingredients, strategic political options of movement organizations, and eventually some judicious political initiatives by state actors. In particular, state actors were very wise in offering environmental groups some kind of institutional status that regulated their public participation in the aftermath of the EC adhesion.

Movement organizations also gained strength and political legitimacy to play the role of the opposition in the field of environmental politics for at least two reasons. First, the ecological mainstream discourse in Portuguese society was still 'radical' enough to highlight practical contradictions of government policy- and decision-making. Second, mainstream opposition parties conceded the field either by inertia, by having been caught unprepared for disputes over environmental issues, or by virtually admitting the non-prioritary policy status ascribed by the government to environmental policy-making. Ultimately, this gives credit to the wise strategy of governmental actors in defining the policy field as an expert, a-political domain.

The institutionalization of environmental politics in Portugal followed an odd, contradictory model. First, it does not draw upon green party politics, but resembles the model of government *versus* opposition splits. Second, the primacy of the role of the state as actor and promoter of development and modernization of the economy virtually prevented the emergence of a 'politics of interests' on environmental issues, both inside and outside the state sphere. In accordance with a long and continuing tradition of state centralism and protectionism in the political culture, the state retained its monopoly over the definition of the policy-making agenda. Eventually movement organizations were consulted, and ad-hoc separate negotiations with industries, local governments, grassroots protesters, and the parliamentary opposition were conducted.

Thus, movement organizations took the form of volunteer, a-political, issue associations and of public interest groups. However, given the deficit of interest intermediation, they had to resort to other forms of participation and mobilization, drawing upon means of communicative action, expertise discourse, field research, lobbying, press releases, educative action. Eventually, they became parallel channels of democratic participation and public interest intermediation. Still environmental agendas continued to be determined by how rigorously governmental actors were able to apply basic environmental policy orientations negotiated with, or dictated by, Brussels.

Contrary to expectations of being similar to green movement careers well-known in other advanced democratic societies, the environmental movement in Portugal followed a particular path in consonance with the nation's own path towards modernity – that is, towards achieving political, economic, and social structures prevailing in other European central nations. Environmental collective action and protest, while furthering the institutionalization of environmental dialogue, also reflects the re-emergence of means of action for civil society. Besides having contributed effectively to the increase of environmental consciousness, it had also stimulated institutional change for a new kind of dialogue between state and civil society actors in Portugal. Moreover, it has definitely contributed to strengthening the role of the public space in policy-making and policy agenda-setting processes, and has undoubtedly pioneered new forms of institutional participation of public interest groups in policy and decision-making structures which to this day have centered upon self-sufficiency of state administration.

REFERENCES

Aguilar Fernández, S. (1993) Corporatist and statist designs in environmental policy: The opposing roles of Germany and Spain in the community scenario, *Environmental Politics* **2**, 223-247.

Almeida, J.F. (1995) Evoluções Recentes e Valores na Sociedade, in E.Sousa Ferreira and H. Rato, *Portugal Hoje*, INA (Instituto Nacional de Administração), Oeiras, pp.55-70.

Amaral, D.F. (1994) Lei de Bases do Ambiente e Lei das Associações de Defesa do Ambiente, in D.Freitas do Amaral and M.Tavares de Almeida, *Direito do Ambiente*, INA (Instituto Nacional de Administração), Oeiras, pp.367-376.

Barreto, A. (1994) Portugal, a Europa e a democracia, *Análise Social* **29**, 1051-1069.

Braga da Cruz, M. (1995) *Instituções Políticas e Processos Sociais*, Bertrand Editora, Venda Nova.

Bruneau, T.C., and Maclcod, A. (1986) *Politics in Contemporary Portugal. Parties and the Consolidation of Democracy*, Lynne Rienner Publishers, Boulder, CO.

Domingos, J.J.Delgado (1978) *Inteligência ou Subserviência Nacional? (Vol. 1) Crise do Ambiente, Crise de Energia, Crise da Sociedade. Alternativas*, Edições Afrontamento, Porto.

Eder, K. (1985) The 'New Social Movements': Moral crusades, political pressure groups or social movements, *Social Research* **52-54**, 869-890.

Eder, K. (1995) Rationality in environmental discourse: A cultural approach, in W. Rüdig, *Green Politics Three*, Edinburgh University Press, Edinburgh, pp.9-37.

Eder, K. (1996) *The Social Construction of Nature. A Sociology of Ecological Enlightenment*, Sage, London.

Eloy, A. (1996) O Ambiente e o Ordenamento do Território, in A. Reis, *Portugal. Veinto Anos de Democracia*, Temas e Debates, Lisboa, pp.331-345.

Eyerman, R., and Jamison, A. (1991) *Social Movements. A Cognitive Approach*, Polity Press, Cambridge, MA.

Fernandes, J.M. (1982) Ecologistas Portugueses. Um labirinto de ideias e de correntes, *Expresso-Revista* 16.01.1982, 14R-16R.

Fernandes, J.M. (1983) Para onde vão os ecologistas portugueses, *Expresso-Revista* 19.03.1983, 19R.

Franco, A.C. (1984) As presidenciais de 1985 e os ecologistas, *Pela Vida - Informação e Coordenação Ecológica - Suplemento do Gazeta das Caldas* **50**, 8-9.

Freire, J. (1985) Deixar o pessimismo para os tempos melhores, *A Ideia* **36-37**, 7-16.

Giner, S. (1986) Political economy, legitimation, and the state in Southern Europe, in G. O'Donnell, P.C.Schmitter and L. Whitehead, *Transitions from Authoritarian Rule. Prospects for Democracy. Part I. Southern Europe*, The John Hopkins University Press, Baltimore, pp.11-44.

Graham, L.S., and Wheeler, D.L. (eds.) (1983) *In Search of Modern Portugal. The Revolution and its Consequences*, The University of Wisconsin Press, Wisconsin.

Jamison, A. (1996) The shaping of the Global Environmental Agenda: The role of Non-Governmental Organisations, in S. Lash, B.Szerszynski and B. Wynne, *Risk, Environment and Modernity. Towards a New Ecology*, Sage, London, pp.224-245.

Jamison, A., Eyerman, R., and Cramer, J. (1990) *The Making of the New Environmental Consciousness. A Comparative Study of the Environmental Movements in Sweden, Denmark and the Netherlands*, Edinburgh University Press, Edinburgh.

Lemos, P. (1988) O Associativismo e a Defesa do Ambiente, *O Verde* **14**, 48-50.

Lopes, J.M. (1984) Ecologistas. Uma decada de "verdes" ilusões, *Expresso-Revista* 21.07.1984, 12R-14R.

Matos, H. (1995) Quem são os ecologistas? *Forum Ambiente* **10**, 17-25.

Matos, L.S. (1991) O Sistema Político Português e a Comunidade Europeia, *Análise Social* **27**, 773-787.

Maxwell, K. (1995) *The Making of Portuguese Democracy*, Cambridge University Press, Cambridge, MA.

Melo, J.J., and Pimenta, C. (1993) *Ecologia e Ambiente*, Difusão Cultural, Lisboa.

Miranda, J. (1994) A Constituição e o Direito do Ambiente, in D.Freitas do Amaral and M.Tavares de Almeida, *Direito do Ambiente*, INA (Instituto Nacional de Administração), Oeiras, pp.353-376.

Opello Jr., W.C. (1983) The continuing impact of the old regime on Portuguese political culture, in L.S. Graham and D.L. Wheeler, *In Search of Modern Portugal. The Revolution and its Consequences*, The University of Wisconsin Press, London/ Wisconsin, pp.199-222.

Os Amigos da Terra (ed.) (1989) 10 Anos de Luta Ecológica, Selection in facsimile of the periodical Pela Vida - Informação e Coordenação Ecológica - Suplemento da Gazeta das Caldas Lisboa, Os Amigos da Terra - Associação Portuguesa de Ecologistas.

Rosa, H. (1990) Ambiente, que te quero verde, *O Jornal* 18.05.1990, 61A.

Santos, B. Sousa (1990) *O Estado e o Sociedade em Portugal (1974-1988)*, Edições Afrontamento, Porto.

Santos, B. Sousa (1994) *Pela Mão de Alice. O Social e o Político na Pós-Modernidade*, Edições Afrontamento, Porto.

Statham, P. (1995) *Political Pressure or Cultural Communication? An Analysis of the Significance of Environmental Action in Public Discourse: A Methodological Technique for Qualitative Data Collection*, Unpublished manuscript, European University Institute.

Valente, S. (1994) *O "verde" e a comunicação social. Os processos de produção da "notícia-ambiente"*, ISCTE, Lisboa.

Vitorino, A. (1995) A Democracia Representativa, in E.Sousa Ferreira and H. Rato, *Portugal Hoje*, INA (Instituto Nacional de Administração), Oeiras, pp.327-350.

Chapter 15

Tourism Policy and Sustainability in Italy, Spain and Greece

A Comparative Politics Perspective

Geoffrey Pridham

1. INTRODUCTION: FOCUSSING ON THE POLITICS OF TOURISM

The politics of tourism is hardly developed as an area of academic and specialist concern. In political science, tourism rarely if ever makes its presence felt as a serious scholarly subject even though the role of government in tourism is vital involving coordination, planning, regulation, stimulation and also that of entrepreneur (Hall 1994: chapter 2). A review of major international journals in policy studies as well as politics and public administration only reveals a shortage of material on the political dimensions of tourism. The same goes for the field of tourism studies where political dimensions have been largely neglected for a concern primarily with problem solution and prescriptive models of policy-planning (Hall 1994: 2, 6). This neglect is clearly surprising in view of the fact that tourism is widely regarded as the world's largest industry and is one that shows all signs of growing in the future. In the three countries under review it has become quite vital to their economic development and strategy.

This obvious gap in work on tourism is variously explained. There is an unwillingness on the part of decision-makers in government and the private sector to acknowledge the political nature of tourism, and therefore a lack of interest in research on it. At the academic level, tourism still suffers from an image of an easy if not frivolous option offering obviously attractive fieldwork; while, on the other hand, there are real and substantial methodological difficulties in conducting research due to a multiplicity of frameworks for

K. Eder and M. Kousis (eds.), Environmental Politics in Southern Europe, 365–391.

analysis, the subjectiveness of work in this area and, not least, the lack of a consensus on what actually constitutes tourism (Hall 1994: 4-7).

Does, however, growing interest in sustainability and the link made between this and tourism policy, as notably in the case of the European Union, mean that a new avenue is now opened for studying the political dimension? If so, it is important to bring a new perspective into the discussion of tourism studies, and one not solely linked to standard approaches in policy studies. For this reason, a comparative politics approach is taken as this allows one to focus on what motivates and affects ideas and action in the tourism area. This cannot resolve the methodological difficulties mentioned above, but it may reveal relevant trends in policy matters, especially through taking countries where tourism is a predominant activity. Whether this gives a new stimulus to academic interest in the politics remains to be seen.

Firstly, the role of tourism policy in sustainability is explained before outlining the approach adopted in the comparative approach. That is based on the consequent need to examine environmental and tourism policy patterns in conjunction, since it is hypothesized that what really drives - or is likely to drive - change in the latter is the overall commitment to environmental concerns on the part of public authorities at both European and national levels. Various questions therefore arise. How far is this true, or does tourism policy have any prospects for becoming more autonomous? What are the likely key determinants - or combination of determinants - of policy dynamism in the tourism sector? Do cross-national trends, taking three countries from Mediterranean Europe, show any new development or perhaps portray a pessimistic scenario or are their differences such as to confirm or question assumptions made in this chapter?

In responding to these questions, we look therefore in turn at (a) the EU and the international environment, (b) the policies and structures of national governments and then (c) at the domestic political arenas, asking respectively the questions: what kind of top-down pressures there are; what sort of political commitment there is; and, what form of bottom-up developments have been occurring. By looking in conjunction at both environment and tourism sectors, the paper will draw on a British ESRC-funded project and will discuss comparatively the three Southern countries of Italy, Spain and Greece.[1] Research methods were those familiar in policy analysis. They involved extensive elite interviews with policy-makers and other actors (e.g. environmental organizations, political parties, relevant journalists) as well as utilizing policy documentation and specialist publications.

[1] The four-year ESRC-funded project (1991-95) on 'Environmental Standards and the Politics of Expertise in Europe' will result in a joint book by Weale et al. (2000).

2. SUSTAINABILITY AND TOURISM POLICY

Tourism policy should potentially benefit from increased commitment in the 1990s to sustainable objectives. The new emphasis on cross-sectoral integration of environmental concerns could mean that the overall drive behind these carry fresh energy into the tourism sector. However, on a cautionary note, the very strategy of sustainable development (SD) has faced many obstacles and some of these are really predictable.

The strategy of sustainability, as represented by the European Union's Fifth Environmental Action Programme of 1992, has faced a multiplicity of problems in carrying through its ambitious approach. These have included resistance from traditional policy approaches, national administrative systems but also economic interests and consumer attitudes.[2] The 1992 Programme had already opted to go down the road of sectoral choice by concentrating on five policy areas: manufacturing, energy, transport, agriculture and tourism. The European Commission's Interim Report of 1994 was cautiously optimistic in seeing the Fifth Programme as marking a turning-point in European approaches to SD, but it also admitted 'there was still a failure to get to the real heart of some of the key issues' (European Commission 1996: 3). It was, admittedly, too soon to expect much change, although the Progress Report of 1996 painted much the same picture - with progress in some respects, but a marked variation over the integration of environmental considerations into the target sectors (European Commission 1996: 3-4). After that, a new Action Plan was considered for the period 1997-2000. This pointed to the need to establish priorities for action in integrating environmental considerations into other policy areas, but also proposals for promoting sustainable consumption and production patterns (Baker 1998).

Of the five sectors, that of tourism has encountered a particular difficulty in applying SD. The Progress Report of 1996 identified the core problem when it noted: 'tourism is a highly fragmented and diverse sector where a very large number of economic and other interests need to be reached before the effects of change can be felt' (European Commission 1996: 4). It is thus difficult to treat tourism as a simple case of 'sector integration', and therefore problems of policy approach and management are more complicated than in other sectors. At the same time, its own cross-sectoral linkages argue strongly for the kind of policy integration approach advocated by Brussels. This was recognized in the 1992 Action Programme which argued: 'Tourism represents a good example of the fundamental link which exists between

[2] For a discussion and categorization of the problems of implementing EU environmental legislation in Southern Europe, see Pridham (1996a).

economic development and environment, with all the attendant benefits, tensions and potential conflicts', for 'if well planned and managed, tourism, regional development and environment protection can go hand in hand' (European Commission 1992: 7).

The concept of sustainable tourism (ST) emerged in the late 1970s. While initially focusing on the impacts of tourism, the debate gradually widened beyond environmental issues to include economic, social and cultural matters as well as questions of power and equity in society. Aiming to sustain but also diversify the resource base on which tourism depends, it seeks to reduce tension created by the complex interaction between the tourism industry, visitors, the environment and the communities which act as host to holiday-makers (Bramwell and Lane 1993).

Since this concept is now rooted in the general concept of SD, is the basic difficulty faced in the tourism sector a result of the latter's intrinsic problems perhaps magnified by the diffuse nature of this sector? Or, insofar as the Progress Report of 1996 indicated more sustainable practices from Northern than from Southern member states (Pridham 1996b: 19-20), was the problem rather one of less efficient systems among the latter? Alternatively, was the basic difficulty simply due to tourism policy in the Southern countries having become too deeply rooted in their overriding priority accorded economic development since the 1960s - and somewhat earlier in Italy's case - given the supreme importance of this sector in these countries' wealth creation?

This chapter, while recognizing there is a truth in all three hypotheses, focuses on the political dynamics of policy-making in the sectors of the environment and tourism. So much academic discussion of SD has been dominated by economic research or ecological concerns or planning approaches, with little attention to what drives a policy forward.[3] But, as Hall argues, tourism is not the result of a rational decision-making process for it is a product of complex, interrelated economic and political factors (Baker et al. 1997: 3). The reason for focusing on the political dynamics of sustainability relates broadly to its radical policy implications, involving a basic change in approaches and procedures among policy-makers but also reflecting changes in public attitudes and behavior (European Commission 1992: sections 3.3). In other words, such changes may be facilitated by improved administrative procedures, for example, but the real push for change has to come from persistent political pressure at different levels.

An approach is developed here that draws on comparative politics methods. It is inspired by three notions which are complementary. Firstly, Put-

[3] A first attempt to look at this dimension is Baker et al. (1997).

nam's idea of 'push' and 'pull' characteristics relating to changes in political attitudes and behavior suggests that policy commitment is not purely or simply a matter of ideological conversion - as here, to the concept of SD - but may arise from situational dynamics (Putnam 1973). Secondly, Hanf identifies three sets of pressures that could work to bring about a political commitment to a strategy of SD. These are: politically relevant groups in societies could 'demand' it with or without support of the main economic actors; political leadership might embrace the idea and attempt to convince relevant domestic actors of the need to integrate both the environmental preconditions and consequences of economic activity into a country's developmental strategy; and, this commitment becomes imposed upon policymakers by a relevant actor-system outside the country such as the EU (Hanf 1995: 5). Thirdly, the specialist literature on ST has highlighted a range of different obstacles including: pressure from demand in tourism growth as well as the increasingly hedonistic philosophy of many people (Müller 1994: 134); the problem in de Kadt's words of 'changing the ingrained habits of the consumer society's consumers' (de Kadt 1990: 22); and, the intense competitiveness and short-term view of tour operators. In addition, there is the slow and deficient response of governments as well as traditional public attitudes, which are often seen as the most intractable problem (Croall 1995: chapter 4).

This chapter adopts the view that, if national governments are indeed lethargic, then a combination of top-down pressures from Brussels and bottom-up pressures from the domestic arena are likely, albeit gradually, to produce change if these pressures grow and are sustained. It also hypothesizes that sustainability is a very elite concept that may not necessarily be 'bought' down below, among national and especially local publics. However, it is relevant to look at that level for developments that may facilitate sustainable objectives without being as such conceptually driven by this strategy.

3. THE EUROPEAN UNION AND THE INTERNATIONAL CONTEXT

The history of EU policy on the environment is generally well-known, from its modest origins in the early 1970s with the First Action Programme of 1973. There subsequently occurred a broadening of its concerns so that by the mid-1980s the view had emerged in Brussels that environmental protection was a fundamental part of economic and social policies and not merely a moral imperative (Hildebrand 1992: 25-26). From then on, European environmental policy acquired a new dynamic, linked to the plan for the Single European Market, which created various specific pressures including tighter decision-making procedures in the Council of Ministers, the harmonization of national laws and new legal provisions defining environmental policy. An increased momentum occurred while the Maastricht Treaty introduced qualified majority voting for most environmental matters and upgraded the European Parliament's say over environmental standards (Hildebrand 1992: 28-39). Just as the Fifth Action Programme of 1992 embraced the most ambitious approach so far to environmental concerns at the EU level, so the UNCED summit in Rio of the same year highlighted the big issues. One could say that national governments were persuaded through this international framework to embrace a more principled position. This had, however, diffuse effects, although as indicated it was really only in the EU that the institutional framework provided a consistent pressure on the environment and that was a fairly recent development.

The question nevertheless remains whether this intensification of EU environmental policy activity has been as effective as intended. During the 1980s, there was a dramatic rise in infringement cases over the non-implementation of EU environmental legislation. The reasons for this pattern were various and included the problem of adapting complex national legislation to the requirements of EU law, the variety of national and sub-national structures involved in implementation, differences of national legislative culture but also problems in interpreting concepts contained in many EU environmental directives (Collins and Earnshaw 1992: 217). It goes without saying that the usual EU procedure of directives leaves much to be desired especially in the environmental field. EU directives have often been unskillfully drafted and over-complex, a problem magnified by the invariably technical content of environmental matters. Imprecision in regulations has sometimes allowed member states to delay or make exceptions. One should not forget that not only has the method of direct regulation, so far favored here, run up against legislative cultures, but the use of the directive in the environmental area allows for flexibility in the means of implementation, thus creating scope for avoiding commitments (Pridham 1996a: 50-51). All the same, it is clear that the EU role over the environment has been much expanded and consolidated over the past decade and more, and that its Action Programme on sustainability represents a major challenge to national governments to change their policy ways and approaches.

Somewhat by contrast, the EU role in the tourism sector is quite new and is distinctly modest. Brussels has generally placed a low priority on tourism policy, except over coastal zones, basically leaving responsibility in this sector to national and, where applicable, regional governments (Montanari 1992: 33). Its relevance for the markets of member states dictated a potential interest, and this began to develop in conjunction with the Single Market, so that in 1992 the Council of Ministers decided there should be an EU-level policy to strengthen tourism with the first Action Programme for Tourism. But it was only with the Fifth EAP of 1992 that the environmental dimension to tourism policy was given prominence, and this has led to various programs to promote environmental friendly tourism (Pridham 1996b: 10). However, moves to include tourism among the EU's common policies have been opposed due to differences in national interest in the tourism sector, but also over whether further EU action should concentrate on tourism directly or through related sectors, some of which are more developed as EU concerns (Pridham 1996b: 10-11). As it is, the small tourism unit (formed in 1989) in DG XXIII of the European Commission has not been particularly effective, often finding itself having to react to other DGs' policies.

Certain Southern states like Italy and Greece have urged more of an EU competence in this area. There is a practice of international co-operation over tourism among countries in the Northern Mediterranean, through the UN's Environment Programme (UNEP) and its Blue Plan for the Mediterranean as well as intermittent conferences of government ministers from the whole of the Mediterranean region. This is not surprising since no less than a third of the world's tourist market concentrates on this region, with projections of a massive increase in the tourist trade over the next decades (Pridham 1996b: 6). Tourism has been of major importance to the economic development of the three countries being examined in terms of balance of payments, national income and GDP as well as, obviously, employment. But it has had damaging effects on the environment in different ways, and this is now generally more recognized at governmental level and in the media. Clearly, such pressure is greatest where environmental quality directly confronts mass tourism, as over the state of beaches. Significantly, there has been a gradual change in consumer demands with tourists insisting on a clean environment (Pridham 1996b: 11-12). Changing tourist demand forms part of a general cultural change that not merely challenges the suppliers of tourism but must, ultimately, have implications for sustainable policies in the future.

In short, therefore, it could be said that the most consistent international pressure comes from the world tourist market and the perceived threat to tourism from growing environmental degradation rather than from the EU as such. The latter's role in the tourism sector has so far been to contribute to the greater cross-national awareness of the link between environmental concerns and tourism prospects. If it does, however, have an effect in changing approaches to tourism policies this is more likely to come from its far more ambitious strategy on the environment with consequences this might have for sustainable tourism practices. So far, judging by progress made under the Fifth EAP, this has not happened to any great extent, and must remain a hope for the future.

4. NATIONAL GOVERNMENTS: THEIR POLICIES AND INSTITUTIONAL STRUCTURES

Given that the EU role in tourism matters is minimal and that its environmental policy role is certainly greater but relatively recent, it follows that looking at the national governmental levels is vital in understanding why policy action does or does not occur. As noted in the introduction, governments tend to perform multiple functions with respect to tourism promotion. According to Hall, there is an almost universal acceptance among governments that tourism is of unquestionable benefit as an economic activity; yet, paradoxically, tourism is usually granted a modest status in national administrations and only in recent times has tourism reached the political agenda of national parliaments (Hall 1994: 23-29). How do the Southern European countries appear in this respect?

4.1 National policies on the environment and tourism

Until well into the 1980s, Italy, Spain and Greece placed a relatively low priority on environmental policy as reflected in their approach to problem-solving, policy outlook and degree of activism in the area. Compared with certain Northern European states, like the Netherlands and those in Scandinavia, these Southern countries arrived relatively late on the environmental policy scene, although Italy is a partial exception. While placing a high priority on economic policy considerations, their approach to environmental matters tended for long to be reactive, dominated by response to crisis or emergency (Pridham 1994: 86).

In recent years, there has been a slow change in policy thinking with more appreciation of preventive approaches. Undoubtedly, a major factor has been the influence of the EU which has in effect contributed the bulk of these countries' legislation on the environment especially in the relatively new member states of Greece and Spain. In Spain's case, policy was not particularly advanced before joining the EU in 1986, following which Madrid took some two years to implement the whole corpus of EU legislation in this area (more than 90 laws) - unlike Greece, which was slow in adopting European legislation after entry in 1981 (and in contrast to Portugal, which deferred this action) (Weale et al. 2000: chapter 5). But this impact of Brussels is also true of Italy despite having introduced her own major laws on environmental matters sooner than the other two countries, for the vast majority of national measures since the late 1960s have been the outcome of EU directives and international treaties (Bianchi 1992: 92-93). However, this tendency to follow the initiative of Brussels - without seeking to influence it very much - has not necessarily produced an efficient policy output. Italian legislation on the environment is known for its abundance of laws, their lack of clarity and the absence of coordination between them (Weale et al. 2000: chapter 5).

Moreover, Italian policy was long marked by the superior economic imperative compared to the environmental one. This rather exclusive priority persisted until the late 1980s, when Ruffolo (Minister of the Environment, 1988-92) attempted to introduce a more long-term perspective through initiating national reports on the state of the environment and pushing for a preventive approach with three- and even ten-year plans. These were a novelty in establishing policy priorities and criteria for legislation, and represented a significant shift in Italian environmental policy. However, it also illustrated how much policy initiative depends on committed individual ministers. This casts some doubt on policy continuity given the relative brevity of Italian coalition governments and therefore regular changes in the environment portfolio (which Ruffolo held for an unusually long period) (Weale et al. 2000: chapter 5).

This dependence of policy strategy on individual ministers is also seen in Greece, where Tritsis was instrumental in the first half of the 1980s in giving the environment a new prominence - another example being Laliotis in the mid-1990s. In Athens, an official in the Greek ministry complained: 'the government might change, maybe there won't be this minister but another minister who'll change everything or the priorities will be reversed by another politician'.[4] In general, the dominant concern in the ministry (YPEHODE) has been its other responsibility of public works. The values of development have carried great weight in policy considerations since the 1960s, when industrialization began, with clientelistic interests playing a powerful backstairs role (Weale et al. 2000: chapter 5). Some leverage for EU influence exists, nevertheless, since Athens regards Brussels as an important source for funding, including environmental programs. However, the EU can also exacerbate conflicts between developmental and environmental

[4] Interview with Maria Kritikou, Ministry of the Environment, Athens, December 1992.

policy concerns, as shown by the experience of the first Integrated Mediterranean Programme (granted to Crete) and the pilot scheme for the Prespa National Park which went wrong environmentally (Weale et al. 2000: chapter 5).

Spain has in the past given a much greater priority to economic growth over environmental protection, seeing the latter as basically an obstacle to the former. Again, as in Greece, public works have predominated in the same ministry. However, EU influence has gradually been felt and since 1992 - the year of the Fifth EAP and Rio - there has been a sustained policy effort to try and respond to the arguments of sustainability using strategic plans and introducing new institutional procedures. The Spanish government acknowledged that the Rio summit obliged it to 'resign its model of development...in the form of a national strategy for the environment and sustainable development' (Ministerio de Obras Publicas, Transportes y Medio Ambiente 1993: 27). Moreover, this change of direction has not been so dependent on committed individual ministers being in place at the right time, for this policy approach has been better coordinated in Madrid than in Athens or Rome (Ministerio de Obras Publicas, Transportes y Medio Ambiente 1993: 27).

It is clear that the concept of SD has indeed penetrated the corridors of power in these three Southern capitals in the course of mid-1990s, for it became rhetorically respectable for governments to proclaim the values of sustainability following Rio .[5] It does not, however, follow that the concept is perfectly understood or even consistently so when comparing the three countries. Moreover, the idea of how to put the concept into practice has not radically altered in recent years. There is only limited evidence that the approach of policy integration has been carried very far in promoting an environmental dimension to other policy sectors. Nevertheless, first steps have in recent years been taken with regard to tourism, although these owe far less to the EU action on tourism and more to EU and international pressure for change on the environment or national policy initiative on tourism matters.[6]

Italy in particular has a long history of tourism with mass tourism beginning to take off in the later 1950s (King 1991: 61). But much time passed before sustainability arguments began to affect policy thinking on tourism let alone practice. As late as 1989, a report of the Ministry of the Environment

[5] See Pridham and Konstadakopoulos (1997:132-33). This is based on elite interviews with policy-makers conducted in the three capitals during 1992-95 for the ESRC-funded project.

[6] For a fuller discussion than here on national tourism policies in these three Southern European states, see Pridham (1999).

in Rome admitted: 'For a long time, tourism has been considered a sector with little impact on the environment... the principal reason for this belief has to be sought in the restricted definition of environmental degradation, limiting it to forms of pollution and not considering the consumption of resources' (Ministero dell'Ambiente 1989: 3.57). At first, tourism did not feature prominently in Ruffolo's strategic plans except for the condition of coasts, since their relevance was more indirect through projects in other sectors linked to tourism (Pridham 1996b: 14). Then, in December 1993, the Italian government produced the National Plan for Sustainable Development, with tourism being one of the chosen sectors (it added waste to the five sectors of the Fifth EAP). It defined national objectives to make mass tourism compatible with environmental concerns, with measures including the protection of fragile tourist areas, diversifying the supply of tourism and campaigns to promote ST (Ministero dell'Ambiente 1993). This plan, which found no equivalent in the other two Southern countries, was presented as a direct response to Rio but also to the EU's recommendation of 1990 that 'strong measures' be adopted to help integrate tourism with the environment. Significantly, it was the Ministry of the Environment that coordinated cross-ministerial collaboration to produce the plan. The Ministry of Tourism did not take any lead in parallel to the Environment Ministry (Pridham 1996b: 14).

In Spain, early policy on tourism was based on the simple premise that the coastal variety offered immense economic advantages which should be exploited so measures were introduced to promote tourism without any planning mechanisms. It was not until the mid-1980s that action was taken to combat seasonal and regional intensity of tourism development, but environmental considerations were slow in appearing (Pridham 1996b: 9). It was not until the late 1980s that the environmental dimension to tourism began to be operationalized seriously. The Ley de Costas (Law on the Coasts) of 1988 aimed at better coastal management and provided for natural conservation in coastal areas. It proved effective and was followed by a Plan de Costas for 1993-97 to stimulate environmental regeneration (Ministerio de Obras Publicas, Transportes y Medio Ambiente 1993: 157). It was shortly afterwards that the influence of the EU and Rio was felt, since when Spanish governments have rather more resolutely than those in Italy and Greece pursued sustainability in the tourism sector by: developing a more strategic approach linking it with environmental concerns; programs following the Ley de Costas aimed at coastal improvement; some movement in developing alternative forms of tourism; and, the appearance of regional and local initiatives to counter the threat to tourism from environmental degradation (Pridham 1996b: 15-16).

Greek policy on tourism long failed to establish any link with other sectors of the economy. However, by the early 1980s, some link with regional policy began to emerge although it was not before the 1990s that efforts to integrate the environment with other policies occurred. Undoubtedly, the growing problem of environmental degradation and its threat to tourism was creating a sense of urgency (Pridham 1996b: 9). Unlike Italy, however, there was no overarching approach based on SD and there was less strategic drive compared with Madrid. Instead, the emphasis has been on focused projects motivated by a real concern for tourism interests. And there was a new effort to attract upmarket tourists - a new departure in a country known as a low-price destination for mass tourism - with an expansion of forms of alternative tourism (Pridham 1996b: 18).

The present decade has therefore seen a distinct policy change on tourism, with some interesting differences evident between the three countries, with for example Rome making more of a formal strategic effort but Athens showing more impact in terms of concrete policy action of a microcosmic kind. The link between the two sectors is evident in Italy at the strategic level, while in Spain the emphasis on coastal quality shows how much this link may determine sub-sectoral priorities. It is clear that European and international influences have been at work - and these are readily acknowledged in policy circles in the Southern countries - but a more brutal pressure comes from a growing concern that economic interests in the tourist industries there are seriously under threat and that action has to be taken to prevent this worsening in the near and long-term future.

4.2 National institutional structures

In looking at patterns of policy management in the Southern member states it is necessary to re-examine whether or how far their general reputation for inefficiency is justified. Indeed, it is also relevant to note if any institutional changes or improvements have occurred with respect to the environment and tourism sectors, given European pressure on sustainability.

The creation or presence of an environment ministry may be seen as symbolic recognition of the importance of environmental policy - a new policy area to emerge in the 1970s - although this cannot predetermine actual policy effectiveness in terms of location of important environmental powers and policy coordination (Weale et al. 1996: 260). Somewhat later than in Northern Europe, environment ministries or ministries so part-called were set up in the Southern countries at different stages during the 1980s - a single Ministry of the Environment in Rome in 1986, while in Athens and

Madrid ministries combining environment and public works together with another portfolio were formed in respectively 1980 and 1993. In 1994, Spain acquired a separate Ministry of the Environment. However, in none of these cases did the creation of such a ministry give a real impetus to environmental policy-making. The Italian ministry, for instance, was exceedingly small by Rome standards and it had to struggle to gain any bureaucratic influence in rivalry with the larger ministries which retained important powers in the environmental area (Weale et al. 1996: 267-268). And, as indicated above, the mega-ministries in Athens and Madrid showed a distinct bias towards public works rather than the environment.

Introducing effective environmental management into traditional bureaucracies is very difficult. It involves some radical restructuring of ministerial responsibilities that is bound to encounter bureaucratic resistance because of established interests in the government machine. This is seen particularly in the Greek case where the environment part of YPEHODE has not succeeded in creating any effective role in cross-ministerial coordination between many different ministries which retain a wide range of particular environmental responsibilities. Co-operation across Athens is further inhibited by a lack of clear division of labor and numerous cases of overlapping tasks (Weale et al. 2000: chapter 6). In Madrid, on the other hand, institutional structures in the environmental field are less fragmented. However, the general picture of institutional fragmentation in the South has been acerbated by ineffective coordination at both horizontal (inter-ministerial) and vertical (center-periphery) levels - with Spain, again, being something of an exception in the latter respect. Although the environment has gradually acquired more visibility and, to a limited degree, more bureaucratic influence compared with a decade and more ago, this situation has continued to act as a break on the ability of governments in these countries to respond to the call of SD.

The institutional structures in Southern Europe are rather less fragmented for tourism policy than for environmental policy, although obviously the problems of the latter affect performance in the former. However, the tourism structures tend to lack institutional weight, even more so than the environmental ones, despite the importance of this sector for the national economies of these countries. But there are some differences between them, as we shall see.

In Italy, the state of affairs at the national level has remained unsatisfactory. The existence of a Ministry of Tourism gave visibility to this policy sector, but it was understaffed, underfunded and enjoyed little bureaucratic weight (King 1991: 82). Changes have occurred, not necessarily leading to serious improvement. Powers over tourism planning and administration were transferred to regional governments in 1983, but the division of responsibilities between national and regional levels has remained unclear (King 1991: 81-82). The other major change has been the abolition of the Ministry of Tourism, Entertainment and Sport by referendum in 1993, together with two other ministries as well as the proportional representation electoral system, as a result of a general protest vote. Eventually, a lower status Department of Tourism, attached to the Prime Minister's Office, was created, but that had little effect all the more as other ministries like the Environment and Industry continued to play an important part in tourism policy (Pridham 1996b: 21). Two other central organs exist in the national tourist organization (ENIT) and the government-run travel agency (CIT), but both have been criticized for their antiquated ways of operating although CIT has now improved (King 1991: 82). Undoubtedly, these problems with tourism management contribute to Italy's failure to maximize the economic benefits from its many touristic resources.

In Spain, there is a Ministry of Industry, Trade and Tourism. There is thus a different model, analogous to the structure for environmental management in Madrid until 1994. This makes for greater institutional weight without the same kind of internal conflict between different portfolio interests as in the environmental case. While the Secretariat-General for Tourism in the Industry Ministry includes among its tasks the formulation of national tourism policy and the international promotion of Spanish tourism, important powers have been transferred to the Autonomous Communities (regions) including the bulk of tourism management. But center-periphery coordination in the tourism sector works better than in Italy, with many cases of agreements between the two levels for improving the quality of tourist zones as well as on a range of environmental sub-sectors relevant to tourism (Pridham 1996b: 21).

In Greece, there has been a Ministry of Tourism as such since 1987 and its establishment signified an upgrading of this policy area. Several other ministries have been involved in tourism policy, such as National Economy, Public Works, Culture and Agriculture, reflecting the institutional fragmentation that marks the government structure in Athens. The creation of a separate Tourism Ministry (previously a tourism section under National Economy) was aimed at maximizing tourism's benefits for the Greek economy at a time of increasing international competition (Chiotis and Coccossis 1992: 135). The idea was to improve tourism services and also to take advantage of new EU programs coming on line (Leontidou 1991: 90-91). The principal tasks of the Ministry are to plan tourist development and to supervise the monitoring and control of effluents from tourist installations such as hotels. It has control over different entities such as the Hellenic Organization for Tourism (EOT), the School of Tourism Professions and the Hotel Chamber of Greece. The institutional structure is in one sense simpler than in the other two countries because of the centralized system. There is no substantial sharing of powers with sub-national authorities, although municipalities have some responsibilities in other sectors that affect tourism, e.g. sewage and waste (Pridham 1996b: 20-21).

In conclusion, it is possible to say that policy commitment has increased in these Southern countries, but it is a recent development as of the 1990s and especially the past half-decade. Undoubtedly, the EU has in the environmental area been a significant influence and here the Rio summit provided an added stimulus. It is clear, too, that official concern for ST has also grown but for largely separate reasons relating to national economic interest and new international demand for environmental quality. The EU has practically little influence in this sector except indirectly through its strategy on sustainability where tourism has been one component. Cross-national differences are also apparent in the timing and pace of policy commitment, even though all three countries have in the past subscribed quite distinctly to the economic imperative.

Typologizing these Southern countries as environmental laggards is therefore misleading if not incorrect, although certain (but not all) Northern member states are more advanced. Differentiation is all the more necessary when we compare their institutional structures, although a general picture emerges of fragmentation. While this fragmentation is less true of the tourism structures than the environmental ones, the lack of institutional weight in the former casts doubt on these countries' ability to make a concerted cross-ministerial drive to carry into effect bolder policies in the tourism sector. Spain emerges as rather more effective in this respect than the other two countries. This is evident too when looking at the impact of EU membership on Madrid, where this has prompted greater bureaucratic status for the environment, enhanced coordination with the regions and helped to open up the administration to consultative procedures. By contrast, the Rome administration emerges much weaker despite Italy's longer EU membership (Weale et al. 2000: chapter 6).

5. THE DOMESTIC POLITICAL ARENAS: A GROWING DEMAND FOR SUSTAINABILITY?

The domestic arena is a diffuse entity, the term being collective for a variety of actors and influences that may impinge on policy-making and that may interact with each other to a greater or lesser extent or possibly not to any significant degree. We are concerned here with how far the domestic arenas provide pressures that complement those emanating from the EU and other international sources or alternatively conflict with them. Furthermore, is there any evidence of 'top-down' influences having any dynamic effect on domestic arenas? Any such trend would indicate a potential for sustainability to root itself in member states. We again look comparatively at the three countries in reference to the environment and tourism sectors.

5.1 Economic interests

These play a vital part in formulating policy given the traditional predominant concern with economic development and the now current emphasis on integrating environmental concerns into the planning of production, not to mention the chances of compliance with and implementation of European policies. Traditionally, business interests have been seen as focused on short-term profit so that the longer-term perspective of ST would appear in conflict with this priority. Tour operators, concerned with intense competitiveness, are unlikely to favor a restriction on numbers of tourists in destinations as advocated by sustainable strategies (Croall 1995: 57-61). Transnational enterprises with their considerable resources have considerable power and influence when dealing with public authorities.

In Southern Europe, economic growth concerns have weighed heavily against SD values all the more as business and industry have enjoyed a privileged position and considerable influence in the corridors of power. Deeply-rooted mentalities on the part of industry came to the fore in Italy, for instance, when Ruffolo as a pro-environment minister pushed for a more activist and strategically-focused policy. There was strong lobbying against his idea of a new energy tax. Interests such as the construction industry have exerted a powerful pull over government circles, and clearly this been present in the tourism sector too. Significantly, too, the institutional fragmentation which is rather pronounced in the Southern administrations allows business interests to exploit inter-ministerial rivalry where interests appear challenged by environmental initiatives (Weale 2000: chapter 7).

However, growing pressures from the international market for environmental quality have begun to have some effect, though predictably in some quarters more than others. Most of all, large companies have felt this pressure; and, in particular, large multinationals with subsidiaries in the Southern countries have been at the forefront of this change during the 1990s. In Italy, FIAT has led this approach of responding to green demands. Similarly, tour operators have been showing signs of responding to ST requirements when crisis has erupted and their interests are directly threatened. For instance, the support of the International Federation of Tour Operators for the Majorca scheme for ST (new quality regulations, laws controlling traffic and the building of tourist facilities) was explained by a director of Thomson Holidays in the following terms: 'There is no altruism of any kind involved; it's absolutely straightforward - unless something is done we won't even have a business'.[7] While tour operators have a low vested interest in long-term sustainability, they have paradoxically a high potential to influence tourist numbers and behavior, hence playing a pivotal role in pushing forward ST (Wight 1994: 43). However, there are signs in certain countries that some business interests may be sympathetic towards the long-term benefits of ST but rely on government regulation to help support any adjustment (Forsyth 1995: 229). There is likely to be a similar requirement in Southern Europe in the event of business moving more towards sustainable objectives.

But, seen more broadly, the problem of adaptation to the new politics of pollution is complicated in the South by the structure of industry having a very large SME component. Small businesses have not the resources or skills to respond to new market demands, and the question of environmental costs is now more of an open issue there (Weale et al. 2000: chapter 7). This naturally affects the tourism sector in particular since it is based to a large degree on SME if not family-run concerns. Local tourist businesses may not necessarily follow the logic of tour operators or, if they are affected by ST values, they do not have the resources to respond. In Italy, there were serious problems in implementing the EU directives on bathing water quality not least because of pressures from local economic interests, such as beach operators, who used their contacts with local and regional authorities to put pressure on the national government to demand exceptions and postponements from Brussels (Fondazione G. Agnelli 1990: 120).

5.2 Public opinion

Public opinion is important for estimating patterns of environmental awareness but also for its direct and indirect effects on policy choice as well as the possibilities for compliance. Undoubtedly, increased environmental awareness is likely to be a driving force for innovation in the tourist industry, but it needs to be focused. There are also some intractable problems which suggest that progress will be slow. While tourists are increasingly

[7] *The European*, 30 July - 2 August 1992.

demanding environmental quality - particularly those from certain Northern European countries - there remains a common tendency among tourists to switch off consciences while holidaying abroad (Croall 1995: 55-56, 154-155). Discussion here is confined to trends that might have a bearing on policy-makers and their proneness or not to consider sustainable approaches.

Opinion research has shown a general secular growth in the importance of the environment as an issue in Western Europe during the two decades from the mid-1970s, this being most remarkable in Northern countries but subsequently evident at a later stage too in the South. This occurred somewhat earlier in Italy, while in Spain and Greece it did not emerge as an issue of public concern until the 1980s with signs of this growing during the course of the 1990s. But evidence in Spain, for instance, suggests an almost abstract view of environmental matters with a reluctance to engage with the consequences of policy action such as new taxes (Weale et al. 2000: chapter 7). In Italy, there appears as of the last few years to be less discrepancy between 'theory' and 'practice' following an acceleration of environmental awareness in the second half of the 1980s after the Chernobyl crisis. Recent opinion evidence, such as from Eurobarometer in 1996, shows a high degree of acceptance of the EU as an actor in the environmental field with 94% in Greece and 81% in Germany for instance (Weale et al. 2000: chapter 7).

Tourism does not feature as a big issue of public concern, the conclusion being that support for ST is closely linked with environmental concern. But there is a particular twist to opinion in the Southern countries relating to territorial attachment. A matter of tradition and culture, known in Italy as *campanilismo* (the campanile as symbol of the local community), this comes into play when local heritage becomes threatened by change - which tourism can represent on a major scale. At the same time, in touristically developed locations environmental degradation can present a very real danger to local interests who are aware of abrupt turnabouts in tourist choice, often prompted by press coverage.[8]

Territoriality thus provides a powerful though localized source of mobilization since it usually combines with a concrete perception of environmental problems if they are pronounced. Such attachment is strong in certain parts of Spain. According to a survey conducted by the environment department of the Catalan regional government in the early 1990s, the highest level of environmental awareness was found in small rural communities

[8] A dramatic example was the algae crisis in the Adriatic in 1989 which affected tourism on the North-east Italian coast. The German popular press, especially the famous *Bild-Zeitung*, gave prominent and rather hysterical coverage to the problem thus frightening away a mass of German tourists for several summers.

when there existed an environmental problem: 'the small communities are the most sensitive to an unfavorable change and automatically generate agreement'.[9] A well-known example were the debates that surfaced for some time over the despoliation of the Costa Brava arising from mass tourism. This touched public sensitivity about national parks, but also mobilized regionalist feeling which is rather strong in Catalonia (Morris 1992). Territorial motivation for environmental action may have several effects. There may be issues where local or regional authorities are reinforced by concerned local opinion in pressurizing national governments as a direct form of bottom-up pressure. Or, local authority initiatives on environmental protection can gain momentum in this way. There is a growing pattern of this happening in the South, this sometimes linked with efforts to gain funds from Brussels which looks favorably on such activity.

5.3 The media

The media have certainly a considerable potential for stressing environmental concerns in everyday politics. However, there remains the question whether they act more to reinforce short-termism among policy-makers or contribute to longer-term developments like the growth of environmental awareness. This may well vary according to the type of press (quality, tabloid or environmentalist) and whether television is considered (Hansen 1994).

In Southern Europe, the usual pattern was for long one of sensationalism dwelling on environmental scandals. The environment issue appeared only periodically in the Greek press in the early 1980s, invariably after incidents - which meant urban pollution at Athens rather the environment as a whole (Vlassopoulou 1991: 35). This pattern has continued, and is all the more true of television. Such focused pressure can give an extra push to authorities, especially since high exposure can highlight deficiencies in crisis management. However, alongside that there has been an overall growth in media attention over the past decade. In Italy, this was evident from the mid-1980s, although an informed journalist specializing in environmental matters commented that Italian press coverage had actually declined in the three years after the intense interest shown in environmental information at the time of the Rio summit.[10] Similarly, one close observer of the Greek press noted in 1995 that over the previous few years there had been a decline in quantita-

[9] Interview with Joan Puigdollers, directorate for environmental promotion, Generalitat, Barcelona, October 1992.

[10] Interview with Antonio Cianciullo, *La Repubblica*, Rome, April 1995.

tive but an increase in qualitative coverage (e.g. accuracy of information) against a background of growing environmental awareness at the public level.[11]

Whatever the degree of attention to the environment in the press, the general tendency has been to provide description - often in a very fragmentary way - rather than explanation, thus casting some doubt on the media's capacity for educating the public. EU legislation on the environment does receive mention in the quality press, but it is usually a matter of brief and rather intermittent coverage while the consequences of European legislation are not always clearly linked specifically to Brussels. This is much less the case with the environmentalist press and specialist magazines, which rely frequently on scientific expertise, but mainly reach the converted. There has been some cross-national variation as to type of environmental problem covered, according to habitual national concerns, although this may also vary locally. A survey of the selected local press in Italy during 1989-91 carried out by the Ministry of the Environment revealed that waste disposal was the most itemized issue, that having an obvious relevance to tourism (Ministero dell'Ambiente 1992: 450-451).

The most common environmental issue relating to tourism is by far that of coastal water quality, with the most regular coverage seasonally determined, i.e., during the summer period. In this context, environmental organizations have come to play an important part (see below). Otherwise, it has been found in Spain that the local sections of national dailies are more attentive to environmental issues than the news sections on national matters; while in Italy there is a strong tradition of local newspapers which inevitably affects reporting on the environment.[12] Once again, territoriality comes across in environmental matters.

5.4 Political parties

Parties are central actors in these systems and it is relevant to consider also those not in government as a measure of how far political competition may serve or not to enhance the profile of environmental policy and tourism concerns. It may, however, be said immediately that in these Southern countries parties have not generally been major proponents of the environmental cause as they have become in Germany, for instance, - a possible exception being the much smaller Verdi, which joined the Italian parliament in 1987. But green parties are very weak in the other two countries. Tourism

[11] Interview with Peter Diplas, *Kathemerini*, Athens, March 1995.

[12] Interview with Enrico Fontana, journalist, AIGA, Rome, May 1992.

is not a major party-political issue at the national level, it being subsumed within economic development, although this sector can feature in party activity locally - this fitting with territoriality in public opinion trends.

Some brief points may therefore be made. Inevitably, parties have long been exponents of developmental concerns, all the more when they were linked to relevant interests, as was notably true of Italy's old party system before the major political disruption of the 1990s. This link was sometimes blatantly voiced as in Spain where in the early 1990s it was noted that 'politicians are convinced green measures do not win votes'.[13] Individual parties have made a passing issue of a particular environmental matter, especially when a crisis coincided with an election campaign, as happened several times when the Athens pollution cloud worsened.

But parties have not as a rule been a consistent pressure for a priority to environmental measures (Weale et al. 2000: chapter 7). In Greece, party interest in the environment issue has fluctuated over time after it began to engage attention in the early 1980s, but always as essentially secondary to developmental programs. In Italy, it was noted even at the start of the 1990s that 'the country's main political parties still regard environmental protection as external, peripheral or only partially relevant to the production-distribution-consumption function of society' (Alexander 1991: 106). And, since then, preoccupation with the turmoil in Italian politics has not left much space for green issues to move from the edge of the political stage. However, the Verdi have become more present in local coalitions and in 1996 entered the national government of the broad Left. It follows, therefore, that political parties are - except in certain localities - unlikely to act as a serious pressure on tourism matters.

5.5 Environmental organizations

A far more pertinent actor are these organizations which may operate either as a semi-institutionalized 'insider' exerting influence on government policy or as an 'outside' pressure mobilizing public opinion. In the Southern countries, they have by and large acted more in the latter way, although - significantly - the large ones have in recent years begun to develop 'insider' links with Brussels, combining this with their continued pressure on national and, where necessary, sub-national authorities.

Environmental organizations are fairly numerous in these countries, but their political impact on government has with some variation been rather limited. This is linked to environmental decision-making having been rela-

[13] *The Economist*, 27 April 1991.

tively closed in Spain and Greece in particular, connected to the fact they only became new democracies from the mid-1970s. Matters have improved here mildly in the 1990s, but in Italy there is more of a practice of consultation between the Environment Ministry and the large organizations like the Lega per l'Ambiente, although that has sometimes depended on the minister in office. On the other hand, much energy has been spent on public campaigns over environmental issues on which the different organizations, large and small, concentrate.

In Italy, the Lega has devoted much attention to the quality of bathing water, with obvious implications for tourism. Using its own monitoring facilities, it has annually tested the state of the Italian coast, beach by beach, and has usually contested the official data issued by the Ministry of Health. Its own results have been well publicized. The *Guide to the Clean Sea*, issued in 1992, proved such a success that its 1993 edition appeared also in English, French and German for the benefit of tourists. A Lega spokesman described the guide as 'blowing the whistle on all those local authorities and operators who think they can just hoodwink holidaymakers'.[14] Among Italians themselves, such 'alternative' data from the Lega have been particularly effective against a background of widespread distrust of political authorities in that country. This kind of action has forced governments, already aware of Italy's tourism prospects, to tighten their monitoring of beaches'.[15] That has thus contributed to environmental quality in Italian tourism. But the case of the Lega is somewhat exceptional in Italy and also Southern Europe. We are clearly looking at an environmental organization that is notably well provided for in infrastructure and expertise.

In Spain, environmental organizations developed slowly after Franco but there was a full range of them, including Spanish branches of the main international federations, by the 1990s. In Greece, there is a remarkable diversity of groups with quite special concerns, some of which like the Sea Turtle Protection Society have an evident link with tourism. This diversity, spiced with a touch of Greek individualism, makes it difficult to coordinate national campaigns and maximize pressure on the government in Athens, a weakness enhanced by these organizations lacking party-political links - unusual in Greece among pressure groups (Pridham, Verney and Konstadakopoulos 1995: 261). The large organizations have the most presence, like WWF Greece which places much emphasis on information campaigns, including

[14] *The Sunday Times*, 30 August 1992.

[15] Interviews with Mauro Albrizio, Lega Ambiente, and with Enrico Fontana of AIGA, in Rome, May 1992. See also issues of the weekly *Panorama* for the month of August 1993.

for tourists, and lobbying local governments and international bodies to legislate for protection of the natural environment.[16]

Given some frustration met by these organizations in operating in these national systems, it is not surprising they have sought to develop strong transnational links within the EU. The opportunity has come from the European Commission's need to rely on environmental organizations for concrete information on environmental defaults in member states. Several such organizations in Southern Europe have exploited this channel for influence. The Lega, for instance, has provided the Commission with dossiers on problems of applying the Environmental Impact Assessment (EIA) directive and on the environmental effects of public works in Italy. Some of the larger ones have also participated in various EU research programs on the environment. But this has not stopped these organizations criticizing the environmental damage caused by EU developmental programs (Weale et al. 2000: chapter 7).

6. CONCLUSION

It is clear, first of all, that the 1990s have witnessed qualitative changes with respect to policy approach and political behavior on environmental matters in the three Southern countries of Italy, Spain and Greece. As shown above, these changes have been evident on all three levels - the supranational, the national and the sub-national and societal. That is significant in itself and it does represent a source of hope, though not necessarily optimism, about the future. How that works out in terms of sustainability's progress obviously depends on factors like the course of the world economy. However, the question that finally confronts our analysis is whether the trends that have appeared in the environmental and tourism sectors in the 1990s are likely to continue and possibly deepen.

Some top-down pressure has come from the EU but essentially via the environment sector and therefore with indirect effects on tourism policy. More decisive has been growing international demand for environmental quality in tourism although this is so far a minority concern. Policy outlooks in these three countries in the Mediterranean have certainly been rooted in traditional views of tourism as an opportunity for quick economic growth, but they have had to start adapting. Difficulties have arisen as the tourist industry is itself fragmented and heavily dependent on many small concerns. Also, national institutional structures are not such as to make for effective

[16] *The Athenian*, October 1993, article by G. Valaora, head of WWF office in Athens.

policy action, except possibly in Greece. At the same time, some bottom-up pressures have begun to emerge, especially those linked with locality. Of various actors, it has been environmental organizations more than the media or political parties that have taken up the cause of environmental quality in tourism especially concerning coastal areas. The problems facing these three countries are broadly similar. Some differences are apparent, as over policy style and institutional structures as well as public opinion concern and the influence of environmental organizations, but these are not fundamental.

At the EU level, environmental policy had already before this decade acquired a new dynamic and, even though there are signs of this losing momentum in the 1990s, the greater EU role in this sector has become consolidated. Almost, by contrast though, the EU policy on tourism remains inchoate although the sector has gained some formal recognition. But a direct EU influence on national tourism policy is not particularly noteworthy. What is striking from examining EU policy action is that its push in the environmental area is what carries ST with it. And this dependence of ST on SD is evident at other levels too, whether it be national policy initiative along SD lines (as with the Italian plan of 1993) or the pressure exerted by environmental organizations. However, what counts most of all as a direct push factor is the changing international market in tourism which has already begun to have a quite decisive effect on policy thinking in Southern Europe.

At the national governmental levels, policy thinking is far from being the same as policy action let alone policy implementation. We have seen that in both environmental and tourism sectors new types of measures have commenced, and the policy hand of Brussels is very present in environmental legislation in the South. However, all this does not amount to a radical departure; and caution is further dictated by the nature of institutional structures, which on the environment are very fragmented (though somewhat less so in Spain) and on tourism carry no great bureaucratic weight. Institutional reform has not really occurred, notwithstanding some improvements in coordination of Spanish administration; and, although this is a hot issue in Italy, it has not in any way facilitated the prospects for sustainability in the two sectors. For this reason, it is difficult to speak of serious progress towards policy integration involving tourism and the environment. There remain two many functional obstacles to achieving this.

Turning to the domestic arenas, there is little suggestion that SD values as such - as a form of ideology - cut deep or have been consciously adopted as a motive for action. A possible exception in these countries is environmental organizations, especially large ones, which are indeed plugged into the sustainability debate, not least those ones with good connections in Brussels and elsewhere. Otherwise, sustainability is an elite concept largely imported in these countries from Brussels and Rio. But it is not ideology but rather raw economic interest or cultural factors like territoriality that serve to pull people in effect towards solutions that have an affinity with SD. Some cracks have appeared in the wall of the economic imperative that for long dominated attitudes and behavior in the postwar period, but changes on the part of economic actors, public opinion and the media do not represent an overturning of different obstacles to ST. And, we should not forget that perhaps the most important political actors - namely, parties - have been the least affected by these developments.

All the same, it is significant that various linkages between these three levels have developed in the two policy sectors, including those between the supranational and the domestic levels. That points to a continuing dynamic, but the general picture that finally emerges is one of diffuse, disjointed and distinctly fragmented progress which is nevertheless significant in its own way. Taking Hanf's three sets of pressures, we may summarize as follows: some but not all politically relevant groups in society have begun to demand changes that bear a relevance to sustainability, with however limited and probably not full-hearted support from certain (i.e., large) economic actors; political leadership has rhetorically embraced this philosophy but have, to differing degrees in the three countries, failed to convince as to the compelling need to integrate environmental with economic concerns and demonstrated only limited capacity to alter their ways of policy performance; while, finally, the external actor system of the EU has certainly not imposed a policy commitment but it has managed to half-persuade but also encourage changes in policy approach that may or may not eventually lead to a more sustainable world.

REFERENCES

Alexander, D. (1991) Pollution, policies and politics: The Italian environment, in F. Sabetti and R. Catanzaro, *Italian Politics: A Review*, Pinter, London, pp.90-111.

Baker, S. (1998) Models of sustainable and unsustainable development: making sense of the European Union's practice, *Contemporary Political Studies* **6**, 159-170.

Baker, S., Kousis, M., Richardson, D., and Young, S. (eds.) (1997) *The Politics of Sustainable Development. Theory, Policy and Practice in the European Union*, Routledge, London.

Bianchi, A. (1992) Environmental policy, in F. Francioni, *Italy and the EC Membership Evaluated*, Pinter, London, pp.71-105.

Bramwell, B., and Lane, B. (1993) Sustainable tourism: An evolving global approach, *The Journal of Sustainable Tourism* **1**, 1-5.

Chiotis, G., and Coccossis, H. (1992) Tourist development and environmental protection in Greece, in H. Briassoulis and J. Straaten, *Tourism and Environment*, Kluwer Academic Publishers, Dordrecht, pp.133-143.

Collins, K., and Earnshaw, D. (1992) The implementation and enforcement of European Community environmental legislation, *Environmental Politics* **1**, 213-249.

Croall, J. (1995) *Preserve or Destroy: Tourism and Environment*, Calouste Gulbenkian Foundation, London.

de Kadt, E. (1990) Making the Alternative Sustainable: Lessons from Development to Tourism, Discussion Paper No. 272, Institute of Development Studies Sussex, University of Sussex.

European Commission (1992) *Towards Sustainability. Fifth Environmental Action Programme*, Brussels.

European Commission (1996) *Progress Report on the Implementation of the Programme "Towards Sustainability"*, Office of Publications, Luxembourg.

Fondazione G. Agnelli, (1990) *Manuale per la Difesa del Mare e della Costa*, , Turin.

Forsyth, T. (1995) Business attitudes to sustainable tourism: Self-regulation in the UK outgoing tourism industry, *Journal of Sustainable Tourism* **3**, 210-231.

Hall, C. (1994) *Tourism and Politics: Policy, Power and Place*, John Wiley & Sons, Chichester.

Hanf, K. (1995) Institutional Prerequisites of Sustainable Development, Working Paper No.109, Institut de Ciencies Politiques i Sociales Barcelona, Universitá Autonoma di Barcelona.

Hansen, A. (ed.) (1994) *The Mass Media and Environmental Issues*, The Leicester University Press, Leicester.

Hildebrand, P. (1992) The European Community's environmental policy 1957-1992, *Environmental Politics* **1**, 13-44.

King, R. (1991) Italy: Multi-faceted tourism, in A. Williams and G. Shaw, *Tourism and Economic Development*, Belhaven Press, London, pp.61-83.

Leontidou, L. (1991) Greece: Prospects and contradictions of tourism in the 1980s, in A. Williams and G. Shaw, *Tourism and Economic Development*, Belhaven Press, London, pp.84-106.

Ministerio de Obras Publicas, Transportes y Medio Ambiente (1993) *Medio Ambiente en España 1992*, Author, Madrid.

Ministero dell'Ambiente (1989) *Rapporto al Ministero sulle Linee di Politica Ambientale a Medio e Lungo Termine*, Author, Rome.

Ministero dell'Ambiente (1992) *Relazione sullo Stato dell'Ambiente*, Author, Rome.

Ministero dell'Ambiente (1993) *Piano Nazionale per lo Sviluppo Sostenibile in Attuazione dell'Agenda 21*, Author, Rome.

Montanari, A. (ed.) (1992) *Il Turismo nelle Regioni Rurali della CEE: la tutela del patrimonio naturale e culturale*, Edizioni Scientifiche Italiane, Napoli.

Morris, A. (1992) A sea change in Spanish conservation: with illustrations from Gerona province, *Journal of the Association for Contemporary Iberian Studies* **5**, 23-30.

Müller, H. (1994) The thorny path to sustainable tourism development, *Journal of Sustainable Tourism* **2**, 131-136.

Pridham, G. (1994) National environmental policy-making in the European framework: Spain, Greece and Italy in comparison, in S. Baker, K.Milton, and S. Yearly, *Protecting the Periphery: Environmental Policy in Peripheral Regions of the European Union*, Frank Cass, London, pp.80-101.

Pridham, G. (1996a) Environmental policies and problems of European legislation in Southern Europe, *South European Society and Politics* **1**, 47-73.

Pridham, G. (1996b) Tourism Policy in Mediterranean Europe: Towards Sustainable Development? Occasional Paper No.15, Center for Mediterranean Studies Bristol, University of Bristol.

Pridham, G. (1999) Towards sustainable tourism in the Mediterranean? Policy and practice in Spain, Italy and Greece, *Environmental Politics* **8**, 97-116.

Pridham, G., and Konstadakopulos, D. (1997) Sustainable development in Southern Europe? Interactions between European, national and sub-national levels, in S. Baker, M. Kousis, D. Richardson and S. Young, *The Politics of Sustainable Development: Theory, Policy and Practice within the European Union*, Routledge, London, pp.127-151.

Pridham, G., Verney, S., and Konstadakopulos, D. (1995) Environmental policy in Greece: Evolution, structure and processes, *Environmental Politics* **4**, 244-270.

Putnam, R. (1973) *The Beliefs of Politicians: Ideology, Conflict and Democracy in Britain and Italy*, Yale University Press, New Haven, NJ.

Vlassopoulou, C.A. (1991) *La Politique de l'Environnement: Le Cas de la Pollution Atmosphérique à Athènes*, Working Paper, University of Picardy.

Weale, A., et al. (1996) Environmental administration in six European states: secular convergence or national distinctiveness? *Public Administration*, 255-74.

Weale, A. (ed.) (2000) *Environmental Governance in Europe: An Ever Closer Ecological Union?* Oxford University Press, Oxford.

Wight, P. (1994) Environmentally responsible marketing of tourism, in E. Cater and G. Lowman, *Ecotourism: A Sustainable Option?* John Wiley & Sons, Chichester, pp.39-55.

Conclusion

Is There a Mediterranean Syndrome?

Beyond the North-South Divide

Klaus Eder and Maria Kousis

1. THE SOCIAL FUNCTIONS OF ENVIRONMENTAL POLITICS

Environmental policy-making will change the way in which societies organize their economic relationship with nature. The implications of environmental politics for the re-organization of work and leisure have become obvious to people everywhere. The environment creates jobs different from those associated with industrial society. It also changes our habits as consumers by providing choices regarding consumption; that is to say, we, i.e., European societies, live increasingly in a world where starvation is no longer a constraint on survival and feeding one's family is no longer the mechanism that shapes the life-course decisions of most individuals. Most of us live in a world in which making a living is a normality. We distinguish ourselves through something beyond simply making a living: conspicuous consumption, free time, investment in an dwindling number of children.

This emerging type of society is mobilized through concern for the environment. The concern differs, depending on the type of environmental impact which human action has upon nature. This is the reason why the environmental impact of tourism is an important theme in countries – especially those of the Mediterranean – which depend heavily on tourism.[1]

[1] Tourism is a major economic sector in Greece, Spain and Italy and is expected to grow much further in the next few decades (Pridham and Konstadakopoulos 1997: 146). The Mediterranean countries (especially the European) host 30% of international tourism. According to different Blue Plan scenarios, the 135 million international and national tourists visiting only the coastal regions of the Mediterranean in 1990 could become 235 to 350 million by 2025 (Plan Blue, Mediterranean Action Plan, UNEP, MCSD, 1998).

K. Eder and M. Kousis (eds.), Environmental Politics in Southern Europe, 393–406.

There is a Mediterranean syndrome which is, first and foremost, nothing other than the effect of a shared concern for tourism's use of the environment. There is no equivalent effect in the North in terms of size and impact. The centrality of this concern emerges in several of the papers. Yet it is not the only concern. The chapters on Italy, Spain, and Greece mention other concerns such as the agricultural sector, the informal sector, the industrial sector – e.g. concerns about chemical plants. What is common to all these concerns is that they affect people's lives. Most importantly they affect people's interests, but often they also affect their identities, and it is the interplay of interests and identities which accounts for the diversity of reaction of the people, ranging from individual rational action to collective mobilization in the name of collective identities. To bring some analytical order into this array of phenomena we will make some concluding proposals.

2. THE TRANSFORMATION OF THE EUROPEAN CLEAVAGE STRUCTURE

Our argument at the outset has been that the old cleavage structure in Europe which historically separates East and West, North and South, so well described and analysed by Stein Rokkan (1975, 1983) and taken up by Flora (1999), is undergoing significant transformations in the course of the political changes that have been set in motion by the process of European integration. Viewed in this context, environmental policy-making is, at first glance, a special case of analysing the political and economic effects that European integration has on the structure of the European social landscape.

The environmental issue, however, is more than just a special case; it is a key to understanding the transformation of the old European cleavage structures insofar as environmental issues have joined classic social policy concerns in shaping the economic cleavages in present-day European societies. This argument targets one side of the coin. On the other side, cleavage structures in Europe are affected by environmental policy issues not only in terms of economic interests, but also in terms of collective identities. Environmental issues mobilize people in much the same way as social issues did prior to the creation of modern welfare states in Europe. Thus, we assume a structuring effect of environmental issues in the organization of the social landscape of Europe.

Thus, the impacts on the environment are a most important issue for its continuation (Lanquar et al. 1995).

Over the course of the discussions presented above, this perspective has led – unforeseen! – to a new sociological appraisal of terms such as the subnational, the local, or even the people. It has drawn attention to what can be called (1) a social life-world, which organizes people's lives independent of national and supranational impositions on these life-worlds, and (2) the social effects of economic life which transcend political control and regulation. In these terms, the making of a European society in the process of European integration has so far been neither registered nor researched in social-scientific analysis.[2]

The theoretical categories used to understand what happens in Europe have instead been dominated by political categories which function well as long we as we can assume an identity between the people and the nation, between society and state. This identification, constitutive of the nation-state as invented in Europe, allows one to disregard the difference of polity and society.

Making such a distinction is necessary in order to understand the complex changes that Europe is undergoing. This is not to consider the difference in terms of a normative idea that pits civil society against the state, but rather in terms of the idea that any polity rests upon a social basis, upon people who are governed, who eventually do not like to be governed, and who only in rare (and pathological) cases long to be governed. This tension between political order and social life has been lost in much of the social research on Europe. It is the paradoxical effect of studying the implementation of environmental directives in Southern Europe that 'society' re-appears in empirical social research.

Within the dominant conceptual model, which understands Europe in terms of supranational, national, and subnational levels, the 'social' appears only as the subnational, as the lowest element in an hierarchical order ranging from the complex and the universal to the non-complex and particular. Such an image has a quasi-paradigmatic role in research on Europe, and it is still virulent in much of what is said in the chapters above. The empirical studies, however, show that the 'subnational' or 'local' element breaks the confines of such models and forces us to think of these phenomena in different terms.

[2] There are as usual exception to this rule. Historically oriented social scientists have worked on such issues. A good example is Michael Mann (1988, 1993, 1999). Comparativists have begun such work (Therborn 1995; Crouch 1999).

3. THE OTHER SIDE OF THE MEDITERRANEAN SYNDROME

One major outcome of the studies presented in this volume has been to uncover the driving political forces which are fuelled by and react to environmental politics and policies in Europe. To what extent does environmental policy-making 'made in Brussels' provide an opportunity structure for environmental politics? To what extent does the environmental politics that emerges from such an opportunity structure shape the collective action aimed at the social organization of an emerging society in Europe? The contributions to this volume have identified and compared the respective collective actors, their political as well as their economic opportunities and constraints, the ways they organize their social space, and the way they view and frame sustainability.

What emerges from such discussion is a highly differentiated picture of the North-South divide. Given such diversity, the discourse of a North-South divide turns out to be a counterfactual discourse. Its function is not to describe reality but to classify it. The policy-analytical frame of a 'Mediterranean syndrome' distinguishes between the member-states and these distinctions structure the transnational political field as it is constituted in the process of European integration. It is, therefore, first of all a convenient tool for making politics in Europe. Such classificatory distinctions are discursive devices which are used to construct an interaction order on the elite level. Policy discourses on sustainability are especially well suited to reinforce such a classification since the concept of sustainability is part of a transnational culture which refers to universalist principles, for example to the equity principle. Several of the contributions above make this point, most explicitly Redclift's article. The strong normative character of this concept makes it an ideal tool for classification: reality can be classified and then evaluated in terms of deviation from, or concordance with sustainability principles.

In what follows, we will first describe some of the real differences between the countries in question and identify the shaky ground on which the discourse on a Mediterranean syndrome rests. This will be done by looking into the specific political and economic opportunities of environmental action as well as the specific structures of the social space for such action. Then we will again take up the question of sustainability and show how this discursive device has a double function: not only does it function to hide particularistic interests behind a universalistic ideology, but it also functions to provide rationales for collective action in the name of protecting the environment. The North-South divide, we claim, is a simplifier of reality which has real consequences.

3.1 Political Opportunities and Constraints

There is a lack of comparative evidence concerning the 'laggard' label as it is applied to Northern as well as Southern countries and regions at a time when the environmental agenda reflects concerns of the North (Redclift; Louloudis et al.; Aguilar; Daoutopoulos et al.). There is more evidence of differences among EU member-states (Pridham; Briassoulis; Louloudis et al.; Boerzel 2000; Kousis 1999; Pridham 1994). To place the 'laggard' label on Southern Europe is misleading, given, for example, the better tourism-related environmental policy in Spain as compared to Italy – an older EU member with more experience in tourism (Pridham). This pattern also applies to the adoption of agro-environmental policies, which vary not only across Southern European countries, but across Northern European ones as well (Louloudis et al.).

At the same time, environment-related institutional reform is weak in the tourism (Pridham) or the agricultural sectors of the countries studied (Louloudis et al.; Daoutopoulos et al.). As regards tourism, the delay in incorporating environmental policies mirrors the sluggish action of the European Union's own weak tourism-related institutions (Ruzza). This in turn reflects the low priority given to tourism on the environmental policy agenda of the North (Pridham).

As concerns agriculture, there have been changes. The resistance to EU policies of the 1980s has subsided in the 1990s, and the prospects for agro-environmental policies have improved since the mid-90s. Given the weak administrative structures as well as the poor coordination between the central state and rural areas, only limited use was made of such policies (Louloudis et al.; Daoutopoulos et al.). This weakness was in part overcome by the utilization of a long tradition of interpersonal relations, as well as by the slow, but steady progress in collaboration with environmental NGOs such as WWF or GOS (Greek Ornithological Society) (Louloudis et al.).

Other characteristics of Southern European areas, especially prior to the 1990s, include a lack of expertise and the gradual assumption of expert roles by NGOs (Aguilar), leftist-nationalism (Barcena and Ibarra), scant legal provision for citizen rights (Jimenez; Redclift), defective coordination, and political corruption which is, however, also to be found in non-Southern European countries (Diani). By contrast, personal relationships have enhanced informal consultations between the state and NGOs on a general level (Diani). In this transitional period, some of the old cleavage characteristics appear to remain while member-states are simultaneously experiencing the influence of EU policies (Triandafyllidou).

Thus, the traditional secretive and statist approach is now giving way to experiences of participation (Aguilar). During the past decade, in all Southern European countries, steady and positive moves have been made towards collaborations between the state and environmental NGOs (Louloudis et al.; Diani; Aguilar; Barcena and Ibarra; Gil Nave; Jimenez). When approached for assistance, Southern European states have also responded positively to grassroots groups (Kousis; Aguilar, Figuereido et al.; Barcena and Ibarra).

Nevertheless, environmental demands are still inadequately met by state actors (Kousis; Figuereido et al.; Barcena and Ibarra; Aguilar; Gil Nave; Schnaiberg et al., forthcoming), be it in agro-environmental policies (Louloudis et al.; Daoutopoulos et al.), in tourism-related environmental policies (Ruzza; Pridham), or in other industrial policy issues (Jimenez).

3.2 Economic Opportunities and Constraints

The extent to which environmental policies will leave a strong mark is greatly influenced by the intensification of global economic competition and the ensuing wariness accompanying regulations that might hinder economic growth (Golub 1998). At its core, the EU's ecological modernization internalizes environmental costs within processes and products, leaving member-states in a competitive position on global markets. Yet, there is little evidence that the idea of economic competitiveness has been revised to reflect sustainability aims (Redclift). Economic liberalization increases the risk of rising environmental degradation and ecological marginalization.

Europeanization and economic liberalization have been the two major processes which have steadily molded the political and economic context of Southern Europe since the 1980s (Lavdas 1996). EU institutional arrangements normally lead states to retreat from economic activities (e.g. national airlines). EU competition policy intervenes in the regulation of mergers and monopolies while simultaneously restricting the use of a widely employed instrument in Southern Europe: the subsidization of national industry.[3]

The shift to privatization in Southern Europe is closely linked to Europeanization (Lavdas 1996: 253). This increasing competitiveness and the

[3] Thus, companies in crises are no longer be supported by the state. The negotiation of exemption regimes from EU control of state assistance for small and medium sized enterprises (characteristic of Southern European economies) is especially critical for their survival (Lavdas 1996: 239).

restructuring of world markets has spurred the growth of the informal sector in Southern Europe and elsewhere, especially during the past three decades (Briassoulis). The state not only tolerates, but may even stimulate informal activities in order to enhance economic growth, resolve social conflicts, and promote political patronage (Briassoulis).

The tension between economic development and environmental protection is visibly exacerbated by the Cohesion Funds, which give priority to economic criteria, as in the case of Spain, where dams were constructed to solve the water problem (Pridham and Konstadakopoulos 1997; Yearley et al. 1994).

As far as environmental policy in relation to tourism is concerned, there is no direct EU influence worthy of note (Ruzza), since the policy has a Northern bias (Pridham). Thus, Southern European states, given their vested interests in the tourism economy, try to push and direct environmental policies in this area, albeit not very successfully. Unlike tour operators, who are not directly touched by principles of long-term sustainability, or small family businesses, which have few resources at their disposal, large tourism multinationals have started to feel the demand for environmental quality on international markets (Pridham). Concerning agriculture, a high dependence has been built up on those core countries which provide the latest technology and inputs for contemporary agricultural production (Daoutopoulos et al.). The lack of integration across policy sectors is evident in EU environmental policies, this being one of its major shortcomings for all member-states. Non-coherent EU policies cannot enhance the promotion of sustainable practices. Thus, horizontal as well as vertical integration is required at the EU and member-state levels (Daoutopoulos et al.).

Based on the information above, we can identify a structural cleavage between the interests of the North and the South that is resolved to varying degrees by the different Southern European countries. It is their position in a system of economic dependence within EU society that structurally determines the responses of Southern countries with vested interests in agriculture and tourism. The Mediterranean syndrome turns out to be a device used to attribute to the 'culture' of the countries in question the environmental consequences of Europeanization and the concomitant liberalization and marketization of economies. The difference between North and South is in fact generated by the structural positions of both within the EU – positions which need to be explained in terms of power and inequality. Simply attributing the responsibility for 'laggardliness' to national deficits instead of taking into account the structural dimension of power and inequality, amounts to a discursive strategy of downgrading the 'irresponsible'.

3.3 Organizing the Social Space

The contributions to this volume also allow one to draw a different picture of Southern European civil societies. They highlight the various forms of a steadily growing environmental movement instead of merely labeling countries as 'laggards' in environmental activism (Ward, Lowe and Buller 1997; Rootes 2000).

The experience of some Southern European countries does not at all portray features of a passive and weak civil society. During the 1960s and 70s, even under the military regime, the Neighbors Movement actively mobilized around social issues and demanded environmental quality in Spanish cities as well (Aguilar). From the mid-70s to the mid-90s grassroots environmental activism made its presence strongly felt in Spain and Greece, and less so in Portugal. In the Basque country, with a strong militant tradition, anti-nuclear mobilizations were marked by intense, violent, and radical activism (Barcena and Ibarra). Mainly through the use of informal networks, local activists mobilized for the protection of the environment, challenging the state, producers, and other responsible groups (Kousis; Aguilar; Figuereido et al.; Pridham). Yet, as elsewhere within and outside of the EU, grassroots environmental activism is not necessarily linked with halting ecosystem degradation. Thus, for example, in agricultural or tourist regions there is only limited protest against these activities as sources of ecosystem damages (Figuereido et al.; Kousis 1999, 2000). This limited social demand for environmental protection is in part related to economic dependence on these activities, as well as a simultaneous lack of other resources with which to confront the environmental issue (Kousis forthcoming).

During the 1990s, after the consolidation of the new democracies and their entry into the EU, the stage for environmental politics changed, and so did the chances for the development of the environmental movement. For the Basque ecologist movement this meant a decrease in violence, radicalism, and nationalism, alongside a simultaneous growth in non-violent, localist, grassroots activism. The movement, having distanced itself from its traditional left-wing nationalist allies, now pursues a more institutional route to resolve the issues (Barcena and Ibarra). While localist encounters were present in the 1990s (Jimenez; Kousis 1999), it was environmental associations which made a strong appearance during this decade.[4]

[4] A good opportunity to study environmental activism cross-nationally is provided by the TEA project, covering Britain, France, Germany, Greece, Italy, Spain and Sweden over the years 1988–1997. For a brief description of the TEA project (EC contract no.: ENV4-CT97-0514), see http://www.ukc.ac.uk/sociology/TEA.html .

The relationship between Europeanization and the environmental movement is of special interest, since environmentalists quickly realized the opportunities which European cooperation offered both inside and outside of European institutions (Rootes 2000). Outside of European institutions, environmentalists can coordinate or synchronize their efforts via campaigns and put their demands to the EU, as did Spanish and other European environmental NGOs who succeeded in modifying the rules governing the concession of EU funds so as to ensure that environmental prerequisites were upheld (Jimenez). This expanded opportunity structure has also been utilized by local environmental activists who have approached EU agencies for assistance, especially since the late 1980s (Kousis; Figuereido et al.). Power, although limited in this context, has been shifting away from the nation state and towards the supra-national state. Other activists have blamed the EU for projects it funds that are damaging to the ecosystem (Pridham).

For Southern European states, especially those belatedly entering the EU, the incorporation of EU environmental policies has also fostered the institutionalization of rapidly growing environmental associations (Kousis 1999). While in the 1980s participation of environmental NGOs in environmental policy circles was very limited (Jimenez; Diani; Aguilar), a different experience characterized the 1990s. Relations between environmental NGOs and the state became more cooperative and less confrontational, with the latter becoming increasingly involved in various types of collaboration with institutional actors (Diani; Jimenez; Gil Nave; Louloudis et al.). At the same time the incorporation of EU environmental laws not only provided for the formal legitimation of ENGOs, but it also contributed to the institutionalization and de-radicalization of these associations. Also, business actors started to be influenced by EU environmental policies. In the agricultural sector this led to informal links with the state that facilitated the implementation of agro-environmental measures (Louloudis et al.). Yet, economic interests appear to take priority and direct any such initiatives (Diani; Pridham; Redclift).

Information about social mobilization and its support within emergent European society remains scattered. However, the role of the informal economy and the role of popular protest show that there are transnational social effects of the Europeanization of environmental politics. What we observe is that the response to Europe is the self-organization of the people in Europe. The process of self-organization of people in Europe is increasingly transnational: farmers across national borders experience common constraints; the middle classes discover economic opportunities which no longer depend on national markets. These experiences nourish potential collective action. Such forms of a transnational social field are still more of an objective potential than a real collective habitus. It is the latent, transnational classes which discover their action potential as European politics evolve. Seen from this angle, environmental politics is an important part of an ongoing process.

Whether the North or the South contributes more to the evolution of a European society is still a speculative question. But the forms of protest-oriented collective action, i.e., popular protest (Kousis), and self-interest-oriented collective action, such as the informal economy (Briassoulis), point to a hypothesis for further research: the social structural characteristics of societies in Southern Europe, in the so-called late-comer nation-states, provide more opportunities for the evolution of a transnational society in Europe than the closed and state-oriented societies in the North, in the core-lands of the nation-state.

3.4 Sustainability Discourse

Finally, the contributions to this volume show that there is a wide variety of sustainability discourses in Southern Europe. Even though the political issue at stake is usually the environment (Diani), environmental NGOs often highlight other issues as well. These issues have different implications for the environment (Louloudis et al.; Barcena and Ibarra; Aguilar; Gil Nave; Jimenez) and help to explain variations in the practical meaning of sustainability. Moreover, the environmental views and practices of economic actors also influence the discourse on sustainability. While large firms are ready and able to initiate ecological modernization (Redclift), small to medium sized enterprises (many of which are family-owned and lack the appropriate resources) do not appear to follow this discourse (Diani; Pridham), especially if they belong to the informal sector (Briassoulis). Finally, the top-down effect of the EU has influenced the views of state actors on sustainability. Thus, in the 1990s, with the incorporation of the EU's environmental policies and laws, they adhere to a more conventional definition (Diani; Jimenez; Gil Nave).

Competing claims on the environment are thus evident in Southern Europe. On the one hand, environmental activists seek a more sustainable use of local ecosystems. On the other hand, producers and state actors usually pursue vested interests and economic growth (Kousis). In this context, and also as a result of EU environmental policies, different conceptions of sustainability have arisen in Southern Europe since the 1980s which do not allow one to assume there to be a cultural deficit in Southern support for the idea of sustainable development.

In terms of national discourses, an ambivalent self-image can be identified in Southern European member-states. Increasingly, national media have come to reflect the competing views that characterize the national discourse on sustainability: these views oscillate between the state's focus on partnership, the voluntary approach adopted by business and the strictly environmentalist orientation of environmental groups. All of these views relate to sustainability (Triandafyllidou). Thus, sustainability becomes a catchword of defining national goals through which divergent interests are channeled and coordinated.

Such national forms of coordinating interests by referring to sustainability discourses vary widely between Southern European member-states. This variation cannot be explained by some national trait as implied in the notion of a Mediterranean syndrome, which puts some countries (the Northern ones) in the box of the 'goodies' and others (the Southern ones) in the box of the 'baddies'. It rather requires an analysis of sustainability in terms of a socially embedded discourse which varies with market position, political possibilities, and the mobilization of people's interests. When we find Southern countries lacking a strong notion of sustainability, then the reason is to be found not in their national character, but in some constellation of factors particular to a given country.

What is common to all countries in the EU is a Europeanization effect, i.e., the constraint placed on reorganization of the rules of the political game. Legal directives, cultural principles, and national interests have to be coordinated; the result is what we call a national political culture (Aguilar; Diani; Triandafyllidou). Instead of being an explanans, national political culture is an explanandum. Thus, the explanatory strength of a theory that, like the thesis of a Mediterranean syndrome, takes as its starting point any kind of shared political culture amongst a group of historically widely divergent political cultures, must necessarily vanish from the outset.

4. RE-CONCEPTUALIZING THE SOCIAL IN EUROPE

The proposal which we present as a general conclusion is to conceptualize collective action 'below' the level of politics and to conceive of 'reactions' to things political as the expression of a genuinely social field of collective action. This brings us back to the theory that society is the basis of all things political. This idea changes the meaning of the 'sub' in terms like subnational: it leads one to conceive of the subnational as the social which founds 'from below' the political.

Such a re-conceptualization replaces the implicit model which shapes the distinction between supranational and subnational. The dominant model is that of a funnel in which the supranational level narrows to the national level, which in turn dwindles further to the narrowest subnational level. This model explains the disregard shown toward the 'subnational' in much literature, except that which is motivated by a more romantic idea of 'small is good'. Our counter model proposes to turn the funnel on its head. It transforms the 'subnational' into the social, which, as society, serves as the basis for the rest. The national is a reduction of this social world: its institutional containment being, in a political form, the nation-state. The supranational represents a further reduction, not only in spatial (Brussels, Luxembourg and Strasbourg in Europe), but also in social terms: it consists of several thousand employees in addition to different kinds of interest brokers. Such a theoretical conceptualization leads to the rediscovery of society – society which has been there throughout the process of European political integration and which changes according to its own logic as it is Europeanization during this integration process.

This theoretical proposal also sheds a different light on the institutions that have emerged in the wake of collective environmental action on the part of elites and of the people. National and supranational institutions allow for the coordination of interests on these levels. However, the participation of non-governmental actors in such institutional forms – which has become a distinctive feature of a series of institutional innovations on these levels (Ruzza, Aguilar, Pridham in this volume) – introduces a social basis which may no longer coincide with the political boundaries that constitute these institutions, namely national and supranational boundaries. Instead, transnational forms of social self-organization will emerge and will disturb the boundary markers of national and supranational institutions. To the extent that such transnational forms of collective action even begin to act on the political level, socially determined institutional change will become prominent.

What role will the Mediterranean countries play in this process? Will they be 'laggards' as defined by the dominant political discourse? Or will they (or some of them) be leaders? This last, counterintuitive question is a provocation. Contrary to the discourse on the 'Mediterranean syndrome', it opens up the theoretical horizon beyond the good old stereotypes inherited from bygone European cleavage structures. It will force us to reconsider and re-evaluate the processes which create transnational social spaces through worker migration, the tourist industry, and the strong anti-statist cultural traditions which happen to be characteristic traits in Southern Europe. As a result, Southern European countries might even become leaders in making a European society capable of containing the political institutions that so far have developed in a society-free realm in Europe. Here environmental politics turns out to be a mechanism for fostering this kind of process. The irony of the story is that the supposed 'laggards' might be the non-intentional 'leaders' in the making of a European society.

REFERENCES

Börzel, T. (2000) Why there is no Southern Problem. On Environmental Leaders and Laggards in the European Union, *Journal of European Public Policy* **7**, 141-162.

Crouch, C. (1999) *Social Change in Western Europe*, Oxford University Press, Oxford.

Flora, P. (ed.) (1999) *State Formation, Nation-Building, and Mass Politics in Europe: The Theory of Stein Rokkan*, Oxford University Press, Oxford.

Golub, J. (ed.) (1998) *Global Competition and EU Environmental Policy*, Routledge, London.

Kousis, M. (1999) Sustaining local environmental mobilisations: Groups, actions and claims in Southern Europe, in C. Rootes, *Environmental Politics, Special issue: Environmental Movements: Local, National and Global*, pp.172-198.

Kousis, M. (2000) Tourism and the environment. A social movements perspective, *Annals of Tourism Research* **27**, 486-489.

Kousis, M. (in press) Tourism: Development or environment friendly? Comparing Corsica, Sardinia, Sicily, and Crete, in Y. Apostolopoulos, D. Ioannides and S.F. Somnez, *Mediterranean Islands and Sustainable Tourism Development*, Cassell Academic Publishers, London.

Lanquar, R. et al. (1995) *Tourisme et environnement en Mediterranée*, Plan Bleu, Sophia-Antipolis-France.

Lavdas, K. (1996) The political economy of privatization in Southern Europe, in D. Braddon and D. Foster, *Privatization*, Dartmouth, Aldershot.

Mann, M. (1988) European development: Approaching a historical explanation, in J. Baechler, J.A. Hall and M. Mann, *Europe and the Rise of Capitalism*, Blackwell, Oxford, pp.8-19.

Mann, M. (1993) Nation-States in Europe and other continents: diversifying, developing, not dying, *Daedalus* **122**, 115-140.

Mann, M. (1999) Is there a society called Euro? in R. Axtmann, *Globalization and Europe*, London, pp.184-207.

Plan Blue, Mediterranean Action Plan, and UNEP, Mediterranean Commission for Sustainable Development (1998) *Synthesis Report of the Working Group: "Tourism and Sustainable Development in the Mediterranean Region" Monaco 20-22 October*, Unpublished Manuscript.

Pridham, G., and Konstadakopulos, D. (1997) Sustainable development in Southern Europe? Interactions between European, national and sub-national levels, in S. Baker, M. Kousis, D. Richardson and S. Young, *The Politics of Sustainable Development: Theory, Policy and Practice within the European Union*, Routledge, London, pp.127-151.

Rokkan, S. (1970) *Citizens, Elections, Parties*, Universitetforlaget, Oslo.

Rokkan, S. (1975) Dimensions of state formation and nation-building, in C. Tilly, *The Formation of National States in Western Europe*, Princeton University Press, Princeton, NJ, pp.562-600.

Rootes, C.A. (2000) The Europeanisation of Environmentalism, in R. Balme, D.Chabanet and V. Wright, *L'Europe des interêts: lobbying, mobilisations et espace européen*, Presses de Science Po, Paris.

Schnaiberg, A., Pellow, D.N., and Weinberg, A. (in press) The treadmill of production and the environmental state, *Organization & Society*.

Therborn, G. (1995) *European Modernity and Beyond. The Trajectory of European Societies, 1945-2000*, Sage, London.

Yearly, S., Baker, S., and Milton, K. (1994) Environmental policy and peripheral regions of the European Union: An Introduction, in S. Baker, K. Milton and S. Yearly, *Protecting the Periphery: Environmental Policy in Peripheral Regions of the European Union*, Frank Cass and Co, Essex, pp.1-21.

Index